Charles,
With best wishes,
Ron

# Analysis of SAR Data of the Polar Oceans

Springer
*Berlin*
*Heidelberg*
*New York*
*Barcelona*
*Budapest*
*Hong Kong*
*London*
*Milan*
*Paris*
*Santa Clara*
*Singapore*
*Tokyo*

C. Tsatsoulis · R. Kwok

# Analysis of SAR Data of the Polar Oceans

## Recent Advances

Springer

Professor Dr. Costas Tsatsoulis
The University of Kansas
Department of Electrical Engineering
and Computer Science
Snow Hall
Lawrence, Kansas 66045-2228
USA

Dr. Ronald Kwok
Jet Propulsion Laboratory
California Institute of Technology
4800 Oak Grove Drive
Pasadena, California 91109-8099
USA

ISBN 3-540-62802-9 Springer-Verlag Berlin Heidelberg New York

Library of Congress Cataloging-in-Publication Data
Tsatsoulis, C. (Costas), 1962-
Analysis of SAR data of the polar oceans : recent advances / C. Tsatsoulis, R. Kwok.
p. cm. Includes bibliographical references and index.
ISBN 3-540-62802-9 (hardcover)
1. Sea ice--Polar regions--Remote sensing. 2. Synthetic aperture radar.
I. Kwok, R. (Ronald), 1955- . II. Title.
GB2595.T73 1998 551.34'3'0911--dc21 97-37831



Cover Design: E. Kirchner, Springer-Verlag, Heidelberg
Production: ProduServ GmbH Verlagsservice, Berlin
Typesetting: MEDIO Innovative Medien Service GmbH, Berlin
SPIN: 10534645 32/3020-5 4 3 2 1 0 - Printed on acid -free paper

# List of Authors

Dr. David G. Barber
Centre for Earth Observation Science
Department of Geography
University of Manitoba
Winnipeg, Manitoba R3T 2N2, Canada
*E-mail: dbarber@mccs.umanitoba.ca*

Dr. Scott Beaven
Naval Command, Control
and Ocean Surveillance Center
Research, Development
Test & Evaluation Division D855
53560 Hull Street
San Diego, California 92152-0001, USA
*E-mail: sbeaven@nosc.mil*

Cheryl Bertoia
U.S. National Ice Center
4251 Suitland Road, FOB 4
Washington, D. C. 20395, USA
*E-mail: cbertoia@natice.noaa.gov*

Frank Carsey
Jet Propulsion Laboratory
California Institute of Technology
4800 Oak Grove Drive
Pasadena, CA 91109, USA
*E-mail: fdc@pacific.jpl.nasa.gov*

Ze Cheng
Senior System Analyst
EquiFax Company
1000 Alderman Drive
Alpharetta, GA 30005, USA
*E-mail: zecheng@hotmail.com*

Mark R. Drinkwater
Jet Propulsion Laboratory, MS 300-323
4800 Oak Grove Drive
Pasadena, California 91109, USA
*E-mail: mrd@pacific.jpl.nasa.gov*

John Falkingham
Canadian Ice Service
373 Sussex Drive, Block "E"
Ottawa, Ontario K1A OH3, Canada
*E-mail: john.falkingham@ec.gc.ca*

Florence Fetterer
National Snow and Ice Data Center
CIRES Campus Box 449
1540 30$^{th}$ St, University of Colorado
Boulder, Colorado 80309, USA
*E-mail: fetterer@kryos. colorado.edu*

Professor S. P. Gogineni
Radar Systems and Remote
Sensing Laboratory
Department of Electrical Engineering
and Computer Science
University of Kansas
2291 Irving Hill Drive
Lawrence, Kansas 66045, USA
*E-mail: gogineni@eecs.ukans.edu*

Richard Harding
Alaska SAR Facility
Geophysical Institute
University of Alaska
Fairbanks, AK 99775, USA
*E-mail: dharding@images.alaska.edu*

Benjamin Holt
Jet Propulsion Laboratory
California Institute of Technology
4800 Oak Grove Drive
Pasadena, California 91109, USA
*E-mail: ben@pacific.jpl.nasa.gov*

Dr. Ron Kwok
Jet Propulsion Laboratory, 300-323
California Institute of Technology
4800 Oak Grove Drive
Pasadena, California 91109, USA
*E-mail: ron@rgps1.jpl.nasa.gov*

Prof. Shusun Li
Geophysical Institute
University of Alaska Fairbanks
P.O. Box 757320
Fairbanks, AK 99775-7320, USA
*E-mail: sli@images.alaska.edu*

Dr. Antony K. Liu
Oceans and Ice Branch
NASA
Goddard Space Flight Center
Greenbelt, MD 20771, USA
*E-mail: liu@neptune.gsfc.nasa.gov*

Prof. David G.Long
Department of Electrical
and Computer Engineering
Brigham Young University
Provo, Utah 84602, USA
*E-mail: long@ee.byu.edu*

Tim Papakyriakou
Laboratory for Earth Obsevations
Department of Geography
University of Waterloo
Waterloo, Ontario N2L 3G1, Canada
*E-mail: tim@glenn.uwaterloo.edu*

Dr. Kim C. Partington
National Ice Center
4251 Suitland Road
Washinton, D.C. 20395, USA
*E-mail: kpartington@natice.noaa.gov*

Chich Y. Peng
Oceans and Ice Branch
NASA
Goddard Space Flight Center
Greenbelt, MD 20771, USA
*E-mail: peng@neptune.gsfc.nasa.gov*

Dr. Anthony J. Sephton
Space Systems Finland
Keilaranta 8
FIN-02150 Espoo, Finland
*E-mail: tony.sephton@ssf.fi*

Dr. Leen-Kiat Soh
Department of Electrical Engineering
and Computer Science
The University of Kansas
Lawrence, KS 66045, USA
*E-mail: lksoh@eecs.ukans.edu*

Andrew Thomas
Noetix Research Inc.
265 Carling Avenue, Suite 403
Ottawa, Ontario, K1S 2E1, Canada
*E-mail: andrew.thomas@noetix.on.ca*

Prof. Costas Tsatsoulis
Department of Electrical Engineering
and Computer Science
The University of Kansas
Lawrence, KS 66045, USA
*E-mail: tsatsoul@eecs.ukans.edu*

Carl Wales
Alaska SAR Facility
Geophysical Institute
University of Alaska
P.O. Box 757320
Fairbanks, AK 99775-7320, USA
*E-mail: cwales.@images.alaska.edu*

Prof. W. F. Weeks
6533 SW 34th Avenue
Portland, OR 97201-1077, USA
*E-mail: willy@imagina.com*

Professor Dale P. Winebrenner
Polar Science Center
Applied Physics Laboratory
University of Washington
1013 NE 40th Street
Seattle, Washington 98105, USA
*E-mail: dpw@apl.washington.edu*

# Contents

*We dedicate this editorial work to our children.*

CHAPTER 1

# Recent Advances in the Analysis of SAR Data of the Polar Oceans

C. Tsatsoulis and R. Kwok

## Contents



## 1.1 Sea Ice and Climate

Studies of the behavior of the world's sea ice cover are particularly important because sea ice, with its associated snow cover, serves as an insulative layer separating the frigid Arctic atmosphere from the relatively warm underlying ocean. In fact, the heat fluxes through typical thick multiyear ice are two orders of magnitude less than fluxes observed under equivalent meteorological conditions through newly formed ice leads. Even though the areal coverage by thin ice is small (several percent) in the Arctic, the integrated flux through this area would be comparable to that through thicker ice. The presence of sea ice cover also changes the surface albedo from a value of 0.15 (for open water) to 0.84 (dry snow on the surface of sea ice). Therefore, decreases in sea ice extent and thickness caused by climatic warming would result in a pronounced positive feedback amplifying the observed change. A warming trend would be expected to result in a decrease in the extent and thickness of sea ice, in turn causing further atmospheric and oceanic warming and thus contributing to even further decreases. It is this sea ice feedback that contributes to the increases in mean annual temperatures that have been projected in simulations changes in global climate (Manabe and Stouffer 1980). In the Greenland Sea, where Arctic sea ice exported through the Fram Strait encounters the warmer North Atlantic water, there is net ice melt. It appears that deep convection - the mixing of surface water with deep ocean water - in the Greenland and Iceland Seas is conditioned by freshwater export from the Arctic Ocean. This melt process may be important in maintaining the stratification of the ocean in this area, thus preventing significant overturning (Aagaard and Carmack 1994). It was hypothesized that at some level of increased export of ice, convection would cease; the northward transport of Atlantic water would decrease; and the climatological ice cover would spread southward. Conversely, if the export were to decrease, a northward movement of the ice cover would be expected due to a weakening of the stratification in the Eurasian Basin.

Analysis of SAR Data of the Polar Oceans
Edited by C. Tsatsoulis and R. Kwok

Based on such analyses, sea ice extent and thickness would appear to be sensitive indicators of climate change, and this has resulted in considerable interest in developing procedures for monitoring the extent and behavior of the sea ice cover. It is the intent of this book to provide a survey of the procedures currently being developed for remote sensing of the sea ice cover with synthetic aperture radar.

## 1.2 Synthetic Aperture Radar Remote Sensing of the Sea Ice Cover

In the early 1970s, the launch of polar orbiting satellites allowed sea ice studies using data from visible and infrared imagery. However, sea ice data analyses based on visible and infrared wavelength imagery are limited by cloud cover and the lack of solar illumination during the polar winter. Extensive cloud cover during the summer months obscures the ice pack a large percentage of the time. These limitations have led to increased reliance on sensors from the microwave portion of the electromagnetic spectrum.

The first sensor to provide day/night large scale remote sensing of sea ice cover was the Electronically Scanning Microwave Radiometer (ESMR), which flew from 1973 to 1976. This sensor was replaced by the Scanning Multichannel Microwave Radiometer (SMMR) which provided an almost continuous record of observations from 1978 to 1987. Current passive microwave observations are provided by the series of Special Sensor Microwave Imager (SSM/I) instruments operated by the Defense Meteorological Satellite Program. These passive microwave sensors are of fairly coarse resolution and are affected by atmospheric water vapor and cloud water content. Analyzed sea ice products of ice concentration and ice types are routinely produced on a 25-km grid.

In 1978, the polar orbiting SEASAT satellite carrying an L-band synthetic aperture radar (SAR) produced the first high resolution imagery for observation of the Earth's surface from space. SEASAT imaged a 100-km-wide swath at 25-m resolution. Although it operated for only 90 days, SEASAT collected a sizable volume of data which contributed to furthering the understanding of SAR and its applications. With its moderate frequency, the L-band waves penetrated relatively deeply into the older, low salinity ice, so that volume scattering from the inhomogeneities gave a significant return. This provided a marked contrast between the backscatter of first-year and multi-year sea ice, allowing investigators an alternate view of the ice cover to that provided by passive microwave observations. It was also shown that high resolution ice motion could be observed in successive SAR observations. This SEASAT dataset clearly demonstrated the potential of SAR for both research and operational applications in ice-covered waters.

In 1983, the Soviet Union launched the first KOSMOS-1500 OKEAN satellite with an X-band side looking radar (SLR, sometimes referred to as real aperture radar, RAR). Although the resolution was relatively coarse (2.5 km along-track, 1.3 km cross-track), its 450-km-wide swath provided the first true synoptic scale radar images of the earth. There have been several OKEAN satellites, with a continuous series of improvements to the radar complement. Ukraine has taken over responsibility for the OKEAN series, and the latest satellite offers X-band SLR with two 700-km swaths and spatial resolution of about 1 km at the near range. The Soviet Union launched KOSMOS-1870 carrying a SAR that operated in S-band in 1987. It had a very narrow 30-km-swath and featured 25-m spatial resolution. This instrument evolved into ALMAZ-1, which was launched in 1991. ALMAZ-1 used the same S-band (10 cm) wavelength, but had an improved spatial resolution of 11–18 m (depending on swath width)with a much wider swath coverage that could be adjusted from 40 km to 280 km.

On July 17, 1991, the European Space Agency launched the first European remote sensing satellite (ERS-1) carrying a C-band VV (vertically polarized receive and transmit) SAR. ERS-1 was the first spaceborne SAR mission to provide a sufficient quantity of data with enough reliability to be useful for larger-scale applications. The Alaska SAR Facility in the US was constructed to receive ERS-1 data with the purpose of providing SAR data of the polar regions to the US science community. ERS-1 data has a nominal resolution of 25 m but with a limited swath width of 100 km. ERS-2, a copy of ERS-1, was launched in April 21, 1995. At the time of writing, the two satellites are operating in tandem mode which provides 1-day repeat observations of regions covered by the sensor.

The Japanese launched JERS-1 (Japanese Earth Resources Satellite) in 1992. JERS-1 was primarily intended for research purposes and returned to the original L-band of SEASAT and a relatively narrow 70-km swath. JERS-1 suffered a partial failure of its power system shortly after launch but it was still delivering data in 1996.

After a lengthy development program, Canada launched its RADARSAT on November 4, 1995. One mission objective of RADARSAT is the operational monitoring of the sea ice cover. Of primary significance for this role is its ScanSAR mode which provides a 500-km-wide swath with 100-m resolution. This wide swath provides a high repeat imaging capability, essential for monitoring of ice-covered waters. At this time, large volumes of image data are being acquired by RADARSAT.

The future of spaceborne imaging radars for continuous monitoring ice conditions is promising with several missions well into their planning and development stages. Canada plans to launch RADARSAT-2, a copy of its predecessor, around the year 2000. Investment in the development RADARSAT-3, a new generation SAR satellite scheduled for launch about 2005, has commenced. The European Space Agency intends to launch ENVISAT in 1998 with a C-band "advanced" SAR (ASAR) that will have a swath width of approximately 400 km and resolution similar to that of the ERS and RADARSAT radars. NASA has funded the initial design studies of LIGHTSAR with the intent to demonstrate the feasibility of using the latest technology for implementation of an advanced SAR system. The exact configuration and mission objectives are being defined at the time of this writing.

## 1.3 Analysis of Sea Ice SAR Data

If the full RADARSAT data collection capability is realized, starting in 1997, investigators of the polar regions and of global climate as well as the operational community will begin receiving routine high resolution data on the Arctic and Southern Oceans. RADARSAT could acquire more than 60 min data daily (approximately 60 gigabytes of image data). To effectively utilize these data sets and to generate useful conclusions, automated or semiautomated data analysis tools and algorithms are necessary for reduction of these image datasets into geophysical fields which are more useful for the general science community: manual analysis of this volume of data will no longer be affordable or feasible because of the sheer size of the data sets.

High resolution SAR data provide observations of a number of sea ice variables: ice motion, ice type/concentration and surface changes. Procedures for interpretation of SAR data to obtain fields of these variables are in various stages of maturity. Algorithms for tracking common features in sequential SAR imagery are quite well-developed within the perennial ice zone, but better algorithms are required to increase our confidence in the results from the ice margins and coastal regions. The advantage of ice motion measurements is that a thorough understanding of backscatter signatures of sea ice is not required. We can observe ice motion at the 5-km scale with the resolution of the SAR data that is currently available. Extraction of ice type and concentration from SAR data is a more involved matter: the classification of the principal ice types (first-year, multiyear) in SAR data has met with varying degrees of success. In situ validation of ice type algorithms is difficult due to the extent of coverage and the variability of ice types within the ice cover. Overlapping signatures of new ice and old ice confound most algorithms which expect some invariance of backscatter. In summer, the contrasts between ice types are lost due to melt and the appearance of liquid water on the surface. Further work in these areas is required. Perhaps one of more simple, yet useful, procedures is the one that detects the onset of melt and freeze-up. The changes in the ice backscatter during these seasonal transition are quite dramatic. Intensive validation of these algorithms, however, would require a fairly long time series of SAR observations and supporting data. The importance of correlative datasets (e.g., temperature, wind, etc.) for improvement and validation of SAR data analysis algorithms cannot be overemphasized. Some investigators have moved in the direction of fusion of datasets from different remote sensing instruments. It is the intent of this book to survey some of the algorithms and procedures used to measure of these sea ice variables.

## 1.4 This Book

The chapters in this volume contain the results of recent research on algorithms for analyses of SAR data on sea ice. All algorithms and analyses in this book are aimed at the measurement of specific sea ice parameters: the chapters discuss the significance of the polar geophysical phenomena that are being measured and analyzed. In addition to chapters describing SAR data analysis, one chapter discusses the needs and capabilities of the operational community which supplies time-critical SAR analyses to the maritime industry, and another one describes the US processing and archival SAR center, the Alaska SAR Facility, which will produce raw SAR data and high level geophysical products. A brief summary of each chapter follows.

The first half of the book (Chaps. 2–8) is dedicated to SAR data analysis and the second part of the book (Chaps. 9–12) is dedicated to SAR polar systems and facilities.

Chapter 2, by L.-K. Soh, B. Holt, and C. Tsatsoulis, describes an algorithm that segments SAR images of the marginal ice zone and identifies the ice floes present in them. The algorithm implements a novel technique that combines mathematical morphology with statistical segmentation to identify floes and separate them when they are touching. Next, the chapter discusses how measurements of the floe size distributions can be used as a proxy for climatological and geophysical processes, and presents a large study of floe measurements in the Arctic.

Chapter 3, by D.G. Barber, A. Thomas, and T.A. Papakyriakou, presents our current understanding of how SAR may be used in estimating components of the surface energy balance within the marine cryosphere, and describes various statistical linkages between the time series evolution of the microwave scattering coefficient and components of the surface energy balance. These statistical relationships are then exploited utilizing a variety of image processing approaches.

Chapter 4, by S. Li, Z. Cheng, and W.F. Weeks, discusses how intermediate regional scale sea ice deformation can help bridge the gap between the sea ice behavior on the floe scale, which is measurable in situ, and the one on the climate scale, which is used as a direct input to climate models. The chapter describes the algorithm that allows tracking of sea ice motion using SAR data and a method for screening large volumes of SAR products and extracting significant ice deformation events without having to look at either the SAR images or the SAR-derived ice motion products on which no significant ice deformation occurred.

Chapter 5, by S. Beaven, and S.P. Gogineni, presents an approach that integrates SAR analysis with an SSM/I sea ice concentration algorithm in an attempt to generate more accurate sea ice type fractions. Dynamic thresholding is used on SAR sea ice data to generate an estimate of multiyear ice, and this value is used by the SSM/I classification algorithm as an initialization parameter. The authors show how the results of the integrated algorithm improved concentration measurement and matched better with in situ observations.

Chapter 6, by A.K. Liu, and C.Y. Peng, discusses uses of the wavelet transform on sea ice SAR images to provide spectral decompositions via the scale concept. The wavelet analysis is used for ice feature tracking, for estimating the ice edge in the marginal ice zone and in polynyas, and for identifying ice floes.

Chapter 7, by D. Winebrenner, D. Long, and B. Holt, demonstrates the use of automated methods to detect and map the seasonal transitions of Arctic sea ice using SAR data. The chapter discusses how in situ and laboratory observations of the behavior of SAR backscatter led to the development of algorithmic methods to identify and map melt onset and freeze-up of Arctic sea ice. The authors also discuss possible future uses of scatterometry data to complement SAR data for detection of transitions.

Chapter 8, by M. Drinkwater, discusses how in the last few years scientists have for the first time had access to wide, repeat coverage of the Antarctic using SAR and scatterometer data. The main result of analysis of these data is an understanding of how microwave signatures and their time-varying properties can be used as a basis for understanding geophysical changes in the Antarctic sea ice cover. Furthermore, the chapter describes algorithms that have been applied to these data and ultimately lead to data products including ice kinematics fields and area flux information, with which to address mass- or freshwater-balance questions in the Southern Ocean.

Chapter 9, by F. Carsey, R. Harding, and C. Wales, describes the Alaska SAR Facility (ASF). ASF, established in 1985, is the facility that has collected, processed, and archived ERS-1, ERS-2 and JERS-1 data in the past, and which will fulfill all data system aspects for the US Government allocation of data from RADARSAT. ASF compiles and submits data acquisition requests, schedules and receives data directly at antennas at Fairbanks and McMurdo as well as receiving data by courier from other network ground stations around the world, archives raw data, processes raw data into image data, generates a number of higher level products (e.g., sea ice motion), and makes available to the user community software tools to manipulate SAR imagery. ASF is an important resource for scientists interested in the polar regions.

Chapter 10, by C. Bertoia, J. Falkingham, and F. Fetterer, describes the function, operation, and products of the government-supported sea ice services of the Arctic Nations. The chapter discusses the operational needs of the maritime industry and how these needs are satisfied by sea ice centers in the USA, Canada, Japan, the former Soviet Republics, Germany, Norway, and others. The authors describe the current practices of the centers, the sea ice products they generate, the manual and automated techniques they use, and the future directions for tools that will analyze SAR in the timely manner required by operational needs.

Chapter 11, by R. Kwok, describes the RGPS, the RADARSAT Geophysical Processor System, a complete, semiautomated system that will be used to estimate and monitor five sea ice parameters: ice motion, ice age/thickness, date of melt onset/freeze-up, open water fraction, and histograms of backscatter. The RGPS produces a Lagrangian view of the motion field from a sequence of SAR imagery, and uses this field to estimate the ice age and ice thickness histograms of the thin ice fraction of the ice cover in the winter. In the summer the RGPS estimates the open water fraction. The time of melt onset in the spring and freeze-up in the late fall are also estimated by the system.

Chapter 12, by A.J. Sephton, and K.C. Partington, presents a commercial SAR data analysis system that was developed by GEC–Marconi Research Center in the UK. The IPAP system (Ice Pilot Application Project) is described as to the algorithms used to generate sea ice motion and sea ice typing/concentration maps, to identify the ice edge, and to recognize icebergs.

**Acknowledgments.** Cheryl Bertoia and John Falkingham assisted with the historical overview of satellite sensing of the polar regions.

## References

Aagaard K, Carmack E (1994) The Arctic Ocean and climate: a perspective, the polar oceans and their role in shaping the global environment. AGU Geophys Monogr 85

Manabe S, Stouffer RJ (1980) Sensitivity of a global climate mode to an increase of $CO_2$ concentration in the atmosphere. J Geophys Res 85, no C10: 5529-5554

CHAPTER 2

# Identifying Ice Floes and Computing Ice Floe Distributions in SAR Images

L.-K. Soh, C. Tsatsoulis, and B. Holt

## Contents



## 2.1 Introduction

Sea ice that grows in the open seas takes on characteristic forms and shapes that are different from the largely smooth sea ice that grows in calm inlets, for example. Forcing from wind and waves fractures the thickening ice into pieces or floes, which then collide with each other. Over a winter season, the resulting sea ice matrix consists of fairly rounded floes with deformed surfaces composed of ridges and hummocks, separated by open water areas or leads. The new ice that forms in leads cements the floes together into an aggregate floe, which moves as a larger rigid piece composed of many smaller floes. In the summer, aggregated floes disintegrate as the thinner ice "cement" preferentially melts, making the isolated floes easily identifiable.

In studies of the polar regions with satellite imagery, including synthetic aperture radar (SAR) imagery, identification of ice floes and their outlines is important for examining both large- and small-scale processes (Rothrock and Thorndike 1984; Burns et al. 1987; Campbell et al. 1987). Most floe-related issues of interest are in the areas of classification (Haverkamp et al. 1995), registration and motion tracking (Fily and Rothrock 1986, 1987; Vesecky et al. 1988; Banfield 1991; Kwok et al. 1991; McConnell et al. 1991), route planning (see for example a navigation scheme based on object avoidance, Holmes and Jungert 1992), floe size measurements (Hall and Rothrock 1987; Korsnes

Analysis of SAR Data of the Polar Oceans
Edited by C. Tsatsoulis and R. Kwok

1993), and lead analysis (Fily and Rothrock 1990). For these floe-dependent approaches, geometric descriptors of an ice floe such as size, perimeter, orientation, and circularity are used as inputs to label, monitor, register, or classify the ice floe. Thus, such issues necessitate accurate identification and separation of ice floes.

However, adjacent ice floes touch each other forming a network of connected ice floes which conceals separation among ice floes and hinders ice floe-dependent analysis. In the following, we present an innovative concept called the *restricted growing concept*. The general model of this concept is aimed at achieving object separation and is applicable to many kinds of imagery. In our particular application to SAR sea ice imagery, we combined probabilistic labeling and morphological operations as means of implementing the concept. In this paper, we emphasize the application of the restricted growing concept to floe size distribution. We also present the ice floe separation module within an integrated system designed to identify and measure floe size, distinguish open water areas, and separate out remaining forms of ice not identifiable as floes. Preliminary results using these procedures are provided for a time series analysis in the Arctic.

## 2.2 Background

In image processing, objects are separated from the background using image segmentation. Conventional image segmentation mechanisms include thresholding (Gonzalez and Woods 1992), region splitting and merging (Horowitz and Pavlidis 1976), and relaxation (Hummel and Zucker 1983; Rosenfeld and Kak 1982). These techniques have been able to successfully extract objects from the background, and where segmentation was the primary objective, these mechanisms have been sufficient. However, in cases where (1) separation among objects is desirable and (2) objects are known to be contiguous – for example, detection of individual objects in radiographic imagery (Hall *et al.* 1971), recognition of defects in electronic patterns (Ejiri *et al.* 1973), and identification of ice floes in SAR imagery (Hall and Rothrock 1981, Daida and Vesecky 1989) – we need an object separation module designed to accomplish the task.

For ice floe identification in satellite images, Banfield and Raftery (1992) devised a technique that integrated mathematical morphology and principal curve clustering. An erosion-propagation algorithm (EP) was used to select the potential edge pixels and group them into floe outlines. First, the image was thresholded to obtain its object-background segmentation. Second, the EP algorithm was performed iteratively. At the first iteration, if a pixel was ice and any of its neighbors was water, the pixel was eroded and became water. At the second iteration, the same operation was performed on the image resulting from the first iteration, and so on. After certain iterations of the eroding process, separation among objects would be observed. To prevent subdivision within single floes, a method was developed to determine which of the floes identified by the EP algorithm should be merged, based on an algorithm for clustering about closed principal curves (Hastie and Stuetzle 1989). The authors showed that the approach produced reasonably good results – floe identification and separation. However, the EP algorithm suffers from three disadvantages. First, preservation of floe size and shape is not satisfactory. The morphological effects of the EP algorithm allow a piece of ice floe to shrink with its overall shape roughly intact while sacrificing its actual size. Second, small ice floes have to be eliminated in order to achieve necessary separation. For example, for a specified number of iterations, $i$, ice floes with size less than $(2i+1) \times (2i+1)$ would be discarded. For a low-resolution (100 m per pixel) ERS-1 SAR sea ice imagery, if $i = 7$, then ice floes with size less than 15 x 15 pixels, or 2.25 km$^2$, would be lost. Third, the number of iterations needed for each image is different, and its determination requires human visual judgment. This means experiments would have to be performed in order to manually select the optimal number of iterations.

In another study, Korsnes (1993) proposed a mathematical morphology based technique that estimated floe size automatically. A disc-like mask was used to perform morphological operations such as closing and opening on the floes. The area of the opening operation of the union of disjoint discs of different sizes was used to give the integrated area of the discs with certain radius. The distribution of the integrated areas, by their different radii, defined size distributions of objects in the image. However, there are two disadvantages in this scheme. First, the distribution is simply an estimation since it is based on approximating discs. Second, connected floes are misidentified as a piece of large floe.

In the next sections, we describe the restricted growing concept that enables extraction of ice floes individually by creating separation among floes while (1) preserving the original size and shape of the floes, (2) retaining small ice floes as small as a few pixels, and (3) requiring minimal human intervention.

## 2.3 The Restricted Growing Concept

The main idea of the restricted growing concept is that growing a shrunken version of an object within the boundary of its original version facilitates separation and preserves size and shape. The shrinking process provides separation among neighboring objects. Once this separation is established, we grow the objects back to their original sizes and shapes restricted by the boundary of the original objects. Conventional region growing or morphological dilation schemes do not restrict the process of growing objects such that the acts of achieving both separation and size preservation become contradictory.

According to our definition, an object is a feature in the image and its identification and localization, as well as its size and shape, are desired. Should only the identification and localization or global interpretation of the image be of interest, a basic image segmentation technique will suffice to accomplish the task, ignoring the requirements of accurate size and shape. The object must have definitive shape and size in order to be useful for further processing. This definition serves as the basic objective of the restricted growing concept. By this requirement, objects that are fuzzy (composed of different segment pixels where the object segment dominates), disintegrated (composed of adjacent, disconnected parts due to lost details during the image segmentation or capture process), and connected (linked by pixels misidentified as part of objects during the preprocessing of the image) are not desirable. Hence, an object identification scheme should be able to produce definitive shape and size of the objects encountering these three problems. Figure 1 shows examples of fuzzy, disintegrated, and connected objects.

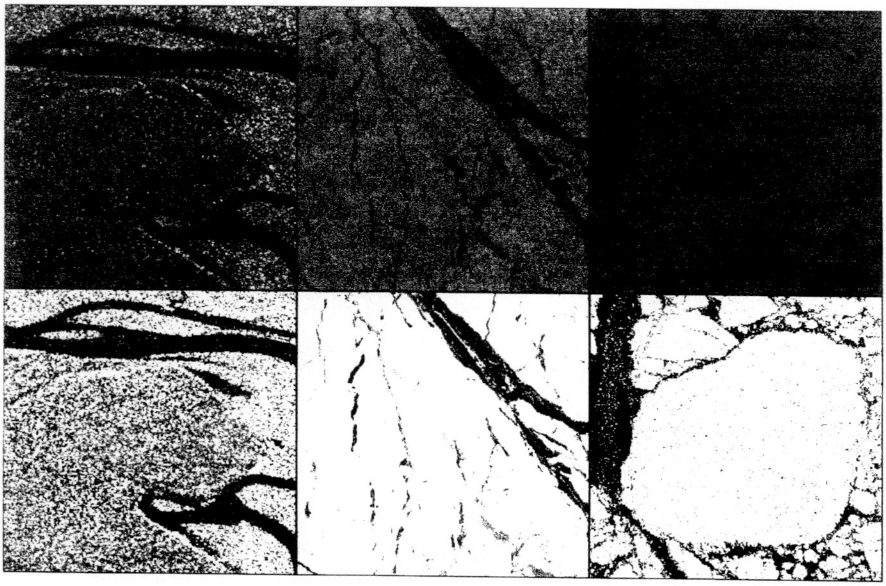

**Fig. 1.** Examples of fuzzy, disintegrated and aggregated objects. Images in the *top row* are the original images and those in the *bottom row* are corresponding intensity thresholded images. *Column 1:* JERS-1 SAR Amazon imagery where the classes of objects are river, swamp and forest. The result of segmentation shows that each class is a fuzzy composite of all three classes, only differing in proportion. (MITI/NASDA retains ownership of JERS-1 data) *Column 2:* ERS-1 SAR sea ice imagery where open-water areas or leads in the segmented image are disintegrated. (Copyright ESA) Column 3: STAR-2 SAR sea ice imagery where a large ice floe is abridged in a network of neighboring floes, inseparable through common image segmentation technique

### 2.3.1
### Initial Segmentation

The restricted growing concept consists of three components: (1) core image, (2) skin image, and (3) restricted growing algorithm. A core object is reduced in size but usually maintains the overall shape of its original version. An image with core objects is a *core image*. This image has two functions. First, it keeps the information of the spatial relationships among floes and of the separation among floes. Second, its core objects are seeds for region growing. Skin objects are usually interconnected and can encompass one or more core objects. These are objects whose original sizes and shapes have been preserved. An image with skin objects is a *skin image*. Usually, the direct object-background segmentation result suffers from fuzziness which provides no definitive boundary information for necessary decision making during the object growing process; on the other hand, a skin image of objects well-defined by their compactness is well-suited for the following growing algorithm. The innovative part of the concept lies with the restricted growing algorithm. This algorithm grows core objects within the boundaries of their corresponding skin objects while preserving the existing separation among the core objects. It was designed in reference to the skeletonization technique by Zhang and Suen (1984). In skeletonization, an object is thinned while maintaining its connectedness whereas in restricted growing, an object is grown while maintaining its separation from other objects. Our algorithm consists of five sequential tests as follows:

**Test 1: Potential Growing Test.** If a pixel in the core image is a nonobject pixel and its corresponding pixel in the skin image is an object pixel, then the pixel is qualified for further tests.

This test selects only nonobject pixels for potential growth. Note that pixels that are nonobject in both core and skin images are deemed as true nonobject pixels without any potential for becoming object pixels.

**Test 2: Isolation Test.** If the pixel in the core image does not have an object pixel in its core image 8-neighborhood, then the pixel is disqualified from the growing process.

The isolation test avoids erroneous separation within an object. For example, small dark specks in an object would be eroded to nonobject pixels during the generation of core image. If allowed to grow, these nonobject holes could become individual object regions within the object that enclosed them. With this test, however, the nonobject holes would be consumed by the growth of the encroaching core objects.

**Test 3: Connectedness Test 1.** If the pixel in the core image has seven or eight object pixels in its core image 8-neighborhood, then the pixel is grown.

If the 8-neighborhood of a pixel has seven or more object pixels, that means all object pixels in that neighborhood are connected. Hence, the growth of that pixel from nonobject to object does not damage the existing (or nonexisting) separation.

**Test 4: Connectedness Test 2.** If the pixel in the core image has less than two "role transitions" in its core image 8-neighborhood, then the pixel is grown.

A role transition is a switch from object to nonobject while following the neighborhood in the sequence shown in Fig. 2a. If less than two role transitions are found, then all the object pixels are connected and further growth does not damage the separation.

**Test 5: Connectedness Test 3.** If the pixel in the core image has its core image 8-neighborhood matching one of the four corner patterns, then the pixel is grown.

The *corner patterns* are shown in Fig. 2b. Any of these patterns could have two or more role transitions yet have all its object pixels connected, qualifying the pixel for the growing process.

The restricted growing algorithm performs these five tests sequentially. At each iteration, the algorithm moves in a raster manner, top to bottom and left to right, until the whole image is scanned. Any growing (a pixel from nonobject to object) is implemented immediately after each pixel has been inspected. To delay the growth effect on the

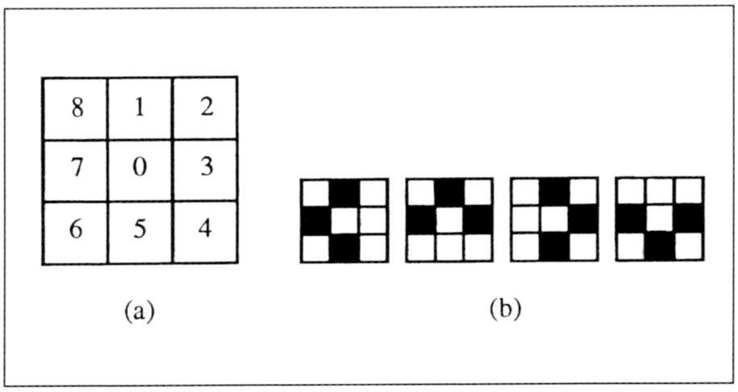


**Fig. 2.** **a** The sequence followed by the connectedness test 2 of the restricted growing algorithm. The center pixel is the current pixel. The sequence starts at pixel number 1, then 2, ..., and ends at pixel number 8. **b** Four corner patterns examined during the connectedness test 3 of the growing algorithm. *Dark pixels* denote object pixels; *unshaded pixels* denote "don't care" pixels

current pixel from propagating to the next pixel, the algorithm skips the next pixel and moves to the one after the next once a growth has been done on the current pixel. The algorithm iterates until the growth rate converges to a pre-specified asymptote such as 0.01% of the initial number of grown pixels.

### 2.3.2
### Probabilistic Labeling

We use probabilistic labeling to analyze the neighborhood of a pixel to decide whether the pixel in the core or skin image is an object pixel. The concept of our probabilistic labeling is similar to that of relaxation, which has been used in a variety of image processing work such as edge detection (Rosenfeld and Smith 1981; Geman et al. 1990), classification (Zhang et al. 1990), and image convergence (Terzopoulos 1986). A basic relaxation algorithm first assigns to each pixel an initial classification or labeling, then computes confidence of each pixel, relaxes the current classification to another classification either stochastically or deterministically, and iterates until the image converges to a satisfactory classification scenario.

In our implementation, we use a threshold slice as the environment in which the confidence or probability of a pixel's being an object is assessed. Instead of relaxing the classification over a number of iterations performed on the image, we use a set of threshold slices and accumulate the probabilities at these different slices. A threshold slice, $S_t$, is an image thresholded at intensity $t$. As we move to a higher $t$, the corresponding $S_t$ imposes a stricter environment on a pixel's neighborhood – the probability of a pixel's having object pixels as neighbors is lower. This translates into a smaller number of object-class neighbors for a pixel at higher $S_t$. For the core image, each object-to-be pixel must have a more constrained set of threshold slices to ensure that each surviving pixel is a qualified seed for the restricted growing. On the other hand, each pixel in the skin image has a less demanding requirement since the objective is to identify and eliminate pixels causing fuzziness in the image. Therefore, we have created two different sets of threshold slices, $\Omega_c$ and $\Omega_S$, for the core and skin images, respectively. Each set of threshold slices, $\Omega(T,I,N)$, is a three-tuple, where $T$ is the starting threshold, $I$ is the interval between successive threshold slices, and $N$ is the number of threshold slices. In our implementation,

$$\Omega_c = \Omega(t(i,j),2,5) = \{S_{t(i,j)}, S_{t(i,j+2)}, S_{t(i,j+4)}, S_{t(i,j+6)}, S_{t(i,j+8)}\}$$

$$\Omega_s = \Omega(t(i,j),2,3) = \{S_{t(i,j)}, S_{t(i,j+2)}, S_{t(i,j+4)}\},$$

where $t(i,j)$ is the threshold computed at pixel $(i,j)$ by the segmentation process. Note that the choices of the interval and number of threshold slices were obtained experimentally. These numbers allow the generation of good quality core and skin images and they accomplish the task computationally fast. In addition, these numbers were set as constants throughout our application of the restricted growing concept to every single image.

To combine the results of each $S_t$ together, we first define a neighborhood mask at pixel $(i,j)$ as

$$\Omega(i,j) = \{(i,j+1), (i,j-1), (i+1,j), (i+1,j+1), (i+1,j-1), (i-1,j), (i-1,j+1), (i-1,j-1)\}$$

The chance of a pixel at $(i,j)$ being an object at threshold slice $S_t$ is

$$c_t[(i,j)=object]=\sum_{m,n\in W(i,j)} Conf(g(m,n),t(m,n)),$$

where $g(m,n)$ denotes the gray level at pixel $(m,n)$, $t(m,n)$ denotes the threshold of the current slice at pixel $(m,n)$, and $Conf(a,b)$ is a step function that returns 0 if $a<b$ and 1 if $a\geq b$. To correlate the chances of a pixel at $(i,j)$ of all threshold slices, we calculate the sums as:

$$C_c(i,j)=\sum_{t\in\Omega_c} c_t[(i,j)=object]$$

$$C_s(i,j)=\sum_{t\in\Omega_s} c_t[(i,j)=object]$$

Finally, to label each pixel in the core or the skin image, we compare $C_c(i,j)$ or $C_s(i,j)$ to a pre-specified threshold, $T_c$ or $T_s$. These two thresholds were obtained experimentally to be $T_c = 0.90$ and $T_s = 0.75$. Every pixel in the image is an object pixel in the core image if and only if $C_c(i,j)$ is greater than $T_c$, and every pixel at $(i,j)$ is an object pixel in the skin image if and only if $C_s(i,j)$ is greater than $T_s$.

### 2.3.3 Morphological Cleaning

Both core and skin images as presented in the previous discussion are noisy. To eliminate the noise effects while preserving the shape of the objects effectively, we use two operators from mathematical morphology, *dilation* and *erosion*, which have been employed to smooth contours and remove noise effects in various imagery, including sea ice imagery (for example, Chou *et al.* 1994). The formal definitions of these two operators and extensions are provided in Matheron (1975) and Serra (1982).

- Dilation: The value of a pixel is replaced with the object class value if one of its 8-neighbors is an object pixel; otherwise, it is set to nonobject. This is analogous to an *OR* operation.
- Erosion: The value of a pixel is replaced with the object class value if all of its 8-neighbors are object pixels; otherwise, it is set to nonobject. This is analogous to an *AND* operation.
- Opening: This is a combination of first erosion and then dilation on an image.
- Closing: This is a combination of first dilation and then erosion on an image.

We further define one complex operation as follows:

- Cleaning: This is a combination of first closing and then opening on an image.

Opening generally trims off peninsulas, breaks bridges, and solidifies channels. Closing generally closes inlets, absorbs islands, and fills lakes. Two examples of closing and opening are shown in Fig. 3. Given the core and skin images, we perform cleaning on each before executing the restricted growing algorithm that grows the core objects within their corresponding skin objects.

Figure 4 shows a series of images depicting different stages of the execution of the restricted growing concept. Figure 4a shows a portion of a STAR-2 SAR sea ice imagery. The image consists of bright ice floes and dark leads. Figure 4b shows the segmented image of Fig. 4a. The image segmentation technique used was binary global thresholding. Note that the ice floes are connected. These networks are inseparable under conventional segmentation. Figures 4c and 4d show the core and skin images, respective-

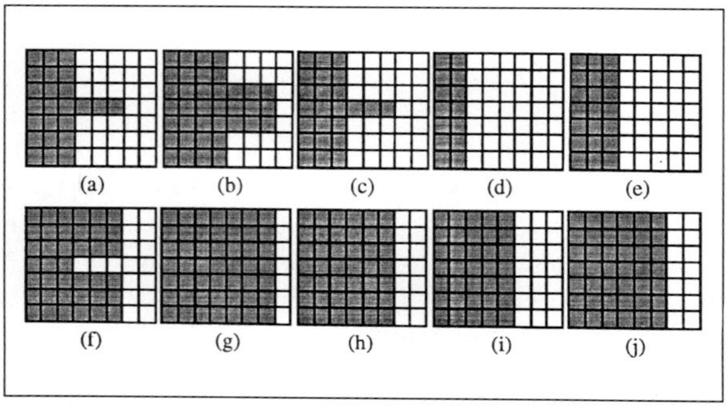


**Fig. 3.** **a** Original image of a peninsula. **b–e** Closing (dilation and erosion) and opening (erosion and dilation). **f** Original image of an inlet. **g–j** closing (dilation and erosion) and opening (erosion and dilation). Note that the peninsula and the inlet are eliminated while the shape of the object region is still preserved. The operator window is 3 × 3. *Dark pixels* denote object pixels; *unshaded pixels* denote nonobject pixels

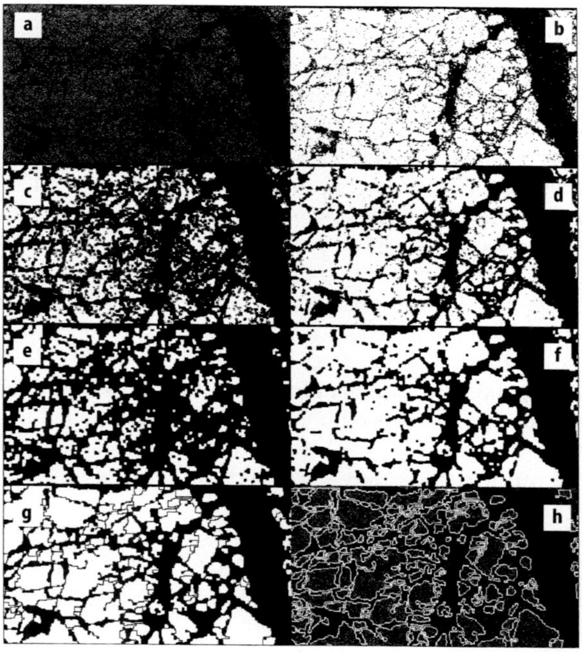


**Fig. 4a-h.** Execution of the restricted growing concept. **a** Portion of an original STAR-2 SAR image (May 1987), **b** segmented image, **c** core image, **d** skin image, **e** cleaned core image, **f** cleaned skin image, **g** final grown image output, **h** superimposition of the floe boundaries on the original image

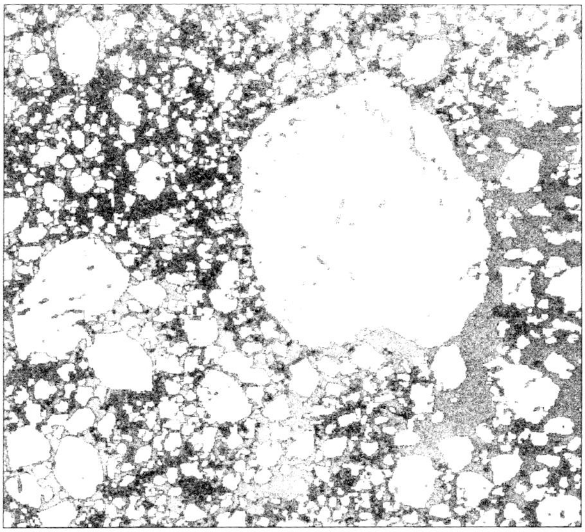

**Fig. 5.** Preservation of size and shape by the restricted growing concept. Floes denoted as white pixels regaining original sizes and shapes superimposed on the original ERS-1 SAR image taken on August 21, 1991 at 75.0°N, 145.3°W (Copyright ESA)

ly. Comparing the two images, one can see that separation has been established among core objects while fuzziness is eliminated from skin objects. Figures 4e and 4f show the morphologically cleaned core and skin images, respectively. Note that most noise effects such as peninsulas, lakes, inlets, and islands have been removed. Figure 4g shows the grown image. Note that the ice floes have been fully restored to their original shapes and sizes, and separation has been established. Figure 4h shows the floe boundaries superimposed on the original image. These images exemplify the difference between the core and skin images, the effects of morphological cleaning, and the relationships among the grown final output, the core image, and the skin image.

Our primary objective in sea ice floe identification is to establish separation among ice floes while preserving the original sizes and shapes. Figure 5 illustrates the extraction of floes that are separated and whose size and shape is preserved. The original image is an ERS-1 sea ice image taken in August of 1991 near the marginal ice zone at 75.03°N, 145.31°W. Ice floes are denoted as white pixels superimposed onto the original image. The grayish regions are water and background-matrix – a mixture of ice and water. Note that the size and shape of each floe are fully restored while separation remains established. For example, the large piece of ice floe in the center has been disconnected from its touching neighbors by destroying about a dozen bridges around the outline of the floe.

## 2.4 Ice Floe Algorithm and Results

In the following, we present in detail an algorithm or series of procedures for determining floe size distribution in SAR sea ice imagery. Our technique consists of six stages: (1) image enhancement, (2) image segmentation, (3) floe extraction, (4) floe filtering, (5) separation of open water from the remainder of the ice cover not identified as floes, termed *water/ice mixture extraction*, and (6) computation of floe size distribution and coverage statistics. Discussing stage 5) further, after floe extraction and filtering, much of the ice cover is not identified as distinct floes. This primarily includes pieces of ice that are on the order of the SAR resolution (variable but generally 100 m or less ) and new ice forms. We have termed this remaining ice *water/ice mixture*[1] for convenience. With the knowledge of the floe, water/ice mixture, and water distributions, we are able to compute complete coverage statistics within a SAR sea ice image. Section 2.5 below discusses an application for examining summer ice where knowledge of the quantities of identifiable floes and their sizes, open water, and the remaining water/ice mixture are important.

**1. Image Enhancement.** To ensure well-defined floes and reduction of radar speckle noise effects (Ulaby et al. 1982), we use median filtering (Huang et al. 1979). Median filtering enhances the image by eliminating maxima or minima in a moving mask (of size 3 x 3 in our case) resulting in more homogeneous objects. (We preferred median filtering for its relative simplicity and effectiveness in case of multiplicative noise.)

**2. Image Segmentation.** Once the image has been smoothed, it is subjected to a segmentation process that separates floes from the rest of the image. This segmentation technique is based on local dynamic thresholding (Haverkamp et al. 1995), which has been proven to be able to detect local details of the segmentation and be adaptive to different SAR sea ice imagery where intensities of similar ice classes vary. Figure 6 shows

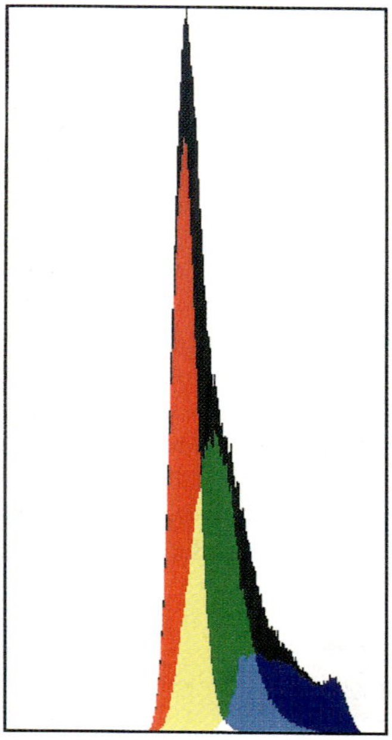

**Fig. 6.** Histograms of an ERS-1 SAR sea ice image. *Black:* The original image intensity. *Red:* The intensities of floes. *Green:* The intensities of water/ice mixture regions. *Blue:* The intensities of water regions. *Yellow, cyan, and white:* Intersections of two or three different regions. The figure clearly shows that intensities of different regions overlap.

1 We will use "substrate" and "water/ice mixture" interchangeably for labeling purposes in our figures.

an intensity histogram of an ERS-1 SAR sea ice image taken on August 19, 1991, at 74.43°N, 144.29°W. The three smaller modes of histograms are generated by collecting pixels of floes, water/ice mixture, and water regions after the image segmentation and water/ice mixture-water distinction processes. It can be seen that the intensities of all three classes overlap.

The local dynamic thresholding technique in (Haverkamp et al. 1995) has been modified to accommodate SAR sea ice imagery composed of floes, water/ice mixture, and water. An image is first divided into smaller, overlapping regions such that each region can be analyzed locally. A histogram is computed for each region, and only regions with high variance histograms will be further processed. The basic concept is to obtain the bisector of each region by examining its histogram. We accomplish this by approximating the histogram with a binormal Gaussian distribution. Given the parameters of the fitting curve, an optimal threshold that dissects the curve into two Gaussian distributions is computed using maximum likelihood, and verified if its histogram passes a bimodality test. Then, a standard deviation minimization technique is used to divide all computed thresholds into two groups. Finally, region and point interpolations are executed to ensure all regions and pixels each have two thresholds. Of the two sets of thresholds, we choose one to segment the image into floes and nonfloes. This decision requires a user input to specify which set of thresholds to be used. If floes are brighter in the image, the set of higher thresholds is used; if floes are darker in the image, the other is used. Figure 7 shows the result of our local dynamic thresholding technique.

3. **Floe Extraction.** For floe extraction, we use our implementation of the restricted growing concept (Sect. 2.3 above). Figure 7c shows the results of our floe extraction phase.

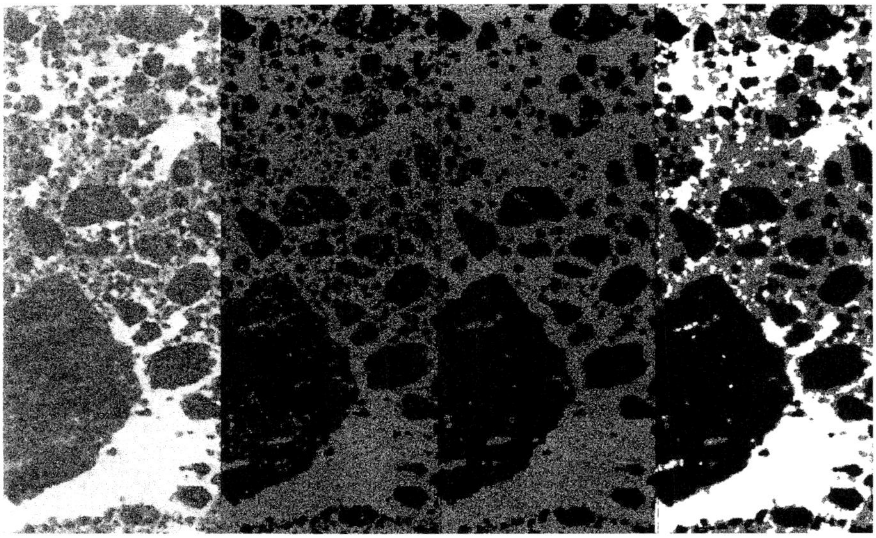

**Fig. 7.** Part of an ERS-1 SAR sea ice image in process for ice coverage. From left to right: **a** The original image (copyright ESA), **b** the segmented image, **c** the floe-extracted image, **d** the resultant image from the water/ice mixture extraction process. *Bright regions* denote water; *gray regions* denote water/ice mixture; *dark regions* denote floes

**4. Floe Filtering.** This stage filters out features that have low confidence as floes. For example, ice conglomerates detected due to extremely poor image quality will not be considered in the floe size distribution calculation. The filter first computes geometric descriptions of each feature in the floe class and eliminates irregular features from the set of valid floes. The irregularity of a feature $f$, $I_s$, is defined as $p_f / a_f$, where $p_f$ is the length of the perimeter of feature $f$, and $a_f$ is the area of feature $f$. If the ratio is high, that means the feature structurally consists of elongated, linear branches, and thus the feature is not a floe. If the ratio is low, that means the feature is compact and thus valid as a floe.

**5. Extraction of Water/Ice Mixture from Non-Floe Regions.** To extract water/ice mixture from nonfloe regions, we have designed a Laplacian-based local dynamic thresholding technique – a variant of Laplacian filtering thresholding (Weszka et al. 1974) and gradient thresholding (Watanabe et al. 1974). Floe regions are masked from further processing. Next, for each pixel $(i,j)$, its Laplacian value (Marr and Hildreth 1980) is computed as

$$L(i,j) = 4 \cdot g(i,j) - g(i-1,j) - g(i+1,j) - g(i,j-1) - g(i,j+1),$$

where $L(i,j)$ is the Laplacian value, and $g(i,j)$ is the intensity level of pixel $(i,j)$. Third, Laplacian histograms of local regions are generated and only histograms with high "edginess" value qualify for further processing.

"Edginess" is defined as follows: Let $H$ be a Laplacian histogram where the bin of the histogram is $b_k$, and $k$ ranges from 0 to $N$–1, $N$ being the number of intensity levels of the image, and where the frequency of each bin $b_k$ is $f(b_k)$. Thus, the histogram of a region $m$, $H_m$, is defined as having its

$$f_m(b_k) = \sum_{(i,j)\in image} \lfloor (L(i,j)) \cdot Count(g(i,j), b_k) \rfloor,$$

where *Count* is a function that returns 1 when its two arguments are equal and 0 otherwise. The edginess of a histogram $H_m$ is defined as

$$E(H_m) = \frac{\sum_{k=0}^{N-2} \left| f_m(b_k) - f_m(b_{k+1}) \right|}{\sum_{k=0}^{N-1} f_m(b_k)} .$$

If a region is homogeneous, then its frequencies of Laplacian values will be close and thus its $E(H_m)$ will be low. On the other hand, if the region is edgy, its $E(H_m)$ will be high.

For each qualified histogram, a threshold is computed by selecting the intensity level that maximizes the accumulative probability that its location is in the middle of an edge transition. This is accomplished by selecting the intensity value $b_k$ such that $f_m(b_k)$ has the largest value among all $f_m(b_i)$ for $i$ ranging from 0 to $N$-1 in the histogram $H_m$. Region and point interpolations are executed to assign a threshold value to each pixel. Finally, a binary decision is performed to segment the processed region into water/ice mixture and water. Essentially, this extraction stage has the same process as the segmentation phase, but instead of using Gaussian approximation to find the optimal threshold, a Laplacian-based edginess measure is used. Figure 7d shows the results of the water/ice mixture extraction phase.

**6. Floe Size Distribution and Coverage Statistics.** After the SAR image is segmented, a program is used to generate two types of numerical statistics: (1) distribution of designated floe size bins of the image, by grouping similar floe sizes into the same bin, and (2) ice coverage – distributions of floes, water/ice mixture and water.

Figure 8 displays a floe size distribution and ice coverage image. The original image is an ERS-1 SAR sea ice image taken in August of 1991 at 74.43°N, 144.29°W. Table 1 shows the distribution of designated floe size bins for this image. The ice coverage percentages are floes 42.05%, water/ice mixture 40.43%, and water 17.52%.

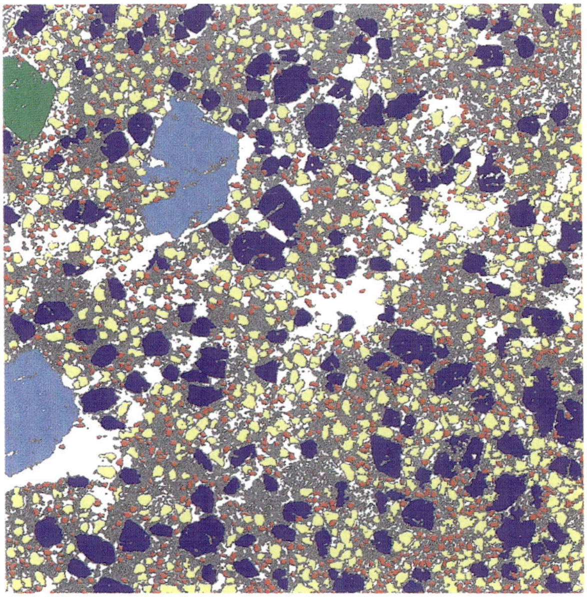

**Fig. 8.** Ice floe size distribution. *Red floes* are floes within bin 0 (see Table 1): each with an area of less than 100 pixels; *yellow floes* are floes within bin 1; *dark blue floes* are floes within bin 2; *green floes* are floes within bin 3; *light blue floes* are floes within bin 4, in which each floe has a size between 12000 and 30000 pixels. Note that there are no floes in bin 5 (floes larger than 30000 pixels)

**Table 1.** Floe size distribution in numerical values for Fig. 8

| Bin | Size (Pixels) | No. of Floes | Area Occupied (%) |
|---|---|---|---|
| 0 | $x \leq 100$ | 1365 | 6.11 |
| 1 | $100 < x \leq 500$ | 685 | 14.29 |
| 2 | $500 < x \leq 5000$ | 141 | 16.58 |
| 3 | $5000 < x \leq 12000$ | 1 | 0.94 |
| 4 | $12000 < x \leq 30000$ | 2 | 4.10 |
| 5 | $30000 < x$ | 0 | 0.00 |

To interpret: For bin 0, which includes all floes with size less than 100 pixels, there are 1365 floes. The total area occupied by the floes in the 976 × 1024 image is 6.11%.

## 2.5 Preliminary Results of an Arctic Summer Time Series Study

In this section we describe initial results from an in-depth study of the floe size algorithm applied to two time series of ERS-1 SAR images during the summer of 1992 in the Beaufort and Chukchi Seas (Figure 9). The thrust of the overall study was to determine whether the floe size distribution varied from region to region and how the distribution evolved during a summer melt period. For example, Vinje (1977) found, using Landsat imagery to measure only those floes greater than 10 x 10 km passing through the Fram Strait, that there was a decrease in the number of both small and large floes during early summer. Here we focus on how the ice characteristics and varying environmental conditions affect the radar backscatter and hence performance of the algorithm.

The Arctic ice cover melts during the summer due to short-wave radiation being absorbed by the ice and the open water between floes. Heat absorption by the wet ice and melt ponds results in thinning of the ice cover. Similarly, absorption by open water causes horizontal or lateral melt of ice floes (Maykut and Perovich 1987). Coupled with melt, ice decay may be increased by mechanical forcing that breaks up the ice cover into smaller pieces which melt rapidly. Using limited field and remote sensing measurements of floe size (Rothrock and Thorndike 1984; Hall and Rothrock 1987), models of ice and of the upper ocean (Maykut and Perovich 1987; Steele 1992) have shown that lateral melt (and therefore ocean heating) is higher in areas where there are many small (less than 200 m) floes than in an equivalent ice/water concentration composed only of large floes. Thus, descriptions of floe size distribution by season and region may provide a more complete picture of the upper ocean summer heat budget from the central pack to the marginal ice zone.

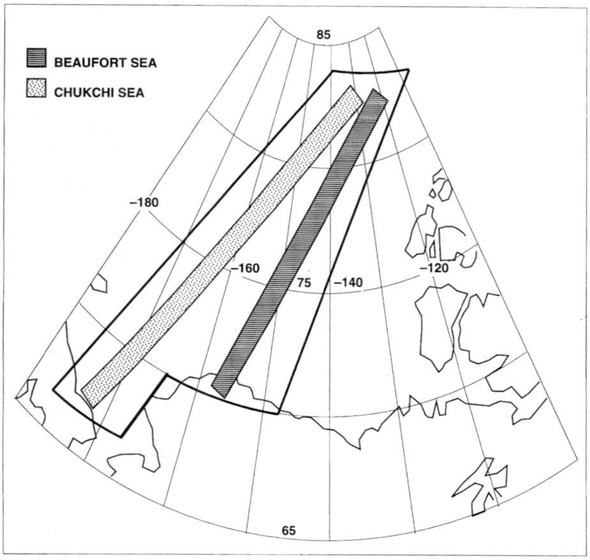


**Fig. 9.** General ERS-1 SAR coverage for in-depth study of floe size distribution for the summer of 1992. Analyzed data were centered around the Beaufort and Chukchi Sea tracks shown

**Table 2.** Floe bin size used in time series study

| No. | No. of Pixels | Maximum area (km²) | $L$(m) |
|---|---|---|---|
| 1 | 1 | 0.01 | 123 |
| 2 | 2 | 0.02 | 174 |
| 3 | 3–4 | 0.04 | 256 |
| 4 | 5–8 | 0.08 | 348 |
| 5 | 9–16 | 0.16 | 492 |
| 6 | 17–64 | 0.64 | 985 |
| 7 | 65–256 | 2.56 | 1969 |
| 8 | 257–1600 | 16.0 | 4924 |
| 9 | 1601–6400 | 64.0 | 9847 |
| 10 | >6400 | >64.0 | >9847 |

From Rothrock and Thorndike (1984). Area = $axL^2$, where $a$ = 0.66 and $L$ = mean diameter.

The algorithm procedures described in the previous sections were applied to ERS-1 C-band SAR imagery sampled down to about 225 m resolution (100 m pixel size) from the finer resolution (25 m) original data. The output was separated into four classes as described above: measurable floes, floes originally characterized by the algorithm as floes but considered too branchy (labeled discarded), open water, and the remainder of the ice cover (water/ice mixture). The floe sizes or area in square kilometers are broken into ten bins, listed in Table 2. We use the relation of floe mean diameter to floe area determined by Rothrock and Thorndike (1984), area = $a \sum L^2$, where $a$ = 0.66 and $L$ = mean diameter. We have tradedoff finer resolution for smaller image files (64 megabytes reduced to 1 megabyte files) with reduced radar speckle. Lateral melt rates are as high as 10 cm per day (Hall and Rothrock 1987), which would not have been possible to measure with this data set even at the finest resolution. Even though the lower resolution prevented the measuring of floes smaller than 100 m, accurate measurements of water/ice mixture extent would in theory provide some estimate of small floe distributions and we could still determine changes in the distribution within the larger size floes.

Before we discuss example results, we first review the extremely variable seasonal characteristics of radar backscatter over sea ice that have been derived from ERS-1 imagery, as these variations have important effects on the floe results. In the winter months, multiyear ice has a stable and highly contrasting (2–4 dB) signature compared to first-year ice (Kwok and Cunningham 1994; Fetterer et al. 1994). Early stages of ice growth have a wide range of returns which often overlap both multiyear and first-year ice (Steffen and Heinrichs 1994). During the onset of melt and continuing into the summer, the presence of liquid water both in the snow cover and in the resulting surface melt ponds that form dramatically reduces (by 4 dB or more) the signatures in both multiyear and first-year ice, so that during the summer months these two major forms of ice are virtually indistinguishable (Onstott 1992; Winebrenner et al. 1994). Through late summer, ablation continues in both the snow and ice; however, much of the surface water drains from the ice. As air temperatures drop below freezing levels, the remaining ice quickly cools and the returns increase rather quickly to winter-like levels (Winebrenner et al. 1996). Taken together the results of these studies cover nearly an entire

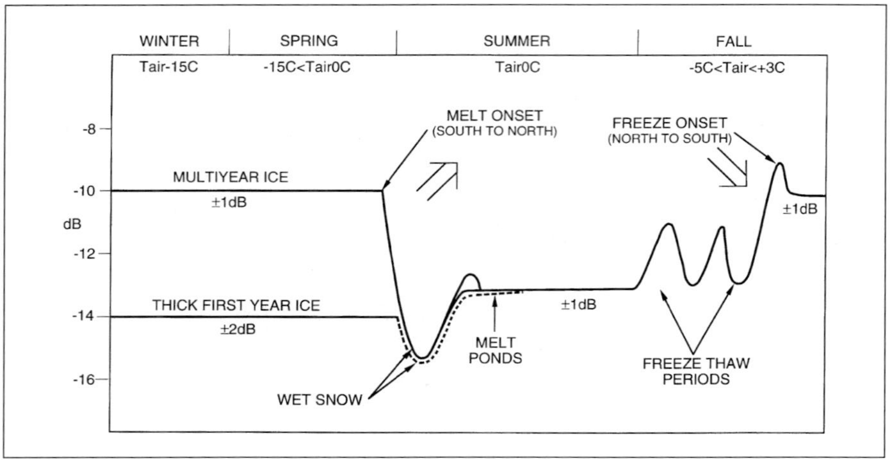


**Fig. 10.** Generalized seasonal ERS-1 C-band VV SAR backscatter of multiyear and first year ice (compiled from Kwok and Cunningham 1994 and Winebrenner et al. 1994, 1996). The approximate ranges of backscatter variations and seasonal air temperatures are indicated.

season and are summarized in Fig. 10. Also indicated in this figure are that the onset of melt and freezing proceed latitudinally over time, with melt occurring earliest in the lower latitudes and freezing in the higher latitudes. Lastly, throughout the year, open water returns are sensitive to wind speed and have a wide range of returns (-20 to -5 dB) that often overlaps all levels of ice signatures. During the summer, however, the contrast between ice (less than -12 dB or lower) and open water (> -8 dB) on windy days (wind speed greater than 5 m/s ) is at its maximum as compared to other seasons.

The best quality floe algorithm results are from SAR images where the floes are dark, the open water and/or water/ice mixture is brighter by several dB, and the floes are visually distinguishable. Two examples are shown in Figs. 11 and 12 (June 18 and June 30, 1992), from comparatively low- and high-latitude regions, respectively. In both figures, the perimeters of the floes in all size ranges are accurately mapped. Note also that the algorithm tends to decompose what appears to be a single large floe into the several smaller floes that actually comprise the larger floe. For the larger floes, it can be seen that large bright linear features contained within the floes are isolated and result in this apparent fragmenting. The long, narrow bright features are segmented into the open water class, but in these cases are most certainly ridges and not open water leads. In this simple segmenting scheme, open water and ridges cannot be separated most of the time, thus making the open water concentration results unreliable. Figure 12 also includes some floes in the discarded class. As can be seen in this figure and later figures, the discarded class are primarily composed of larger floes that appear to touch other surrounding floes. The resulting shape is considered to be too branchy and are thus removed from floe size distribution. This branchiness appears to affect larger rather than smaller floes and thus results in probable undersampling of larger floes.

Examining Figs. 11 and 12 further, it can be seen that smooth gray ice with no identifiable features occupies the space between the darker rounder floes. This is most like-

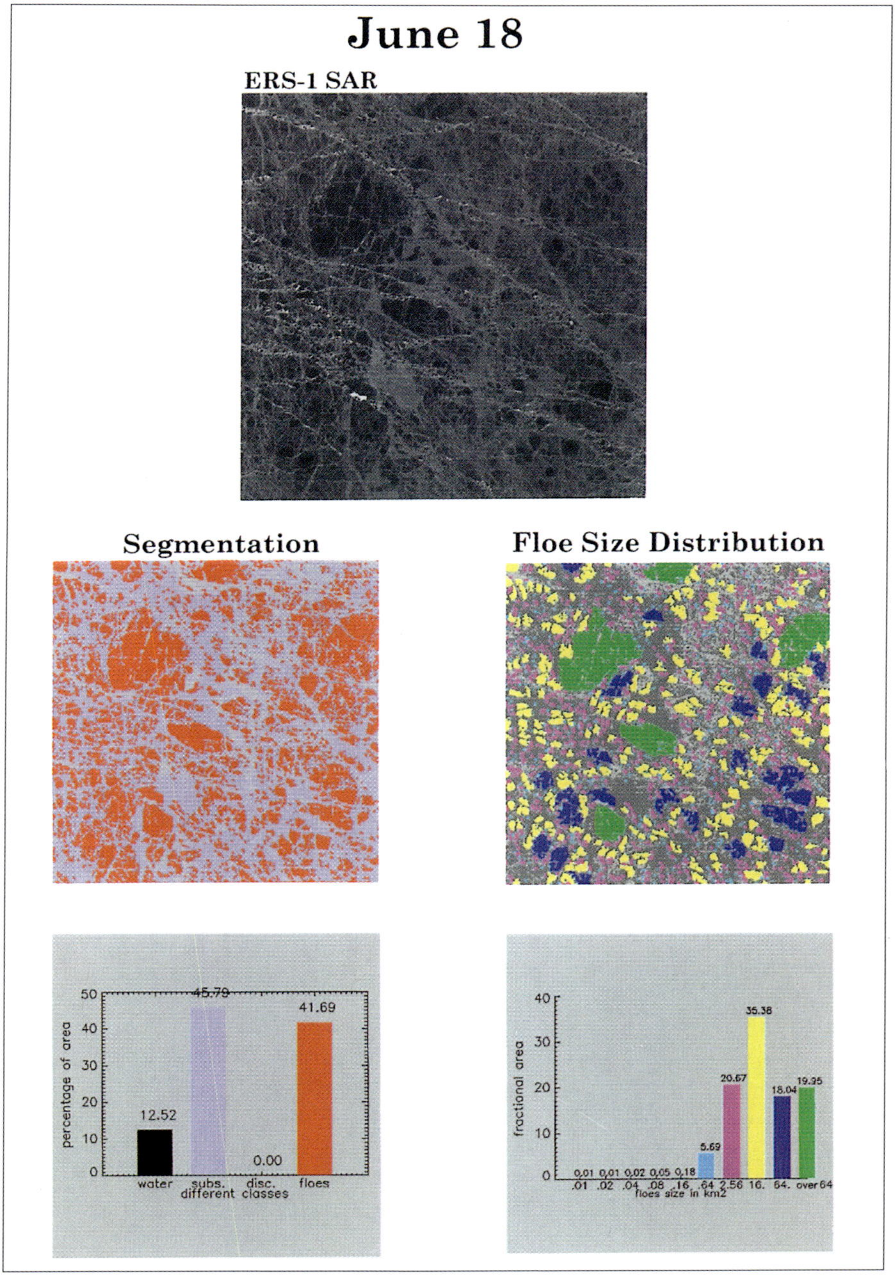


**Fig. 11.** **a** ERS-1 SAR image from June 18, 1992, 76.2°N, 160.5°W (copyright ESA). **b** Segmented image and histogram of measured floes, discarded floes, open water, and substrate. Note that the open water class in the image is *white* and its associated histogram bar *black*.. **c** Image of floe size distribution and size histogram

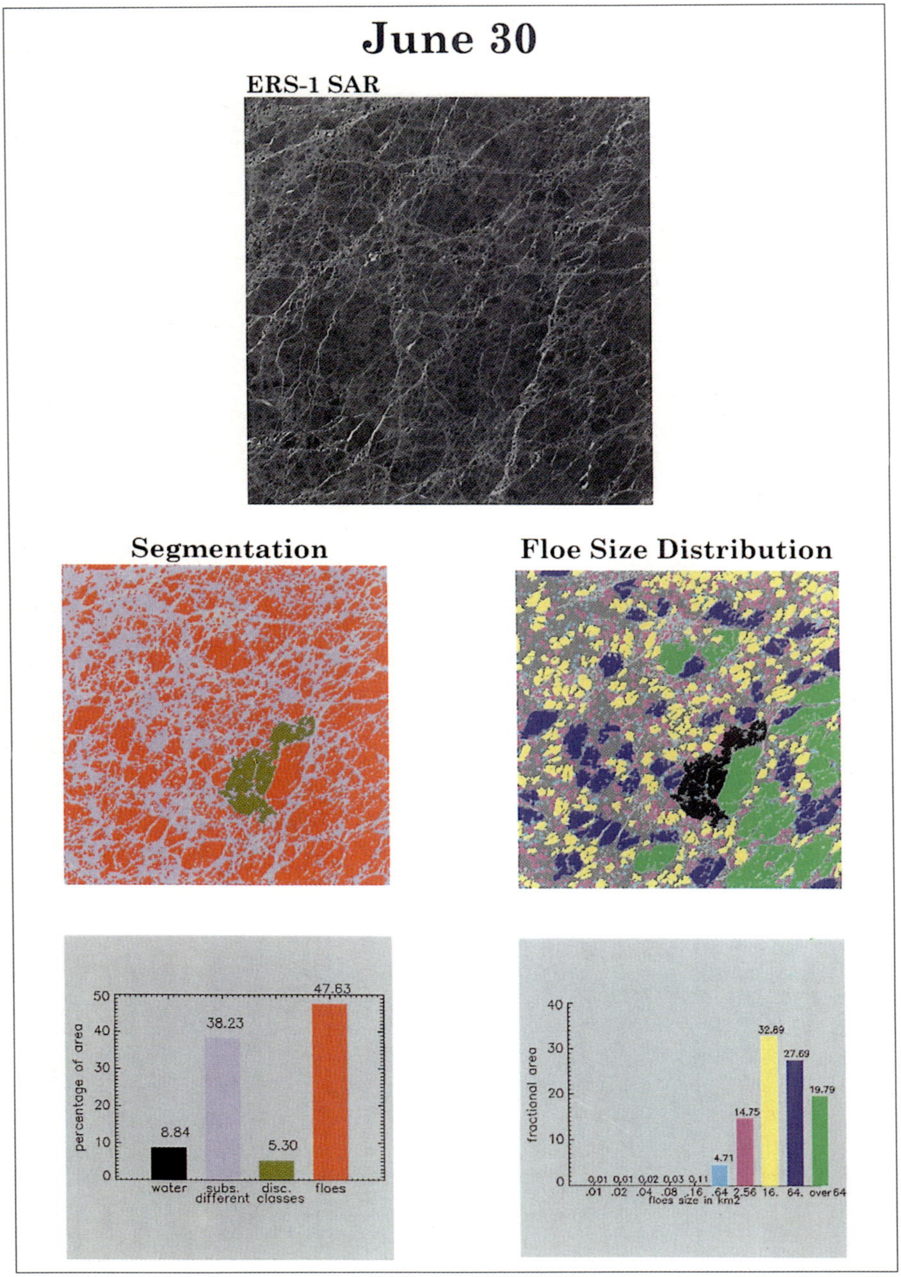


**Fig. 12.** **a** ERS-1 SAR image from June 30, 1992, 82.6°N, 138.4°W (copyright ESA). **b** Segmented image and histogram of measured floes, discarded floes, open water, and substrate. Note that the open water class in the image is *white* and its associated histogram bar *black*. **c** Image of floe size distribution and size histogram

ly first-year ice and is placed by segmentation into the water/ice mixture category. While this may seem confusing or incorrect, it should be borne in mind that this ice is most likely to be thinner than the floes and melt away during the summer months, leaving the thicker, older, more rounded floes intact at the end of the summer. Thus, for comparison purposes with end of summer data, it seems satisfactory that this brighter, nonrounded ice falls into the water/ice mixture rather than floe category. Note that both dates are shortly after melt has begun in the respective latitudes, which was measured by Winebrenner et al. (1994) to have occurred between June 13 and June 20 in 1992.

Figure 13 shows two four-image segments in the high Arctic which cover nearly the same area, one (Fig. 13 a,b) from June 30 (which includes the same data as Fig. 12) and

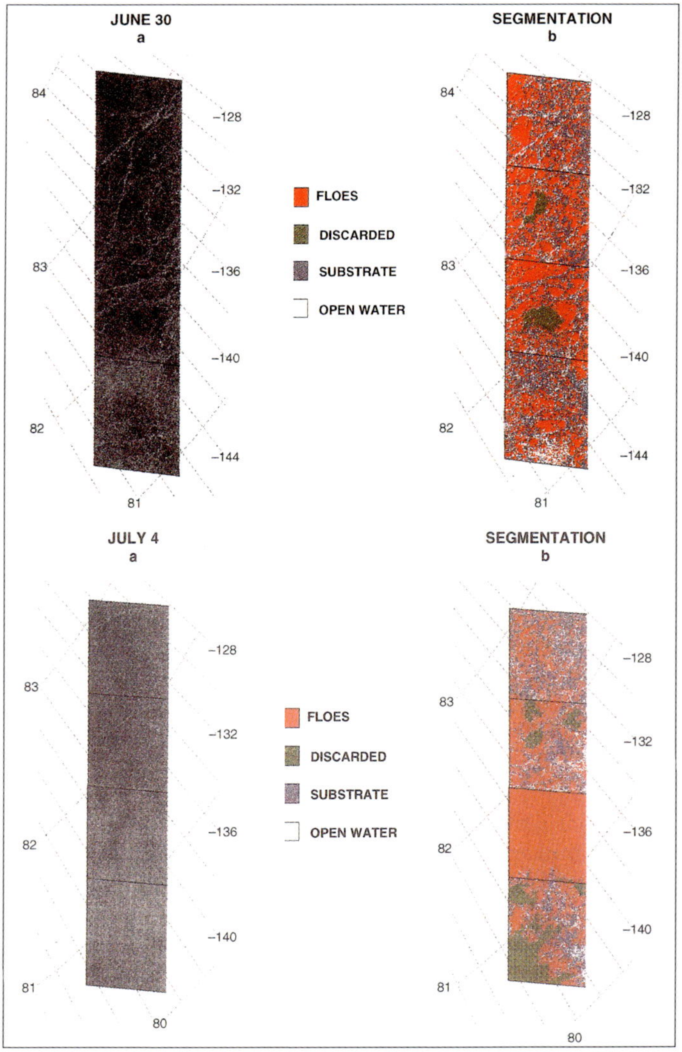


**Fig. 13.** **a** Sequence of four ERS-1 SAR images from June 30, 1992, centered at 82°N, 140°W (copyright ESA). **b** Output display of floes, discarded floes, open water and substrate for **a**. **c** Sequence of four ERS-1 SAR images from July 4, 1992, centered at 82°N, 138°N (copyright ESA). **d** Output display of floes, discarded floes, open water, and substrate for **c**

another data set (Fig. 13 c,d) obtained on July 4. In Fig. 13a, the floes are primarily dark separated by bright ridges, except that there is a small circular region on the border between frames 3 and 4 that is brighter. The floe results in Fig. 13b are largely acceptable, noting that within the brighter region there is a reduction of measurable floes and increased water/ice mixture. On July 4, Fig. 13c shows that the brightened area has extended northward and the floes in frames 1 and 2 are also less dark and distinct. In the brightened areas, there has been an almost complete loss of floe discrimination. We attribute this large-scale change in backscatter in Fig. 13c to one or a combination of these possible explanations: (1) There was an overall cooling trend which has caused refreezing of liquid water in the snow and ice. This could be augmented by the refreezing of water that has surrounded large snow and ice grains, which may form a collection of nodules at the ice/snow interface (Onstott and Gogineni 1985). This explanation is probably not correct since the process was found to occur before extensive melt. (2) A cloud cover is present containing ice particles, which would contribute to the overall scattering levels and reduce the detectability of ice surface features. This has been noted in several ERS-1 investigations and is attributed to the sensitivity of C-band frequency to atmospheric conditions (Johannessen et al. 1994). (3) Massive drainage of melt ponds and surface water has occurred, leaving behind more porous surface ice grains, which would increase scattering (Holt and Digby, 1985). In Fig. 4c of Winebrenner et al. (1994), the buoy closest to these images shows steadily increasing air temperatures above 0 °C after June 18 (melt onset and a steep decrease in backscatter from about -9 dB to -18 dB) and a sharp upturn in backscatter of +4 dB after June 25.

The resulting floe size measurements are dramatically different on July 4 (Fig. 13d) as compared to June 30 (Fig. 13b). Fewer floes are measured in frames 1 and 2, frame 3 indicates one very large floe, and frame 4 shows a mixture of floes, discarded floes, and water/ice mixture. The results of frame 3 are clearly erroneous and are due to the comparatively uniform, featureless backscatter returns. The differing results cannot be attributed to a different field of ice moving into view of the sensor since there are overlapping features between the two dates. Summarizing, large and rapid variations in the ice backscatter in response to changing ice surface and environmental conditions limit the capability of the algorithm to produce meaningful results.

A late summer image and results are seen in Fig. 14 (August 26, 1992). Note in the SAR image the bright open water along one edge. Compared with Figs. 11, 13, the image and output are characterized by fewer large floes, with an increased relative fraction of smaller floes and water/ice mixture as well. This follows the intuitive trend over the summer season of floes becoming smaller with melting, which is also reflected by the relative increase in water/ice mixture or subpixel size floes. Also, the bright open water provides good contrast with the surrounding ice and in this case is also segmented quite accurately.

The algorithm has particular difficulty in at least two situations, when the radar signature of the ice pack is bright, and when analyzing imagery from the marginal ice zone. When temperatures drop steadily below freezing, which occurred during August 29–September 7, 1992, in the Beaufort Sea (Winebrenner et al. 1996), multiyear ice becomes brighter than both first year ice and new ice. In general, results during cold conditions are not satisfactory, even using the 'cold weather' switch in the algorithm, with the floes being largely undersampled. This is likely due to the reduced overall contrast between the multiyear ice and the other ice types, which means that the segmen-

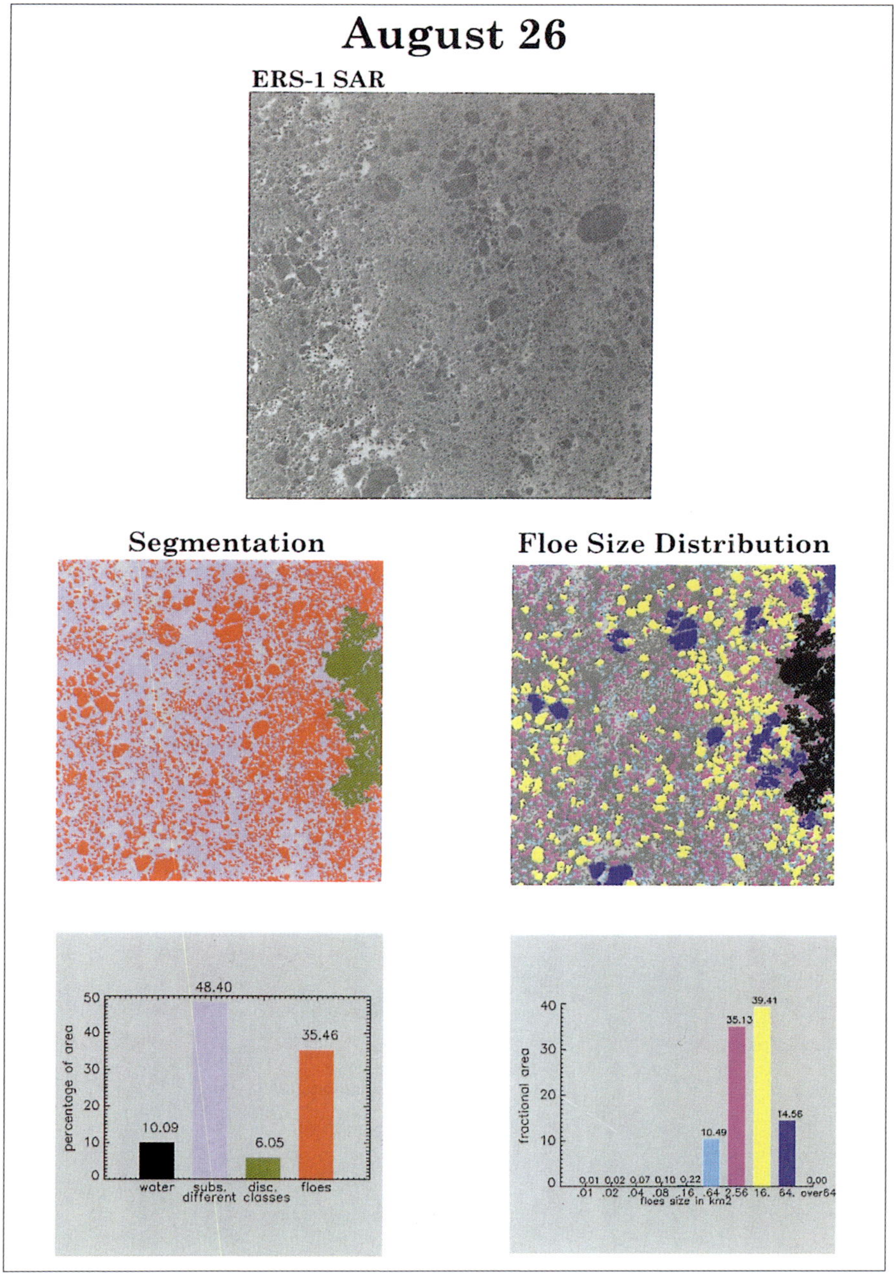


**Fig. 14.** **a** ERS-1 SAR image from August 26, 1992, at 78.2°N, 163.6°W (copyright ESA). **b** Segmented image and histogram of measured floes, discarded floes, open water, and substrate. Note that the open water class in the image is *white* and its associated histogram bar *black*. **c** Image of floe size distribution and size histogram

tation component has difficulty determining the proper categories of each primary grouping. This includes ridges (which are brighter than multiyear ice) often being improperly segmented as open water, and multiyear ice falling into the water/ice mixture grouping.

Within the outermost marginal zones, extreme variations in backscatter from both the ice and open ocean, often within the same image and across short distances, can wreak havoc on accurate segmentation. The ocean variations arise not only from varying wind speeds but from mesoscale eddies, atmospheric boundary layer effects such as windrows, and strong variations in sea surface temperatures. These variations result in a comparatively limited percentage of useful output, especially compared to the optimal conditions of wet-dark floes and windy open water within the central pack.

The floe algorithm is also sensitive to the backscatter fall-off at increasing incidence angles seen in both ice and ocean returns. The ERS-1 imaging beam extends over a range of incidence angles from 20° to 26°. Ocean signatures are strongly dependent on incidence angle, as are ice signatures, although less so. Therefore both the absolute level and the contrast between ice and ocean decrease at higher incidence angles, reducing the ability to isolate floes, and this results in increased apparent branching. The net result is that there are high number of discarded floes at far range, i.e., floes that were originally determined to be floes but were then found to be too branchy. This effect could probably be reduced by increasing the overall backscatter level at higher angles, however, such a boost only at higher angles adds a processing step and would be applied in a nonuniform (or nonlinear) fashion to the calibrated data.

Many of both the positive and the problematic conditions discussed above are shown in Fig. 15, which is a single data take from July 29, 1992. Looking first at the output from the entire data take, the following problematic conditions are readily seen: the discarded class shows up preferentially at the far range (left side) of the descending pass; frames

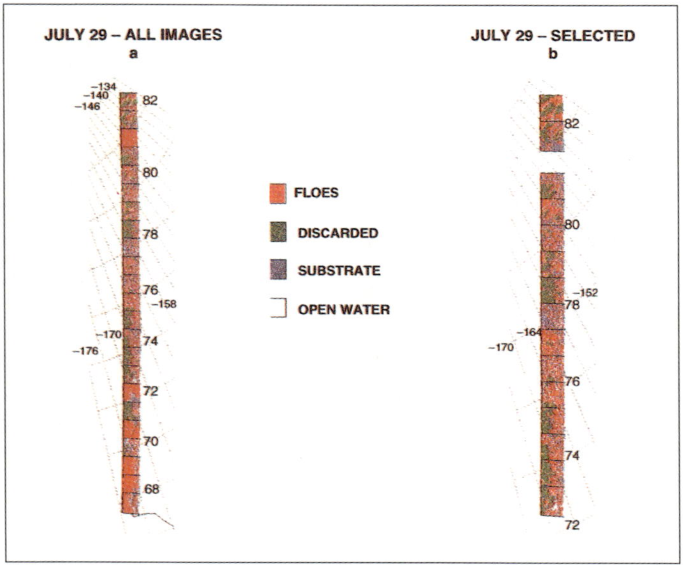


**Fig. 15.**
**a** Sequential segmented output from ERS-1 SAR images from July 29, 1992. **b** Final selection from **a** after frames with unacceptable measured floe results are eliminated

in the marginal ice zone (72°N and southward) have extremely variable and inaccurate segmentation due to the highly variable ice-ocean conditions and resulting backscatter; within the central pack, backscatter variations are localized and strongly affect segmentation. For the latter condition, note the solidly red third frame from the top in Fig. 15a, where the SAR image is basically featureless, and the frame centered at 78°N, where there is a large area of substrate visible, and where the SAR image is comparatively bright and more winter-like than surrounding images. For this data take, the solid red frame was not included in the final selection, but the frame near 78°N was included, since it was felt that the measured floes were accurately segmented. All the frames from 72°N and southward were eliminated from further analysis.

To briefly summarize these preliminary results, the floe algorithm generates good quality accurate measurements of thicker floes of all sizes during the summer when the ice is dark from wetness, the open water is bright, and the floes are readily identifiable to the human eye. Environmental factors which alter the ice and ocean backscatter away from this condition make it problematic for the algorithm to accurately segment the ice/ocean cover.

## 2.6 Summary and Conclusions

We have presented the restricted growing concept algorithm that establishes separateness among objects while preserving their original sizes and shapes. We have also described an implementation based on probabilistic labeling and mathematical morphology, aimed at and tested successfully on synthetic aperture radar (SAR) sea ice imagery. Our implementation is (1) extensible, since it can include any number of classes of objects in an image, (2) flexible, since it can be altered such that the roles of object and background pixels can be switched to focus the user's interest on a certain class of pixels, (3) economical, since it does not sacrifice small ice floes for separateness, (4) fully automated except for the requirement of the user to specify the object class, and (5) computationally efficient.

Moreover, we have designed a technique for computing ice floe size distribution and ice coverage statistics. This technique is able to segment a SAR sea ice image into floes and non-floes, it is able to achieve floe identification and separation using our implementation of the restricted growing concept, it is able to analyze and discard irregular floes, and it is able to distinguish zones of subresolution ice from open water regions. We have shown preliminary results from a study of summer floe size distribution in the Beaufort and Chukchi Seas and how the floe results are strongly dependent on ice and environmental conditions during this dynamic season.

**Acknowledgments:** All the authors acknowledge the support of Robert Thomas, Manager of Polar Programs (Code YSG) at NASA. The work by L.-K. Soh and C. Tsatsoulis was funded in part by NASA grant NAGW-3043. Soh and Tsatsoulis would like to acknowledge the assistance provided by Donna Haverkamp in developing the floe filtering procedure. Holt performed this work at the Jet Propulsion Laboratory, California Institute of Technology, under contract to NASA. Holt would also like to acknowledge the excellent programming support of Remi Roques and Marie-Helene Rio, both graduate students from the Ecole Nationale Supérieure de l' Aéronautique et de l' Espace in Toulouse, France, who performed their end-of-year study projects at JPL.

## References

Banfield J (1991) Automatic tracking of ice floes: a stochastic approach. IEEE Trans Geosci Remote Sens 29: 905–911
Banfield J, Raftery AE (1992) Ice floe identification in satellite images using mathematical morphology and clustering about principal curves. J Am Stat Assoc 87, no 417: 7–16
Burns BA, Cavalieri DJ, Keller MR, Campbell WJ, Grenfell TC, Maykut GA, Gloersen P (1987) Multisensor comparison of ice concentration estimates in the marginal ice zone. J Geophys Res 92: 6843–6856
Campbell WJ, Gloersen P, Josberger EG, Johannessen PS, Guest PS, Mognard N, Shuchman R, Burns BA, Lannelongue N, Davidson KL (1987) Variations of mesoscale and large scale sea ice morphology in the 1984 marginal ice zone experiment as observed by microwave remote sensing. J Geophys Res 92: 6805–6824
Chou J, Weger RC, Ligtenberg JM, Kuo K-S, Welch RM, Breeden P (1994) Segmentation of polar scenes using multi-spectral texture measures and morphological filtering. Int J Remote Sens 15, no 5: 1019–1036
Daida J, Vesecky J (1989) Object-based feature-tracking algorithms for SAR images of the marginal ice zone. Proc Int Geosci Remote Sens Symp IGARSS'89, July 10–14, Vancouver, British Columbia, pp 1121–1124
Ejiri M, Uno T, Mese M, Ikeda S (1973) A process for detecting defects in complicated patterns. Comput Graphics Image Proc 2: 326–339
Fetterer FM, Gineris D, Kwok R (1994) Sea ice type maps from alaska synthetic aperture radar facility imagery: an assessment. J Geophys Res 99: 22443–22458
Fily M, Rothrock DA (1986) Extracting sea ice data from satellite SAR imagery. IEEE Trans Geosci Remote Sens 24: 849–854
Fily M, Rothrock DA (1987) Sea ice tracking by nested correlations. IEEE Trans Geosci Remote Sens 25: 570–580
Fily M, Rothrock DA (1990) Opening and closing of sea ice leads: digital measurements from synthetic aperture radar. J Geophys Res 95, no C1: 789–796
Geman D, Geman S, Grafigne C, Dong P (1990) Boundary detection by constrained optimization. IEEE Trans Pattern Anal Machine Intelligence 12, no 7: 609–628
Gonzalez RC, Woods RE (1992) Digital image processing. Addison-Wesley, Reading, Massachusetts
Hall, EL, Kruger RP, Dwyer III SJ, Hall DL, McLaren RW, Lodwick GS (1971) A survey of preprocessing and feature extraction techniques for radiographic images. IEEE Trans Comput 20, no 9: 1032–1044
Hall RT, Rothrock DA (1981) Sea ice displacement from SEASAT synthetic aperture radar. J Geophys Res 86, no C11: 11078–11082
Hall RT, Rothrock DA (1987) Photogrammetric observations of the lateral melt of sea ice floes. J Geophys Res 92, no C7: 7045–7048
Hastie T, Stuetzle WX (1989) Principal curves. J Am Stat Assoc 84: 502–516
Haverkamp D, Soh L-K, Tsatsoulis C (1995) A comprehensive, automated approach to determining sea ice thickness from SAR data. IEEE Trans Geosci Remote Sens 33, no 1: 46–57
Holmes PD, Jungert ERA (1992) Symbolic and geometric connectivity graph methods for route planning in digital maps. IEEE Trans Pattern Anal Machine Intelligence 14, no 5: 549–565

Holt B, Digby SA (1985) Processes and imagery on fast first-year sea ice during the melt season. J Geophys Res 90, no C3: 5035–5044
Horowitz SL, Pavlidis T (1976) Picture segmentation by a tree traversal. J Assoc Comput Mach 23: 368–388
Huang TS, Yang GJ, Tang GY (1979) A fast two-dimensional median filtering algorithm. IEEE Trans Acoust Speech Signal Process 27:13–18
Hummel RA, Zucker SW (1983) On the foundations of relaxation labeling processes. IEEE Trans Pattern Anal Machine Intelligence 5: 267–287
Johannessen JA, Digranes G, Espedal H, Johannessen OM, Samuel P, Browne D, Vachon P (1994) SAR ocean feature catalogue. European Space Agency Report SP-1174
Korsnes R (1993) Quantitative analysis of sea ice remote sensing imagery. Int J Remote Sens 14: 295–311
Kwok R, Curlander JC, McConnell R, Pang SS (1991) An ice-motion tracking system at the Alaska SAR facility. IEEE J Ocean Eng 15, no 1: 44–54
Kwok R, Cunningham GF (1994) Backscatter characteristics of the winter ice cover in the Beaufort Sea. J Geophys Res 99: 7787–7802
Marr D, Hildreth E (1980) Theory of edge detection. Proc R Soc Lond B 207: 187–217
Matheron G (1975) Random sets and integral geometry. Wiley, New York
Maykut GA, Perovich DK (1987) The role of shortwave radiation in the summer decay of a sea ice cover. J Geophys Res 92, no C7: 7032–7044
McConnell R, Kwok R, Curlander JC, Kober W, Pang SS (1991) Y-S Correlation and dynamic time warping: two methods for tracking ice floes in SAR imagery. IEEE Trans Geosci Remote Sens 29, no 6: 1004–1012
Onstott RG, Gogineni SP (1985) Active microwave measurements of arctic sea ice under summer conditions. J Geophys Res 90, no C3: 5035–5044
Onstott RG (1992) SAR and scatterometry signatures of sea ice. In: Carsey F (ed) Microwave remote sensing of sea ice. American Geophysical Union, Washington, DC, pp 73–104, Geophysical monograph series vol 68
Rosenfeld A, Kak AC (1982) Digital image processing, 2nd edn. Academic Press, New York
Rosenfeld A, Smith RC (1981) Thresholding using relaxation. IEEE Trans Pattern Anal Machine Intell 3: 598–606
Rothrock DA, Thorndike AS (1984) Measuring the sea ice floe size distribution. J Geophys Res 89, no C4: 6477–6486
Serra J (1982) Image analysis and mathematical morphology. Academic Press, New York
Steele M (1992) Sea ice melting and floe geometry in a simple ice-ocean model. J Geophys Res 97, no C11: 17729–17738
Steffen K, Heinrichs J (1994) Feasibility of sea ice typing with synthetic aperture radar (SAR): merging of Landsat thematic mapper and ERS-1 SAR satellite imagery. J Geophys Res 99: 22413–22424
Terzopoulos D (1986) Image analysis using multigrid relaxation methods. IEEE Trans Pattern Anal Machine Intelligence 8: 129–139
Ulaby FT, Moore RK, Fung AK (1982) Microwave remote sensing: active and passive, Artech House, Norwood, Massachusetts
Vesecky JF, Samadani R, Smith MP, Daida JM, Bracewell RN (1988) Observation of sea-ice dynamics using synthetic aperture radar images: automated analysis. IEEE Trans Geosci Remote Sens 26, no 1: 38–48

Vinje TE (1977) Sea ice studies in the Spitsbergen-Greenland Area. NTIS Landsat Report E77-10206

Watanabe S, CYBEST Group (1974) An automated apparatus for cancer prescreening: CYBEST. Comput Graphics Image Proc 3: 350–358

Weszka JS, Nagel RN, Rosenfeld A (1974) A threshold selection technique. IEEE Trans Comput 23: 1322–1326

Winebrenner DP, Nelson ED, Colony R, West RD (1994) Observation of melt onset on multiyear arctic sea ice using the ERS-1 synthetic aperture radar. J Geophys Res 99: 22425–22441

Winebrenner DP, Holt B, Nelson ED (1996) Observations of autumn freeze-up in the Beaufort and Chukchi seas using the ERS-1 synthetic aperture radar. J Geophys Res 101, no C7: 16401–16419

Zhang MC, Haralick RM, Campbell JB (1990) Multispectral image context classification using stochastic relaxation. IEEE Trans Syst Man Cybern 20: 128–140

Zhang TY, Suen CY (1984) A fast parallel algorithm for thinning digital patterns. Commun Assoc Comput Mach 27, no. 3: 236–239

CHAPTER 3

# Role of SAR in Surface Energy Flux Measurements Over Sea Ice

D.G. Barber, A. Thomas, and T.N. Papakyriakou

## Contents



## 3.1 Introduction

The ocean-sea ice-atmosphere interface (hereafter referred to as the marine cryosphere) is a major component of the high-latitude Earth system. Its presence or absence over the polar oceans is an important factor in the overall heat exchange at high latitudes (e.g. Maykut 1978) and is generally accepted as an important component in the climate system because of its potential to amplify small changes through a variety of feedback mechanisms. Energy flows through the marine cryosphere as a function of thermodynamic and dynamic processes, both of which affect, and in turn are affected by, the physical properties of the volume. This results in an integrated system whereby hydrospheric and atmospheric forcings are linked with changes in the geophysical and electrical characteristics of the snow-covered sea ice volume.

Analysis of SAR Data of the Polar Oceans
Edited by C. Tsatsoulis and R. Kwok

In recent years, Earth system scientists have begun to use remotely sensed data to study the planet through complex interactive units. With the advent of the new era of operational microwave remote-sensing sensors (e.g. ERS-1, ERS-2, JERS-1, RADARSAT, SSM/I), it is now possible to obtain a systematic dataset of the marine cryosphere without concern for cloud cover or solar illumination. It is important that an understanding of the relationships between microwave interaction and climate state variables be developed so that fields of radiative and conductive flux conditions may be generated for subsequent use in numerical process models. A basic tenet of this type of observational system is that linkages between the energy regime of the volume and its electrical characteristics may be exploited as a means of estimating the former by the way in which microwaves interact with the latter.

The main problem we face in studying processes operating within the marine cryosphere is the fact that the surface and volume are constantly changing. Ice ablation and accretion are directly related to the variable atmospheric and hydrospheric forcing created by global general circulation patterns. Mass fluxes also affect ice growth and ablation through deposition of a layer which has low conductivity for both thermal and radiative energy. That component of the energy which is transmitted across the upper and lower interfaces is available within the volume to change the physical and electrical characteristics, which then cause changes in the flux proportions being reflected, transmitted or absorbed. This feedback mechanism creates a fundamental link between the physical properties of the marine cryosphere and the nature of electromagnetic interaction with the physical and electrical properties of the interface.

In this chapter we present three sections through which we document our current understanding of how synthetic aperture radar (SAR) may be used in estimating components of the surface energy balance within the marine cryosphere. In the first, Sect. 3.2 (The Marine Cryosphere), we state the scientific rationale for this work and review the important characteristics of a seasonally evolving marine cryosphere from the perspective of its dielectrical, energy balance and microwave-scattering properties. This review sets the stage for Sect. 3.3 (Building Statistical Relationships), in which we present various statistical linkages between the time series evolution of the microwave-scattering coefficient and components of the surface energy balance. These statistical relationships are then exploited utilizing a variety of image-processing approaches in Sect. 3.4 (Generating Image Products). We conclude the chapter with a description of the future directions of this work and a description of how these results are being implemented in large-scale operational processing programs such as the Radarsat Geophysical Processor System (RGPS) and the Ice Information Services of Environment Canada.

## 3.2 The Marine Cryosphere

We consider sea ice as a volume interface between an overlying atmosphere and an underlying hydrosphere; the pertinent physical interconnections can be rationalized within the atmospheric and hydrospheric sub-components of the Arctic system. Hydrologically, sea ice represents a storage volume for fresh water, primarily in the form of ice ridges, multiyear ice and the snow cover on all ice types. The freshwater export of sea ice from polar to temperate latitudes is a significant parameter in the global hydrologic cycle (Prinsenberg 1988).

Radiative and energy balances of sea ice are intimately linked with the spatial distribution of the snow/sea ice/ocean matrix. The alternating presence and absence of sea ice has significant impacts over various temporal and spatial scales. This is particularly important during ice formation and within openings of the Arctic sea ice cover (leads, cracks and polynyas) during winter. It has been found that the small spatial area of cracks and leads has a sufficiently strong enthalpy flux to dominate the total ocean-atmosphere heat flux of the northern hemisphere (Maykut 1978). A snow cover on sea ice acts to decrease the thermal diffusivity of the sea ice, thereby decreasing the heat flux over all ice types. The spatial pattern of snow-covered sea ice and open water also has a strong influence on the radiation balance. The albedo for the ocean surface may range from 3 to 10%, while that for snow-covered sea ice ranges from 80 to 90%.

Ecologically, the radiative, energy and hydrological cycles of the ocean-sea ice-atmosphere interface affect all components of the system. Sub-ice (epontic) primary production within ice-covered regions is known to be light-limited, and higher levels of the trophic system have evolved specific niches within average conditions of ice and snow thickness, growth, ablation, motion and advection. The ecological character and diversity of open water areas, particularly those that occur within the perennial ice cover, may act as sensitive indicators of climate variability and ecological change in the Arctic (Stirling 1980 and Dunbar 1981).

### 3.2.1 Physical and Electrical Properties of Sea Ice

Although the development of the physical properties of sea ice is a continuous process, it is advantageous to compartmentalize unique stages of growth. Following Livingstone et al. (1987), seasonality is defined within the context of microwave remote sensing of a snow-covered sea ice surface into winter, early melt, melt onset, advanced melt, and freeze-up.

*Winter* is the dominant season, lasting from November to May at least. During this season first-year ice types cover a range of thickness conditions from >30 to <200 cm. Younger ice forms such as nilas, shuga and frazil are rare (Maykut 1978). Typically, first-year ice consists of a snow layer covering a primary layer of frazil ice overlaying columnar crystals. Total snow cover increases throughout the winter season, reaching accumulations from 5 cm to over 1.0 m. Maximum rates of snow precipitation occur during the fall and spring seasons when relatively warm air masses exist. When snow is deposited onto the sea ice surface various metamorphic processes begin to operate, creating the "typical" winter structure of the snow cover.

Vertically, snow on sea ice consists of small grains and high density in the surface layers of the snow cover. The grain size and density increase towards the basal layer of the snow volume. Observational evidence (Barber et al. 1993, 1994) shows the existence of a high-brine-volume, low-density, large-grain-size basal layer when snow occurs over first-year sea ice. Immediately beneath the basal snow layer there exists a sea ice frazil layer. This layer is formed in the early stages of ice growth when the ice forms under moderately turbulent conditions. The columnar-oriented crystals occur at deeper levels and are a result of the ice growing under the quiescent conditions of a frazil or shuga layer of new ice. The vertical profile of ice salinity is usually characterized as a "C" shape [5–16 parts per thousand (ppt) at the surface, 4–5 ppt in the centre, and near

30 ppt at the ice-ocean interface]. Ice density is relatively uniform at 0.92–0.96 $g \cdot cm^{-3}$. Liquid brine is interspersed through the ice as intercrystalline brine inclusions and brine drainage channels. The brine inclusions are generally formed in the orientation of growth (vertical) or directed according to the predominant ocean current (Vant et al. 1978). The brine drainage channels are distributed heterogeneously across the ice surface and provide a primary mechanism for brine drainage in the spring (Wakatsuchi and Kawamura 1987).

The electrical properties of snow-covered sea ice can be conceptualized using the complex dielectric constant:

$$\varepsilon^* = \varepsilon' + j\,\varepsilon'', \tag{1}$$

where $j = \sqrt{-1}$

The permittivity ($\varepsilon'$) or real part of the dielectric constant, is a measure of how well a particular frequency and polarization of microwave energy may pass across a dielectric interface. The imaginary part, or loss ($\varepsilon''$) refers to how much of that energy will be absorbed within the volume once it passes across the interface. A high dielectric change across any given interface means that the interaction will be dominated by surface scattering mechanisms. A small change in the dielectric constant across the interface will mean that a large proportion of energy will be transmitted across the interface and will be available for scattering within the volume. The dielectric constant of snow-covered sea ice is highly variable both temporally and spatially.

Multiyear ice typically consists of a surface which has undergone an extensive process of recrystallization during the summer melt period. The surface consists of an array of preferential melt areas (melt ponds) and desalinated low porous hummocks surrounding the melt ponds. The upper layer of recrystallized ice (in the hummocks) has widely varying densities, typically around 0.7 $g \cdot cm^{-3}$. This layer merges into an intermediate layer of slightly higher density which in turn merges into a more solid layer with densities in the range 0.8–0.9 $g \cdot cm^{-3}$. The salinity of both hummocks and melt ponds is considerably lower than that of first-year ice because of brine drainage during the summer. The snow cover is typically deep over melt ponds and shallow over hummocks since the former create natural entrapments for blowing snow. Metamorphic processes occurring within the snow volume tend to favour larger snow grains near the basal layer of the snow cover. There is usually systematic variation in density over the vertical dimension and the snow volume is brine-free. The dielectric constant profile is less variable than in first-year sea ice forms because of the absence of brine.

*Early melt* is a transition period starting with the beginning of snow pack metamorphism and ending when moisture is continuously present in the snow. Metamorphic processes include kinetic energy (wind-induced metamorphosis) and radiant energy. Radiant energy provides for development of ice lenses and ice layers within the snow pack (Miller 1981). Kinetic processes cause metamorphosis from a faceted hexagonal snow flake with intercrystal pore spaces consisting primarily of air, to a more rounded grain state (Colbeck 1982). This can cause density changes from light, newly fallen snow (approx. 0.05 $g \cdot cm^{-3}$) to very dense crustal snow layers (approx. 0.5 $g \cdot cm^{-3}$). Air temperature at the surface rises to 0 °C during some portion of the day. Snow conditions typical of the winter season can change quickly during a warming event. If ambient conditions return to winter norms these changes can be retained within the snow crystal structure retaining a thermal history (Barber et al. 1992).

During *melt onset* free water is continuously present within the snow pack and the ice surface becomes damp at the snow-ice interface. Polycrystalline aggregates form locally at the snow base and superimposed ice layers sometimes form on first-year ice (Holt and Digby 1985). During this stage snow can be characterized as being either in the pendular or the funicular regime. The pendular regime occurs when air occupies continuous paths throughout the pore space and the funicular regime when liquid occupies these paths. This transition occurs at approximately 7% water by volume. In dry snow, grain growth occurs slowly, but with the addition of small amounts of water (2–5% water by volume) grain growth increases markedly (Colbeck 1982).

The melt-freeze cycles create both surface and sub-surface layers (Colbeck 1982). During the early melt season, ice salinities at the surface decrease rapidly. The spatial distribution of these surface salinities is heterogeneous, probably due to the non-uniform distribution of brine drainage channels. The average surface temperature is near freezing. The season ends when the majority of the snow cover is saturated.

During *advanced melt* the snow cover is first saturated throughout its volume and then melts rapidly. During this period liquid is drained from the saturated snow. Tightly packed grain clusters occur after the liquid has drained. Colbeck (1982) reports that grain clusters under these conditions are typically two to four crystal arrangements. These multi-crystal forms can also form aggregates of several hundred crystals, often called polycrystalline aggregates. With drainage of the snow pack a gradient of water volume is set up within the snow with minimum values at the surface and saturation at the base.

The ice surface undergoes melting, often in a cyclical diurnal fashion. The development of drainage networks begins primarily through seal holes, cracks and leads. The salinity of first-year ice decreases. The ice cover decomposes completely in some areas. Transient freezing periods can occur, but the average trend is for melting conditions. Melt waters change the halocline of the local ocean and promote erosion at the interface between the ocean and ice.

At *freeze-up* new ice formation begins with the appearance of frazil, slush or columnar ice crystals developing within the water column. Frazil ice typically occurs under conditions when the ocean surface is roughened by wind. The frazil platelets form within the top few centimeters of the surface and float to the top. These particles aggregate to form shuga or grease ice. Further cooling of this slushy layer results in consolidation of the surface into a dark thin ice type called nilas. This primary ice layer (frozen frazil) is characterized by platelet-shaped crystals in the millimetre size range. Columnar crystals are typical of calm growth conditions and slush ice is a mixture of snow, blown in from surrounding surfaces, and sea water.

As the number of freezing days increases the ice starts to consolidate into the water column. This is termed secondary ice growth and results in ice crystals which have columnar-shaped grains. As the surface continues to cool, brine pockets within the frazil layer are extruded to the surface, resulting in frost flowers (Martin 1979; Hollinger et al. 1984; Drinkwater and Crocker 1988). The salinities of this layer can be considerably higher (45–100 ppt; Drinkwater and Crocker 1988; Martin 1979) than the bulk salinities of the ice. These frost flowers consist of delicate crystalline shapes which typically grow to heights of 2–4 cm above the ice surface (Martin 1979). Brine is also expelled either into the ocean beneath or into brine pockets within the young ice mass as it grows.

The salinity of this ice drops rapidly after the initial freeze-up. After about a week (0.4 m thickness) the desalination stabilizes to a gentler (linear) desalination as the ice sheet thickens. Throughout the growth process, surface and bottom salinities are higher than those in the centre. Bulk salinities range from 5 to 15 ppt (Vant et al. 1978). This period is considered to end when a significant fraction of the young ice has grown to first-year ice thickness (>30 cm).

### 3.2.2
### Energy Balance

The climatic character in the high latitudes is driven by the annual cycle of solar radiation. The spring period marks the transition from polar night to polar day. The seasonal increase in irradiance instigates a reversal in surface net radiation, ablation of snow and ice and a northward shift in the major North American cyclonic trajectories, with accompanying increases in cloud cover, precipitation and advection of southern air masses. The form of the seasonal increase in net radiation is a strong function of its synoptic setting, and therefore subject to interannual variability. It is this dynamic structure which lends support to the notion of estimating surface characteristics and energy flux conditions through the microwave interactions with snow-covered sea ice.

The net radiation of the snow/ice system can be described as follows:

$$Q^* = K\downarrow(1-\alpha) + (L\downarrow - L\uparrow), \tag{2}$$

where $Q^*$ is the net all-wave radiation gain or loss, $\alpha$ and $K\downarrow$ are the albedo and incident short-wave radiation, and $L\uparrow$ and $L\downarrow$ are the upwelling and downwelling long-wave radiation. For energy balance to occur, $Q^*$ must balance against the energy involved in changes in snow structure, conduction of heat within the snow and the turbulent exchanges of latent and sensible heat between the surface and atmosphere. This is compactly described in Eq. (3) (following Oke 1987):

$$Q^* = Q_H + Q_E + \Delta Q_S, \tag{3}$$

where $Q_H$ represents the sensible and $Q_E$ the latent turbulent heat fluxes, and $\Delta Q_S$ is the convergence or divergence of sensible heat fluxes within the snow and ice volume.

Daily averaged net radiation over snow-covered first-year sea ice (Fig. 1) shows considerable interannual variability. While the spring of 1992 was the second coldest of a 46-year temperature record at Resolute Bay, Northwest Territories, each of the three subsequent spring seasons was unseasonably warm (Agnew and Silis 1994). The onset of the rapid increase in $Q^*$ during the spring of 1992 lags over 30 days behind that for 1994. The low net radiative values in 1992 were caused by persistent cloud cover, and frequent snowfall events. In general it appears that net radiation of the entire snow-ice-ocean system ($Q^*$) reverses from being predominantly negative to being predominantly positive early in June or late May (between days 150 and 160) for the first-year ice sites. Once $Q^*$ becomes positive a rapid temporal increase is observed. This is associated with increasing water in liquid phase within the snowpack (Barber et al. 1994).

The onset of this rising limb of $Q^*$ is observed to occur 5–6 days earlier over a multiyear ice surface than over first-year sea ice under the influence of similar atmos-

**Fig. 1.** The evolution of daily average net radiation ($W \cdot m^{-2}$) of the marine cryosphere over first-year sea ice during the SIMMS'92-'95 experiments

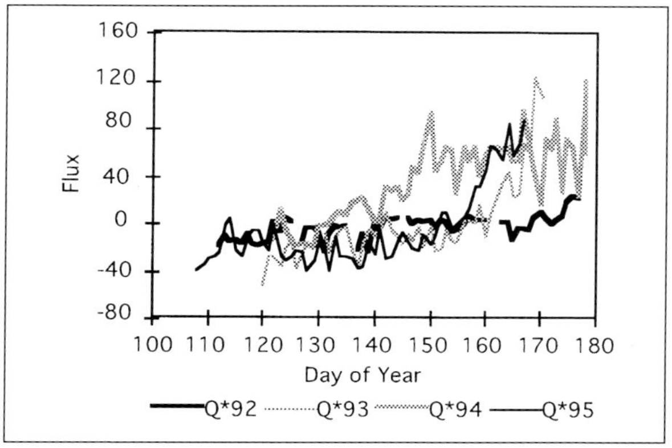


pheres. The rising limb corresponds in time with a decline in surface albedo, an increase in transmitted solar radiation and an increased frequency of non-northerly winds, and is coincident in time when the daily average near-surface air temperature rises above the temperature at the snow-ice interface. The discrepancy between the available energy at the surface, and the available energy to the snow-ice-ocean volume widens as the season progresses because of the increasing transmissivity of the snow volume. The radiation balance at the ice surface ($Q^*_{is}$) remains positive throughout the transition period, owing to the daily contribution of solar radiation received by the ice surface.

The partitioning of net radiation at the surface may be described using the energy balance equation for snow-covered sea ice where $Q^*_{ss}$ is the net radiative flux at the snow surface, $Q_H$ and $Q_E$ are the surface turbulent sensible and latent heat fluxes, and $\Delta Q_S$ is the change in energy storage within the snow. The latter can be expressed using the law of conservation of energy for the snow volume using the following terms: (1) the conductive heat flux at the ice surface ($Q_i$), (2) the conductive heat flux at the snow surface ($Q_S$), (3) absorbed solar radiation within the snow volume ($Q_{abs}$), and (4) energy associated with phase transition ($Q_M$). All fluxes are expressed as $W \cdot m^{-2}$, and energy gains to the surface are denoted as positive values. The change in energy storage encompasses both the divergence and convergence of sensible heat and any latent energy consumed during melt ($Q_M<0$) or gained through increases in the relative volumetric proportion of ice within the snow pack ($Q_M>0$).

Seasonal elements of the energy partitioning during the spring can be summarized by examining computed fluxes over sea ice in the Barrow Strait region of the Northwest Territories (Papakyriakou 1997). The form of the energy balance evolves during the transition toward melt onset. Advanced stages of melt were not observed in the data record during 1992; however, variation in the energy and radiation are apparent during periods of characteristically negative and positive net radiation. Early in the season the surface operates on a net radiative deficit (Fig. 2A). The major heat losses by the surface volume include radiative cooling and sensible and evaporative losses into the near-surface atmosphere. The bulk of the energy losses are made up by ice production associated with ice crystal metamorphism, and superimposed ice growth on

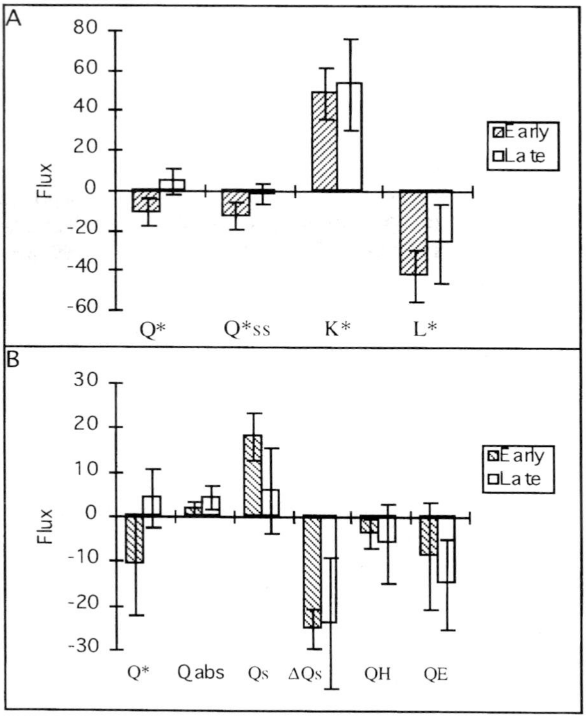


**Fig. 2.** A Radiation balance and B energy balance over a first-year snow-covered sea ice surface during periods in which net radiation was on average negative (early) and positive (late) over the diurnal cycle. Fluxes are representative of pre-melt and the initial stages of snow melt during the spring transition of SIMMS '92

the upper sea ice surface, with secondary contributions through conductive heating at the snow surface, $Q_s$ (Fig. 2B).

Later in the season we observe increased variability in each of the flux terms. The net radiative flux for the snow-sea ice-ocean volume is positive, but the radiation balance at the snow surface, $Q^*_{ss}$, remains negative owing to the transmission of the solar radiation beyond the volume's near surface. Radiation absorbed within the snow volume is comparable in magnitude with $Q^*$. The surface volume is more strongly coupled to the atmosphere, as turbulent losses to the atmosphere represent the major energy sink. Energy losses are predominantly offset by ice production within the snow volume, as conductive heating diminishes in response to the narrowing difference in temperature between the atmosphere and ice surface.

### 3.2.3
### Microwave Scattering

Recent research results (Drinkwater 1989; Winebrenner et al. 1994; Kwok and Cunningham 1994; Barber et al. 1994; Barber et al. 1995; and others) have shown that a distinct pattern exists for the seasonal evolution of the microwave scattering coefficient $\sigma^o$ over various sea ice types. Although these patterns are undoubtedly frequency- and polarization-dependent it would appear that the C-VV manifestation for snow-covered

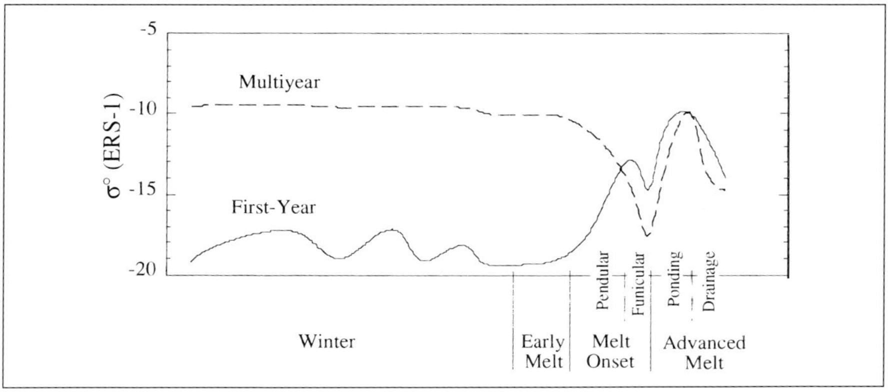


**Fig. 3.** Phenomenological summary of the seasonal evolution of $\sigma^o$ from the ERS-1 SAR for thick first-year and multiyear sea ice over the seasonal periods first proposed by Livingstone et al. 1987. (Adapted from Barber et al. 1995)

thick first-year and multiyear sea ice follows a form approximated in Fig. 3. Following the same seasonal nomenclature used in the preceding section (Livingstone et al. 1987) we can summarize the seasonal evolution of $\sigma^o$ as follows:

*Winter.* Within multiyear sea ice, microwave scattering is very stable during the winter period. The primary scattering mechanism is the bubble structure within the surface hummock layer (Onstott 1992). Older forms of ice appear to have a higher magnitude of scattering, due to the larger and deeper hummock layers which contain larger bubble sizes spaced further apart within the hummocks (Barber et al. 1995). The variability in image texture of multiyear sea ice is due to the spatial pattern of hummock and frozen melt pond surfaces within the multiyear sea ice. The snow cover on the multiyear sea ice is brine-free and of a sufficiently low density and small grain size to be transparent to 5.3 GHz energy.

Within first-year sea ice, microwave scattering appears to oscillate according to changes in the oceanic and atmospheric heat fluxes. During this period a statistically significant inverse relationship, between $K^*$ and $Ts$ with $\sigma^o$, has been determined for a SIMMS'92 case study (Barber et al. 1994). The relationship appears to be related to the effect of heat fluxes on the dielectrics of the first-year ice surface in the very early part of the winter season and then by atmospheric flux effects on the basal snow layer during the warmer part of the winter season. Thin first-year ice types produce a much more complicated seasonal evolution of $\sigma^o$ because of the dynamics of the near surface dielectric properties and the role of the snow cover in altering the scattering volume as the ice matures. Further work is required to detail the location, magnitude and significant scattering mechanisms for thin first-year sea ice in the presence of an evolving snow cover.

*Early Melt.* Within multiyear sea ice, microwave scattering remains fairly stable into the early melt period because hummock volume scattering still dominates. Diurnal differences may be present as small amounts of water in liquid phase become available near the surface of the snow cover at or near solar noon. Because the snow cover is brine-free and the grains are fairly small we do not expect any significant scattering (at 5.3 GHz) from the snow volume during this period.

Within first-year sea ice, microwave scattering appears to be dominated by a combination of basal layer volume scattering and ice surface scattering. During this period we would expect to find a significant difference in solar noon versus solar midnight observations, as small amounts of water in liquid phase would contribute to grain growth and the elevated temperatures would significantly increase the brine volume of the ice surface and snow basal layer. The overall magnitude of $\sigma^o$ will be dependent on the ice surface microscale roughness and its surface brine volume.

*Melt Onset.* In multiyear sea ice, melt onset is denoted by a rapid decrease in $\sigma^o$. The mechanism responsible is the absorption of microwave energy by the water in liquid phase within the snow cover (Winebrenner et al. 1994) and by the presence of water in liquid phase within the hummock structures free of a snow cover. An increase in permittivity and in particular an increase in the dielectric loss effectively mask the volume scattering from the hummocks, thereby decreasing $\sigma^o$. The decrease in $\sigma^o$ proceeds over both the pendular and funicular regimes of snow ablation.

In first-year sea ice, melt onset is denoted by a rapid increase in $\sigma^o$. There are two mechanisms which are likely candidates for the observed increase. At relatively low water volumes (1–3%) the large, wet snow grains in the basal layer may contribute a significant volume scattering term to $\sigma^o$. As the water in liquid phase continues to increase (but is maintained within the pendular regime) it is likely that the snow surface may also contribute a surface scattering term to $\sigma^o$ (Drinkwater 1989; Livingstone and Drinkwater 1991; Barber and LeDrew 1994). It most likely that the snow volume scattering term is the dominant mechanism, since the maximum $\sigma^o$ reached at the first-year ice site was larger than the corresponding one at the multiyear ice site, even though the snow surface roughness was approximately equivalent between the two sites. A distinct dip in $\sigma^o$ at the first-year ice site corresponded with the transition from the pendular and funicular regimes. This transition marks the reduction of brine within the basal layer to near zero, an increase in the water in liquid phase at the base of the snow cover and a reduction of water in liquid phase in the top parts of the snow volume (as the surface begins to drain). These processes could lead to a reduction of both the volume scattering and snow surface scattering hypothesized to dominate the pendular regime conditions.

*Advanced Melt.* In multiyear sea ice, the advanced melt season is denoted by a rapid increase in $\sigma^o$. As surface water forms in the previous year's melt ponds there is an increase in the discontinuity at the air-water interface. The dielectric permittivity and loss ($\varepsilon'$ and $\varepsilon''$) of this fresh water (which has a surface temperature of about +1 °C) is sufficiently large that we can expect a penetration depth on the order of 1 cm (Ulaby et al. 1986). We can expect an increase in scattering if the melt pond surfaces are wind-roughened. A reduction in the water in liquid phase within the hummocks may also cause an increase in $\sigma^o$ by increasing the volume scattering component of $\sigma^o$. Once the ice surface begins to drain there is a pronounced decrease in the multiyear $\sigma^o$. This period coincides with a reduction in the spatial extent of the melt ponds and cyclical changes in the water in liquid content within the hummocks. The exact mechanisms for scattering over multiyear sea ice in the advanced melt season remain largely unexplored and are identified here as a priority for future research.

In first year-sea ice $\sigma^o$ increases over the ponding period and decreases as the ice begins to drain, analogous to that described for the multiyear ice surface. The mechanisms for first year ice scattering during the advanced melt period are also largely unknown.

In this section we introduced the scientific rationale for this work within the context of global climate change and operational ship navigation in ice-infested waters. We then described the average seasonal conditions of the physical, electrical, energy balance and microwave-scattering characteristics of the marine cryosphere within the microwave-scattering seasons proposed by Livingstone et al. (1987). In the next section we investigate the statistical linkages between microwave scattering and various components of the energy balance. Dielectrics and physical properties are then used to explain the physical mechanisms responsible for the statistical relationships. These linkages then constitute a precursor to developing image products (Sect. 3.4) based on the time series evolution of synthetic aperture radar (SAR) scattering coefficients.

## 3.3 Building Statistical Relationships

The research results presented in this chapter have been collected during the Seasonal Sea Ice Monitoring and Modelling Site (SIMMS) experiment. Members of this project team collected data annually over various ice types located in the Canadian Arctic Archipelago between 1990 and 1995. Details of the experiment are available elsewhere (LeDrew and Barber 1994).

### 3.3.1 Univariate Analysis

A time series of ERS-1 data were calibrated and used to produce a summary view of the seasonal evolution of $\sigma^o$ for a first-year sea ice surface (first-year ice site) during SIMMS'92. In this analysis we focus on two time series: the "early" seasonal period (days 105–130, approximately the month of May) and a portion of the "transitional" and "accelerated" periods (days 140–170, approximately the month of June).

The conductive components of the energy balance appear to exhibit a reasonable association with the seasonal evolution of $\sigma^o$ in time series 1. The daily totals of $Qc_o$, show no covariance with $\sigma^o$ until after about day 120 (Fig. 4). The pattern of the $Qc_i$ (Fig. 5) more closely approximates the pattern of $\sigma^o$, suggesting that the snow volume may be having little or no effect relative to the ice surface on $\sigma^o$ during this period. In time series 2 $Qc_o$ follows a similar pattern to $\sigma^o$ while $Qc_i$ appears to be nearly constant. It appears from this relationship that changes in the snow cover both affecting and affected by $Qc_o$ may be detected by the SAR, whereas $Qc_i$ appears to be independent of the observed changes in $\sigma^o$ in time series 2.

In general, as the conductive flux decreases there is a decrease in $\sigma^o$. The similarity in pattern between $Qc_i$ and $\sigma^o$ in time series 1 suggests a slight covariance between the ice surface flux and $\sigma^o$. In time series 2 the covariance of $Qc_o$ with $\sigma^o$ and the lack of association with $Qc_i$ suggests that $\sigma^o$ may be arising (at least partially) from the snow volume. It is important to note that $Qc_o$ began varying in phase with $\sigma^o$ after day 120 and $Qc_i$ stopped varying with $\sigma^o$ after about day 150. During the "early" seasonal period the snow surface is well insulated from the thermodynamics of the underlying vol-

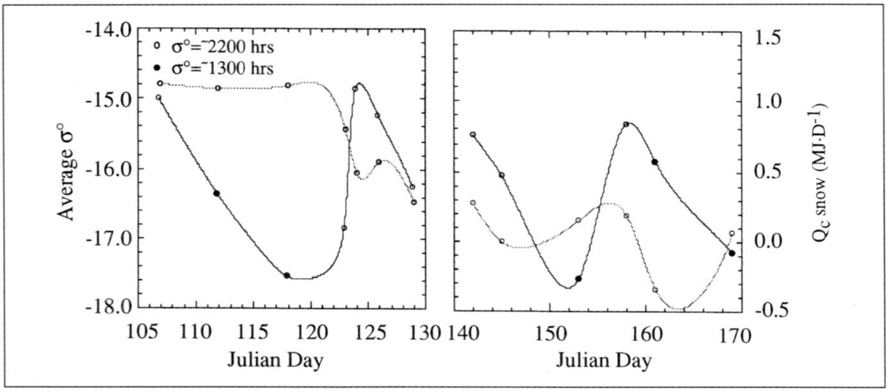


**Fig. 4** Daily total snow surface flux ($Qc_o$) for time series 1 and 2. *Solid lines* denote σ°

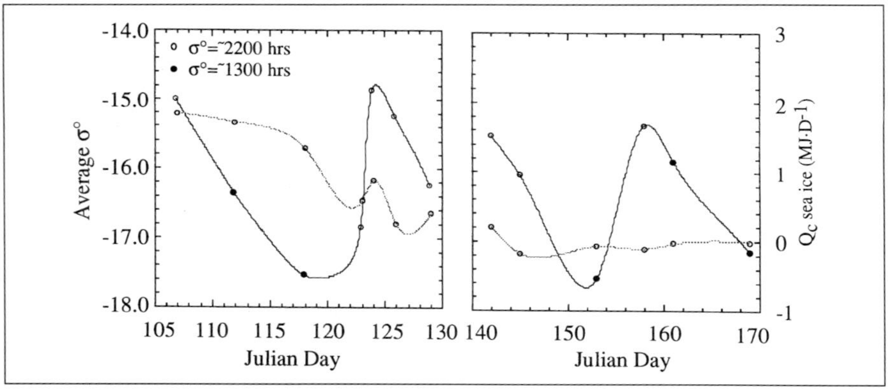


**Fig. 5.** Daily total conductive flux at the sea ice surface ($Qc_i$) for time series 1 and 2. *Solid lines* denote σ°

ume, and therefore $Qc_o$ responds largely to atmospheric forcings. The ice surface on the other hand is insulated from the atmosphere by the snow volume. Early, large fluxes respond to the strong temperature differences between the ice upper surface and the snow base; however, $Qc_i$ is gradually reduced with diminishing temperature gradients. Daily totals of both $Qc_i$ and $Qc_o$ decline after day 120. While $Qc_o$ appears to fluctuate under increasing radiative influence in series 2, $Qc_i$ remains largely unchanged owing to the near isothermal conditions experienced near to the snowpack base.

In time series 1 and 2 the net radiation ($Q^*$; Fig. 6) and net solar radiation ($K^*$; Fig. 7) exhibit a strong out-of-phase association with σ°. The net longwave radiation ($L^*$; Fig. 8) also illustrates a strong association, but one which is in phase with σ°. This is because $Q^*$ balances the net radiative gains ($K^*$) and losses ($L^*$). Since radiative exchange is near-instantaneous, the interpretation of their form is a function of ambient atmospheric conditions and surface geophysical properties. For example the high

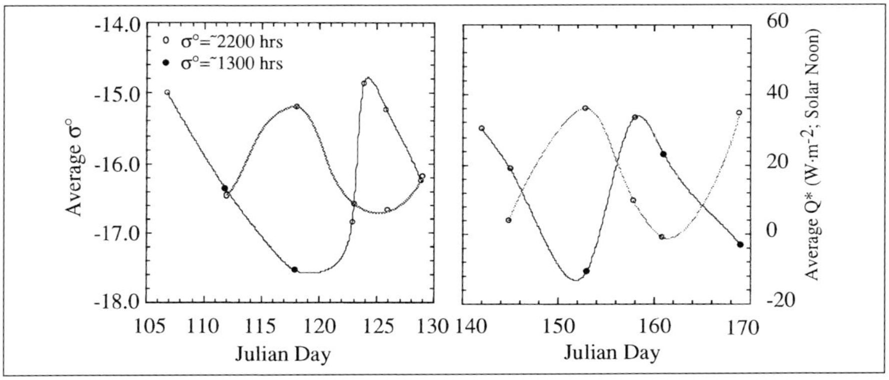


**Fig. 6.** Net radiation averaged for a 30-min period around solar noon over time series 1 and 2 during SIMMS'92

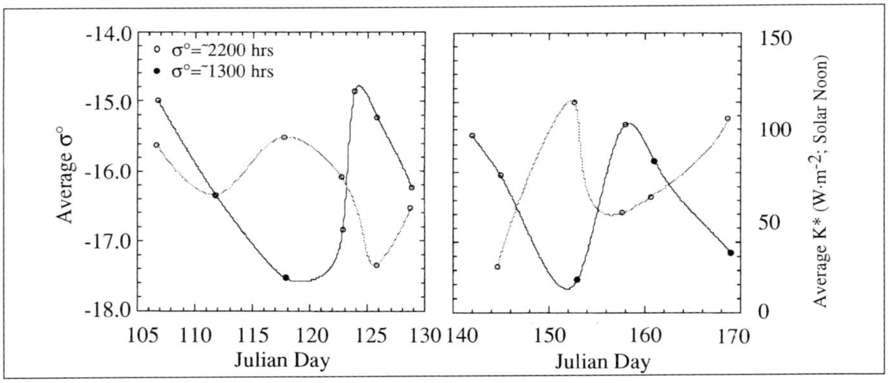


**Fig. 7.** Net shortwave radiation averaged for a 30-min period around solar noon over time series 1 and 2 during SIMMS'92

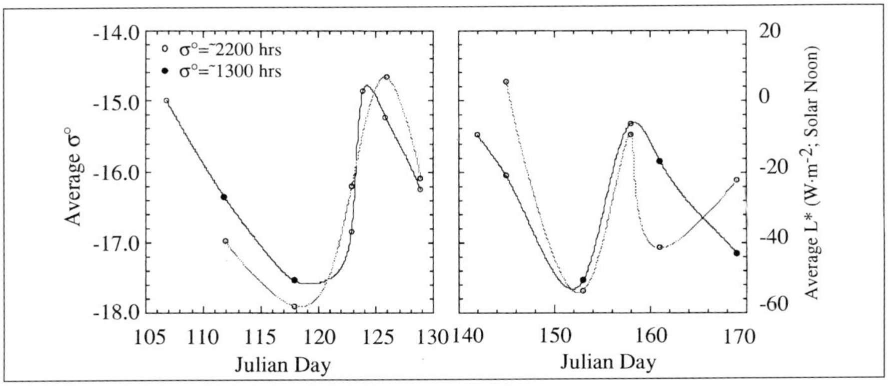


**Fig. 8.** Net longwave radiation ($L^*$) averaged for a 30-min period around solar noon over time series 1 and 2 during SIMMS'92

$K^*$ gains and large $L^*$ losses experienced near solar noon on day 118 indicate clear-sky conditions, whereas the observations on day 126 are attributed to a low-cloud, precipitating atmosphere, with strong surface winds. More detailed analysis of this cloud-cover effect on scattering has shown that changes in the surface dielectrics induced by cloud effects on the longwave flux are detectable within the winter scattering season (Barber and Thomas 1997). Concomitant enlargement of snow grains under temperature gradient morphosis during this period (Barber et al. 1994) served to enhance the radiative gains from $K^*$, further enhancing $Q^*$. Note that $Q^*$ was measured directly, thus deviations in the summation of K* and L* stem from sensor precision and alignment differences. Dielectrically, the sea ice surface and snow volume followed a pattern which appeared to covary with the radiative and conductive fluxes of the snow and sea ice volumes. These forcings can be considered as a causal relationship since the available energy will directly affect the fractional volumes of ice, air and brine and the phase proportions of water within the snow and sea ice, thereby causing a change in the dielectrics.

We used a combined surface/volume scattering model (Barber and LeDrew 1994) to diagnose the scattering characteristics of this condition. In time series 1 the gradual rise in both $\varepsilon'$ and $\varepsilon''$ corresponded to an observed decrease in $\sigma^o$. The downturn in both $\varepsilon'$ and $\varepsilon''$ after day 123 corresponded to an increase in $\sigma^o$ back to a magnitude in scattering similar to the beginning of the time series. The inverse relationship between the complex dielectric constant and $\sigma^o$ suggests that volume scattering $\sigma^o_v$ is occurring (Ulaby et al. 1986). Our diagnostic model predicted a decrease in the volume scattering $\sigma^o_v$ term and an increase in the surface scattering $\sigma^o_s$ term for the range of parameters in time series 1. The decrease in $\sigma^o_v$ was smaller than the increase in $\sigma^o_s$ because of the low magnitude of the dielectric constant in the snow volume, so that the total scattering term $\sigma^o$ tended to increase over the period where the ERS-1 $\sigma^o$ was observed to be decreasing. In time series 2 the same inverse relationship existed between $\varepsilon'$, $\varepsilon''$, and $\sigma^o$, again suggesting that volume scattering was responsible for at least part of the total $\sigma^o$. It is also important to note that the ice surface dielectrics illustrated both a large magnitude and similarity to $\sigma^o$ in time series 1 and that the snow volume did not. In time series 2 there was an overall increase in the magnitude of $\varepsilon'$ and $\varepsilon''$, and the time series patterns also varied in phase with $\sigma^o$.

Determining the exact location and magnitude of the surface and volume scattering contributions to $\sigma^o$ cannot be determined using our diagnostic model since it is not designed to separate ice from snow volume scattering. We can however use the dielectric properties to evaluate the microwave penetration depths, thereby illuminating the important depths over which scattering may occur. It would appear that for dielectric properties and snow densities typical of time series 1, the minimum penetration depth is approximately equivalent to the observed snow depth (approx. 30 cm). Under conditions typical of time series 2 the minimum penetration depth was reduced well under this 30-cm level to values of between 5 and 10 cm. This observation lends support to the supposition that the surface scattering term $\sigma^o_s$ dominated the total scattering cross-section $\sigma^o$ in time series 1 (at least before about day 120), and that the volume term $\sigma^o_v$ contributed a significant amount to the total $\sigma^o$ in time series 2. Given the decrease in $\sigma^o$ with an increase in $\varepsilon'$ and $\varepsilon''$ prior to day 120, it is likely that the brine-soaked, large, elongated snow grains immediately next to the ice surface were acting as volume scattering centers, thereby contributing a $\sigma^o_v$ term to the ice surface $\sigma^o_s$ in specifying the total first-year ice $\sigma^o$.

It is interesting to note that the time series change in snow and sea ice dielectrics followed a very similar pattern. This means that it really does not matter whether the scattering is largely dominated by volume or surface scattering, or whether the scattering originates in the snow volume or sea ice surface. So long as the average volume dielectrics respond directly to the conductive forcings, we can expect a strong linkage between the energy balance characteristics and microwave scattering.

### 3.3.2 Multivariate Analysis

In this section we investigate the contributions of snow temperature and radiative fluxes to the explained variation in the ERS-1 $\sigma^o$ of multiyear ice and first-year sea ice types. We develop multivariate statistical models of various components of the surface energy balance Eq. (2) for the time series of $\sigma^o$ over various years of the SIMMS experiment. The objective here is to determine whether a statistically significant amount of variation in the observed $\sigma^o$ seasonal evolution can be attributed to one or more of the energy balance variables (in the presence of the others). This will provide insights into likely candidate variables to be derived from the time series evolution of $\sigma^o$.

In all the multivariate analyses we used a general linear model where partial multivariate regression coefficients were converted to a standardized form. These standardized partial regression coefficients $\beta'$ are constructed by the simultaneous solution of multiple regression normal equations. This results in a coefficient which can be directly compared (by means of hypothesis tests) both within and amongst analysis results. Further details on the methods are available elsewhere (Barber et al. 1995). These models were developed using a variety of time steps (Table 1). This allowed us to evaluate temporal variations in the observed multivariate relationships

**Table 1.** General categories, variable description and time set allocations for the multivariate analysis at both first-year and multiyear sites. (Adapted from Barber et al 1995)

| Variables | Time sets[a] |
|---|---|
| Temperature partitioned into vertical intervals through the snow layer (*Ts*) in °C | Daily average |
| Temperature partitioned into vertical intervals through the ice layer (*Ts*) in °C | Day Average |
| Net shortwave radiative flux ($K^*$) in $W \cdot m^{-2}$ | 1200 h |
| Net longwave radiative flux ($L^*$) in $W \cdot m^{-2}$ | 2000 h |
| Net all-wave radiative flux ($Q^*$) in $W \cdot m^{-2}$ | ERS-1 coincident |
| Conductive flux at the snow surface | |
| Conductive flux at the ice surface | |

a Daily average: average over 24 h.
Day average: average between 0600 and 1800 h LAT.
1200 h: half-hour averages obtained at the descending ERS-1 overpass time.
2000 h: half-hour averages obtained at the ascending ERS-1 overpass time.
ERS-1 coincident: half-hour averages coincident in time with the ERS-1 overpass times.

Results from the multivariate analyses over three different years of the SIMMS experiment (1992, 1993 and 1994) show that a strong statistical relationship exists between the $K^*$ variable and temperatures within the basal snow layer of both first-year and multiyear sea ice types. In general we found that over both multiyear and first-year ice types $K^*$ explains a statistically significant portion of the seasonal evolution in $\sigma^o$ when the multivariate model includes other components of the energy balance Eq. (2). During the seasonal transition from winter to advanced melt period, as $K^*$ goes up so does $\sigma^o$. The shortwave flux ($K^*$) was found to have a statistically significant $\beta'$ in the presence of all pairwise and the single three-variable set with $Q^*$ and $Ts$. In the presence of $L^*$ the $\beta'$ for $K^*$ was not significantly different from zero in any of the pairwise or three-way combinations of variables.

Variability amongst the time steps (Table 1) indicates that the relationship between $\sigma^o$ and $K^*$ is not highly dependent on the time of day (i.e. the relationship is sufficiently "seasonal" that all time steps produced significant $\beta'$ coefficients for $K^*$). Multivariate testing shows that $K^*$ also explains a significant portion of the observed variability in multiyear ice forms. This relationship is, however, limited to the downward trend in the multiyear ice scattering coefficient within the early melt and melt onset stages (Fig. 3), and the relationship is inverse after the multiyear ice signature rebounds. The same statistical relationships for the basal layer of the snow volume, within a multivariate model, also explained a statistically significant portion of the observed variability in first-year and multiyear microwave scattering. Further details on these statistical linkages are available elsewhere (Barber et al. 1994, 1995; Thomas 1996).

In this section we have shown that there are strong statistical relationships between various components of the energy balance and the time series evolution of the microwave scattering coefficient. The univariate analyses were used to investigate individual components of the energy balance effects on the time series evolution of $\sigma^o$. We also showed that many of the relationships can be explained using a physical scattering model and knowledge of the physical and electrical properties of the evolving marine cryosphere. The multivariate analyses showed that $K^*$ and the basal layer snow temperatures over first-year and multiyear sea ice explain a statistically significant portion of the observed variability in the temporal evolution of $\sigma^o$. Analysis of this relationship is consistent over 3 years of empirical evidence (SIMMS'92–94) and the physical causes of these relationships have been explained using the physical and electrical properties of the snow-covered sea ice.

## 3.4 Generating Image Products

In this section we present research results which exploit the physical and statistical relationships described in the preceding section. We describe how synthetic aperture radar can be used to detect the onset of melt within the marine cryosphere. We illustrate some ongoing work in the field of image segmentation and classification using the time series patterns of the scattering coefficient. We then extend the segmentation work into the broad field of change detection and describe how change images may provide insights into subtle changes in the energy balance of the sea ice surface throughout the annual cycle. In the final section we describe how the temporal evolution of $\sigma^o$ can be used to create climatological albedo classes of the marine cryosphere as the season progresses from winter through to the advanced melt period.

### 3.4.1
### Detecting Melt Onset with SAR

The evolution of the multiyear ice ERS-1 signature offers significant possibilities for the estimation of sea ice physical state changes. Unlike first-year ice, whose signatures are highly variable during the winter period, the multiyear ice signatures are influenced primarily by stable volume scattering from hummock ice (Onstott 1992). The multiyear ice signature is an appropriate indicator of sea ice physical changes since the initial stages of melting are accompanied by a steep, clean drop in $\sigma^o$ caused by liquid water in the snow cover over the ice. The magnitude of the drop is dependent upon the snow depth in which signal absorption occurs. A 2% water by volume in a 10-cm-thick snow cover is sufficient to produce a $\sigma^o$ drop of 10 dB (Winebrenner et al 1994). This response creates an effective means of detecting the onset of melt within the marine cryosphere.

Multiyear ice $\sigma^o$ as an indicator of melt onset has been investigated using a combination of buoy temperature records and ERS-1 data acquired over the Beaufort Sea (Winebrenner et al. 1994). Analysis of multiyear ice $\sigma^o$ frequency histograms reveals significant shifts near the onset of melt from high to low $\sigma^o$ magnitudes due to the absorption of microwave energy in the damp snow cover. Melt onset has been estimated as the date on which the temporally interpolated location of the histogram peak crosses a predetermined threshold from high to low $\sigma^o$ magnitude. The value of this threshold is dependent on the scattering properties of multiyear ice and the thickness of its snow cover. A threshold of -14 dB has been used successfully by Winebrenner et al. (1994) to indicate melt onset from ERS-1 imagery of the Beaufort Sea.

### 3.4.2
### Scene Segmentation and Classification

An ongoing problem in the machine classification of SAR data from sea ice is an effective method of image segmentation and class labelling (i.e. classification). Various researchers have investigated approaches that range from neural networks to texture classification/segmentation approaches (see chapters in this text). Operational use of machine-assisted segmentation and classification are critical if operational sensors such as RADARSAT are to find effective application in areas such as tactical marine navigation (Ramsay et al. 1993) and or global climate change (Kwok and Cunningham 1994).

Time series analysis illustrates that the temporal evolution of the scattering coefficient may be an effective tool for both image segmentation and classification. Although the analyses presented here are limited by the temporal coverage of the ERS-1 sensor, the prospect of segmentation and classification are greatly enhanced with the repeat coverage of RADARSAT ScanSAR products (i.e. daily coverage of the Arctic basin) that are now available operationally.

#### 3.4.2.1
#### *Classification*

During the SIMMS'93 experiment we investigated the utility of the time series evolution of the scattering coefficient for classification of various ice types. Six sites were selected, representing a range of sea ice conditions occurring in the study area (Barber et al. 1995). The three first-year sites (FYI, thin FYI and thick FYI) each show a similar seasonal evolution of $\sigma^o$ (Fig. 9). The variation in the FYI and thin FYI sites between days 50 and 110 arise from the young ice formation and growth. The FYI site consolidated around day 50 and the thin FYI site around day 90. The thick FYI consolidated by mid December of 1992. The oscillating pattern of $\sigma^o$ at the thick FYI site is consistent with observations made over a similar ice site during SIMMS'92 (Barber et al. 1994). The increase in scattering after day 146 corresponds to a period where the water in liquid phase within the snow cover becomes detectable (i.e. 1–3% water by volume), grain size in the basal layer increases, and there is still a significant amount of salinity in the snow volume. The increase in scattering between days 146 and 162 coincides with a pendular snow regime. The water is held in the interstices of the snow matrix and the grain growth is primarily equi-temperature metamorphism with polycrystalline aggregates forming in the basal layer. The drop in $\sigma^o$ between days 162 and 167 coincides with a change in the snow volume to the funicular regime (i.e. water in the snow volume freely drains). This causes a rapid decrease in salinity which increases the absorption by water in liquid phase within the snow cover, thereby decreasing the volume scattering from the large snow grains in the basal layer. The increase in scattering between days 167 and 176 coincided with the complete ablation of the snow cover from a slushy basal layer (days 165–167) to standing water on the ice surface (i.e. melt ponds).

The three multiyear ice sites show a stable $\sigma^o$ evolution during the cold period of SIMMS'93 (Fig. 10). We attribute the relative magnitudes of scattering between the MYI, young multiyear ice (YMYI) and second-year ice (SYI) sites to the fact that the MYI site

**Fig. 9.** Seasonal evolution of the first-year (FYI), thin first-year ice (thin FYI) and thick first-year ice (thick FYI) total relative scattering coefficient $\sigma^o$ for the period day 50-190. The SIMMS'93 field period and the approximate overpass times for the ERS-1 SAR are indicated. *Solid lines* are a spline fit. (Adapted from Barber et al. 1995)

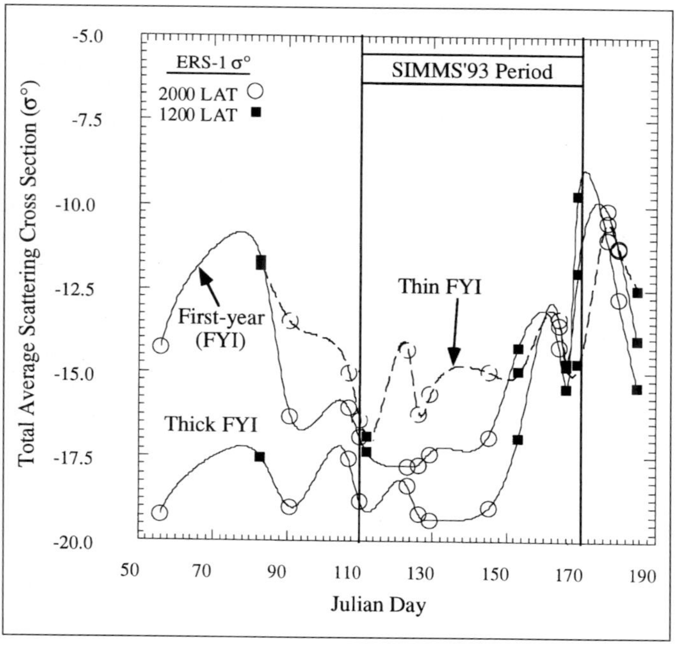

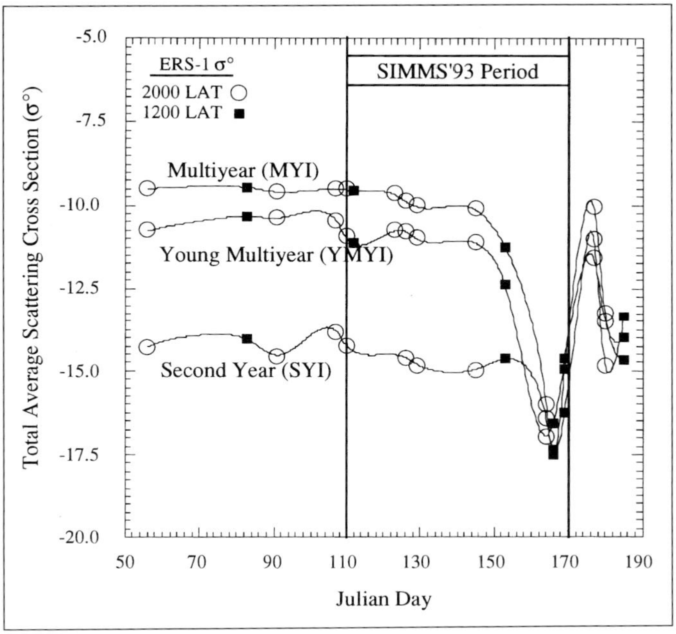


**Fig. 10.** Seasonal evolution of the multiyear (MYI), young multiyear (YMYI) and the second-year ice (SYI) total relative scattering coefficient $\sigma^o$ for the period day 50-190. The SIMMS'93 field period and the approximate overpass times for the ERS-1 SAR are indicated. *Solid lines* are a spline fit. (Adapted from Barber et al. 1995)

had the thickest hummocks with the largest pore spaces of the three forms of multiyear ice. This increases the volume scattering component of $\sigma^o$, thereby causing a higher average $\sigma^o$ for the MYI relative to the YMYI and SYI sites. The reduction in scattering after day 146 corresponds to the increase in water in liquid phase to detectable levels (i.e. 1–3%). The $\sigma^o$ continued to decrease throughout both the pendular and funicular regimes (i.e. up until day 167). Once the snow cover was completely ablated to a melt pond cover the scattering increased (days 167–176). Draining of the surface melt ponds after day 176 corresponded to a decrease in $\sigma^o$.

To evaluate the utility of the time series pattern in the classification of these six sea ice types we subjected the seasonal evolution patterns to a Kolmogorov-Smirnov test. This test ascertains the probability that the two distributions being compared arise from the same parent distribution (i.e. they are statistically indistinguishable). The test assumes a continuous distribution, sampled in an independent, unbiased fashion. The test is sensitive to both magnitude and shape of the distribution (Sokal and Rohlf 1981), making it a good metric for our stated objectives.

Results of this analysis indicate that considerable variability exists between the pairwise contrasts of the seasonal evolution of $\sigma^o$ for the six ice sites tested (Fig. 11). Pairwise comparisons of the first-year ice sites amongst themselves (i.e. "FYI–thick FYI", "FYI–thin FYI", "thin FYI–thick FYI" contrast) is statistically distinguishable ($P$-value = 0.0039). The contrast "FYI–thin FYI" is marginally separable ($P$-value = 0.07) and the "FYI–thin FYI" contrast is statistically indistinguishable ($P$-value = 0.35). Pairwise comparisons of the multiyear ice sites between themselves (i.e. "MYI–SYI", "MYI–YMYI", "YMYI–SYI") shows all pairwise contrast comparisons to be statistical-

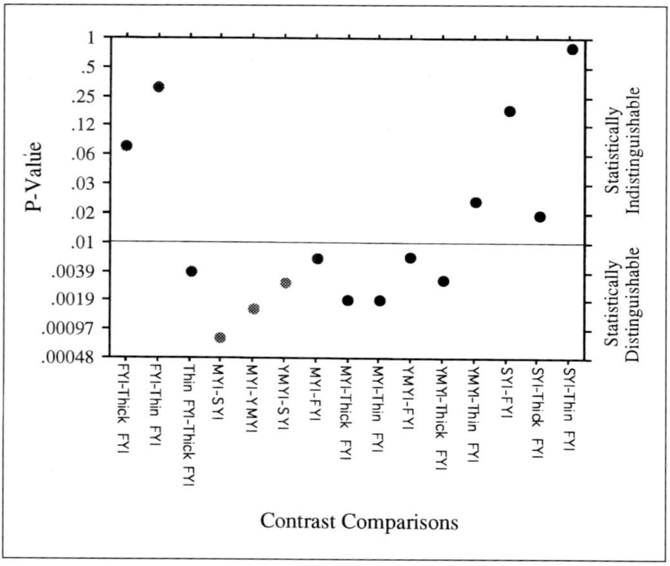


**Fig. 11.** Results of the Kolmogorov-Smirnov non-parametric difference in distribution tests for all pairwise comparisons of the six ice sites discussed in the text. The *P*-values indicate the level of significance in a hypothesis test contrasting the seasonal evolution of $\sigma^o$ between the two ice sites denoted on the *x* axis ("contrast comparisons"). (Adapted from Barber et al. 1995)

ly distinguishable, indicating excellent separation of these ice sites based on the seasonal evolution of their microwave scattering coefficient $\sigma^o$. The remainder of the contrast comparisons indicate that the MYI and the YMYI site were all statistically distinguishable from each of the three first-year ice types (with the exception of "YMYI–thin FYI", which we consider to be marginal with a *P*-value = 0.023). The second-year ice (SYI) was not distinguishable from any of the first-year ice types.

These results illustrate the utility of using the time series evolution as additional information in the classification of sea ice types. Saline ice types (first-year) are easily separated from non-saline types (multiyear) because of the effect of atmospheric forcing on the dielectrics of the saline surface (Barber and Thomas 1997). Separation within multiyear ice sites is possible because of the stability of the seasonal pattern and magnitude shifts due to the age of the multiyear ice. Further details are available elsewhere (Barber et al. 1995).

#### 3.4.2.2
#### *Segmentation*

The time series relationships illustrated in Sect. 3.2.3 (Fig. 3) can be used to separate first year from multiyear forms of sea ice. The approach we are investigating is still in development (Thomas 1996), but the physical mechanisms which allow for this segmentation are well understood (Barber and Thomas 1997). Physical principles, and as a result, segmentation approaches, are different within the winter season and the period spanning the early melt to advanced melt stages of the transitional period.

Within the winter season segmentation is based on the observation that multiyear ice signatures are stable and first-year smooth ice signatures are variable. Ice containing brine in the surface layer and/or within the basal layer of the snow cover can be expected to produce a variable scattering return due to the atmospheric forcing on the surface. Evidence for this relationship has been documented in the preceding sections and is discussed in detail elsewhere (Barber et al. 1994, 1995).

Pairs of ERS-1 scenes within the winter period can be used for ice segmentation using the principle of variable scattering from first-year ice and stable scattering from multiyear ice. One can expect a detectable change in first-year ice scattering with changing atmospheric temperatures and/or presence of cloud cover versus clear skies over the sea ice surface (Barber and Thomas 1997), even when atmospheric temperatures are well below the freezing point.

Difference images over the seasonal transition period also appear to have potential for segmenting the sea ice surface into first-year, multiyear and rubble forms of sea ice. The techniques of change detection outlined above were applied to repeat-pass image pairs spanning the transition from winter to melt onset conditions (Thomas 1996). The images in each pair were mosaics of repeat-pass images covering the SIMMS'95 study area. Each pair is separated by 35 days with an interval of approximately 3 days between image pairs (each successive pair overlapping the coverage of the previous pair by between 40% and 80%). Mosaics were calibrated to a scattering coefficient according to Olmsted (1993). The first image in each pair was acquired during the winter season. Consequently, variation in ice signatures between these winter images is minimal (less than 1 dB for multiyear ice) and the decibel values for multiyear ice can be considered to represent the winter mean.

Once image calibration to decibels had been achieved, change detection images were generated, one for each pair of repeat-pass images covering the melt period of 1995. Image subtraction was used to create change detection images, with the spring period mosaic being subtracted from the winter period mosaic in each case. The resultant change detection images were classified into three change classes: $\Delta\sigma^o > +1$ dB, represented by red; negative change $\Delta\sigma^o > -1$ dB, represented by blue; and "no change:" $\Delta\sigma^o$ between +1 and −1 dB, represented by green (Fig. 12). This three-way classification was chosen as a simple way of illustrating seasonal progression of ERS-1 $\sigma^o$ as a function of ice type.

Bivariate histograms of the two image mosaics in each repeat-pass set were generated to illustrate seasonal $\Delta\sigma^o$ as a function of magnitude in $\sigma^o$. An agreement line bounded on either side by a 1 dB envelope was superimposed over each bivariate histogram (Fig. 12). This provided a graphical representation of $\sigma^o$ changes outside the "no change" $\Delta\sigma^o$ class (−1dB to +1dB).

Difference images and the corresponding bivariate histograms in Fig. 12 show the segmentation classes which are available through different portions of the seasonal evolution (Thomas 1996). The magnitude in the difference classes has been restricted to segmentation of first-year, multiyear and perhaps rubble ice classes. The T1 contrast set (April 24 and May 29, 1995) shows that considerable change occurs in first-year ice forms which have a scattering magnitude less than about −19 dB. The higher-scattering first-year surfaces (i.e. rough) and the multiyear forms show no significant change (difference in scattering is within the ±1dB threshold) (Fig. 12). Unlike first-year ice, significant fluctuations in multiyear ice $\sigma^o$ do not occur during the winter period because multiyear ice $\sigma^o$ is controlled by the bubble structure in hummock ice.

The May 2 and June 6 contrast set (T2) is the first pair where the second image in the repeat-pass pair is clearly within the early melt period. This is evident in the dramatic

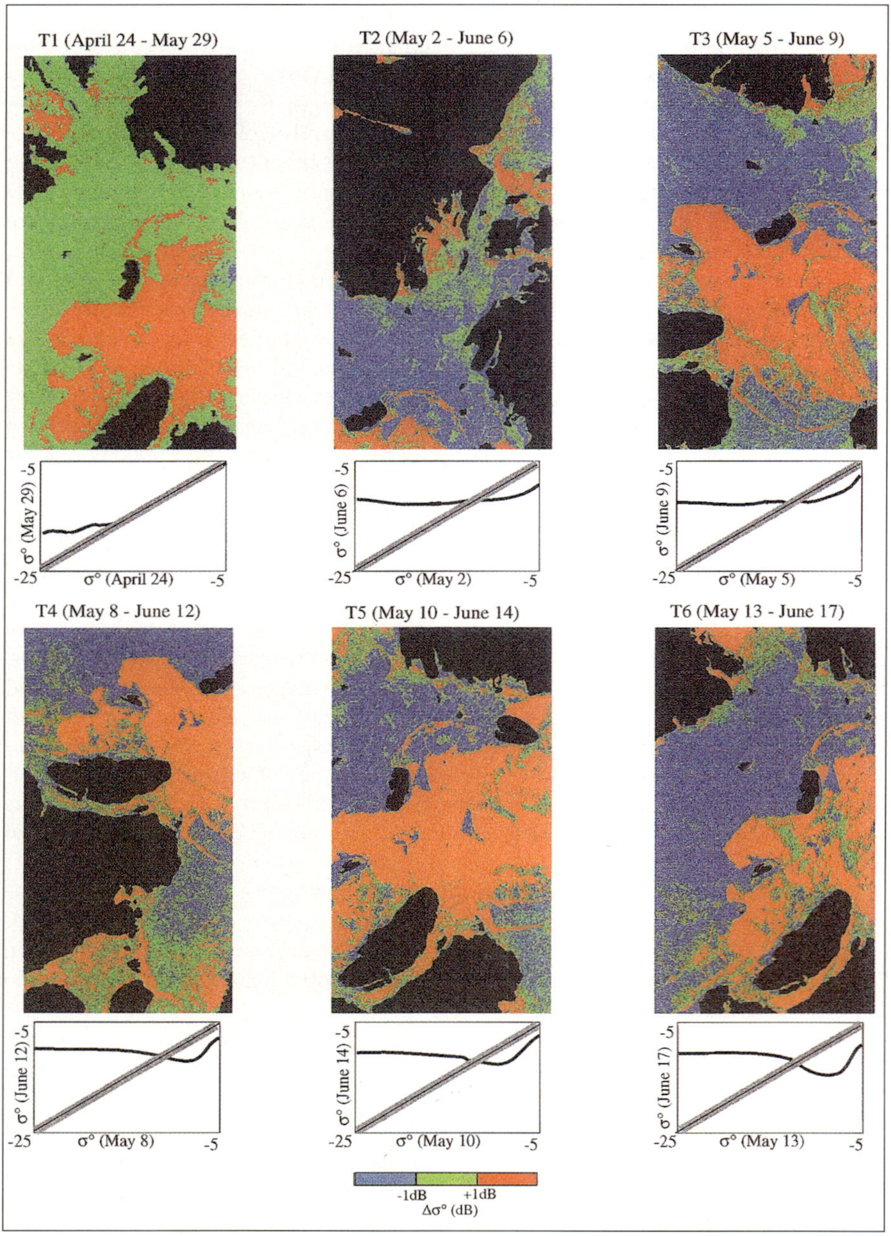


**Fig. 12.** Segmentation of spring transition ERS-1 data utilizing the principles inherent in the effect of water in liquid phase causing scattering differences in first-year, multiyear and rubble forms of sea ice. Bivariate histogram *x* and *y* axes show the winter and spring image $\sigma^0$ components respectively. (Adapted from Thomas 1996)

spread of negative $\Delta\sigma^o$ as microwave energy is absorbed in the deep moist snow cover over multiyear ice. Virtually all multiyear ice $\Delta\sigma^o$ is more negative than –1 dB, whereas stable ice signatures appear to be primarily associated with rubble ice. These changes are also apparent in the bivariate histogram for the two images. Below a $\sigma^o$ magnitude of –15 dB the increase in $\sigma^o$ is inversely proportional to the $\sigma^o$ magnitude of the May 2 image. Negative change in $\sigma^o$ occurs at all $\sigma^o$ above –13 dB, while the range of $\sigma^o$ magnitudes which change less than 1 dB is extremely narrow.

The May 5 and June 9 contrast set (T3) shows an area of conglomerated ice types which display all three change classes. This ice, visible in the southern regions of the image, appears to be a combination of multiyear ice, first-year ice and rubble. The multiyear ice floes are clearly seen as blue ($\Delta\sigma^o > -1$dB). Rubble appears green ($\Delta\sigma^o$ between –1 dB and +1 dB) and first-year ice is visible as veins of red running through the conglomerate. At this point in the time series, it is apparent that the use of $\Delta\sigma^o$ could be used as an effective image segmentation tool. The separation of ice types is preserved throughout the melt in difference images, with the boundaries between positive and negative $\Delta\sigma^o$ change representing first-year ice and multiyear ice respectively.

The T4 and T5 change images are very similar to T3. This indicates that the rate of $\Delta\sigma^o$ or first-year ice and multiyear ice types is stable. Rubble ice $\Delta\sigma^o$ has deviated less than 1 dB from its original winter level. The patterns are consistent with the seasonal evolution illustrated in the conceptual schematic described previously (Fig. 3). The T6 difference image represents the threshold between the melt onset and advanced melt periods when snow melt water begins to pond on the ice surface. T6 indicates the first evidence of a universal increase in $\sigma^o$ for both first-year ice and multiyear ice due to melt water ponding of both first year ice and multiyear ice surfaces. The positive $\Delta\sigma^o$ component displays a magnitude-independent increase from an average of –14 dB to an average of –12 dB between T5 and T6. This indicates the first effects of surface melt water on raising first-year ice $\sigma^o$.

It is apparent from the preceding discussion that change detection images may provide a useful mechanism for segmentation of SAR sea ice scenes. In the winter season, separation of smooth first-year from multiyear and rubble is possible because of responses of the first-year classes to changing brine volumes in the basal snow layer and ice surface (Barber and Thomas 1997). In the seasonal transition period there is a considerable amount of information available on the state of first-year, multiyear and rubble forms of sea ice. Multiyear ice is identifiable by the decrease in its $\sigma^o$ relative to typical winter values and smooth first year ice by its increasing $\sigma^o$. Finally, certain ice types, notably rough ice and rubble, display $\sigma^o$ fluctuations of less than 1 dB between winter and the onset of surface ponding.

### 3.4.3
### Predicting Daily Average Shortwave Albedo ($\alpha$) from $\sigma^0$.

In the preceding sections we have shown that the time series evolution of the microwave scattering coefficient $\sigma^o$ is driven by the electrical and phase proportions of water within snow covered sea ice. We have also shown that the climatological albedo is one of the primary determinants in the initiation of the seasonal evolution of the marine cryosphere. It should also now be apparent that climatological shortwave ($K^*$) and microwave scattering are affected by the same physical changes in snow-covered sea

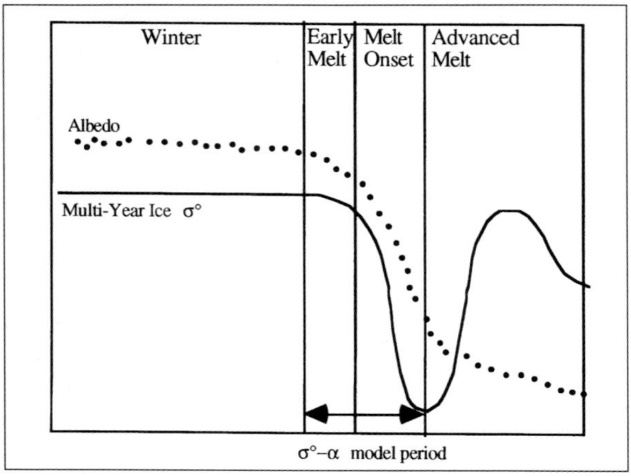


**Fig. 13.** Phenomenological evolution of multiyear ice $\sigma^o$ and $\sigma^o$ over the four SAR scattering seasons proposed by Livingstone et al. 1987. (Adapted from Thomas 1996)

ice. Change detection methods capture this seasonal change in ice type scattering in a consistent and definable fashion (Sect. 3.4.2). It is also evident that the stability of multiyear sea ice scattering in the winter season, followed by a consistent reduction over the early melt season, provides a vehicle by which climatological albedo may be estimated using microwave scattering of the surface. The conceptual relationship between surface albedo and the time series evolution of microwave scattering over multiyear sea ice (Fig. 13) illustrates the general relationship which will be exploited here.

#### 3.4.3.1
#### *Methods*

Statistical relationships between shortwave and microwave interactions were explored using regression analysis between multiyear ice $\sigma^o$ and albedo. Results from detailed analyses (Thomas 1996) indicated that the time period spanning the end of the winter season into the early melt and melt onset periods would be optimal for estimating surface albedo. Due to the limited number of paired data points (multiyear ice $\sigma^o$ with coincident albedo measurement) for each individual SIMMS year, ERS-1 and multiyear ice albedo data from SIMMS'92–'95 were combined, not only to increase sample size, but also to make the model representative of inter-annual variations in multiyear ice types. Both multiyear ice $\sigma^o$ and $\Delta\sigma^o$ were used as independent variables. $\Delta\sigma^o$ was calculated relative to the winter mean multiyear ice $\sigma^o$ for each SIMMS year. This resulted in a series of $\Delta\sigma^o$ values which become more negative the more temporally removed they are from the winter mean. A statistical model was constructed to relate $\Delta\sigma^o$ to albedo over multiyear ice surfaces. Confidence intervals were computed at the 90% level. Computation of the confidence intervals [Eq. (4)] was integral to computing statistically distinguishable albedo classes based on the modelled relationships:

$$\hat{y}_i = \pm \frac{t\alpha}{2S}\sqrt{1+\frac{1}{n}} \pm \frac{n\left(x_i - \overline{x}\right)^2}{n\left(\sum x^2\right)-\left(\sum x\right)^2} \tag{4}$$

These equations were encoded in a computer program and confidence intervals were generated for predicted albedo values. The generated confidence intervals were used as thresholds for image classification. For each 0.1 dB change in the independent variable, the corresponding albedo was calculated with 90% upper and lower confidence limits. Classes were formed whenever the upper limit of the upper confidence interval

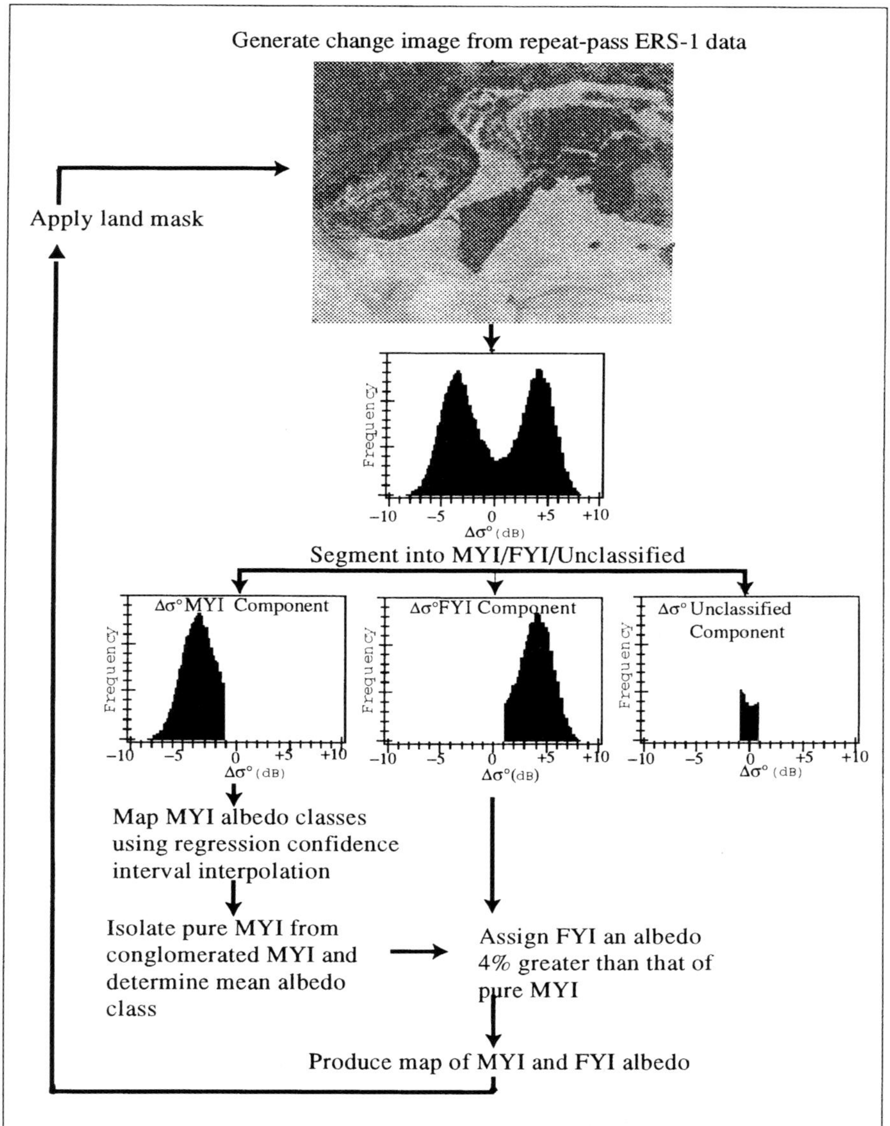


**Fig. 14.** Flowchart of the procedure used to map albedo from MYI $\Delta\sigma^{o}$

fell below the lower limit of the lower confidence interval from the previous class. The mean albedo level for each albedo class was calculated and entered into a table with the upper limit of the upper confidence interval and the lower limit of the lower confidence interval. The range of decibel change associated with each class was recorded and used as the basis for image classification.

The albedo classes were applied to the change detection images generated from repeat-pass pairs of ERS-1 imagery. We used histograms of $\Delta\sigma^o$ for the change detection images as basis of the albedo mapping function (Fig. 14). Negative $\Delta\sigma^o$ histogram values indicate decreasing backscatter from multiyear ice relative to the winter level. Albedo classes for multiyear ice were assigned based on the magnitude of this decrease, with the greatest change corresponding to the lowest albedo.

The winter level of $\sigma^o$ for first-year ice is not sufficiently stable to allow stable $\Delta\sigma^o$–$\alpha$ relationships. As a consequence, first-year ice had to be segmented from multiyear ice and the surface albedo prescribed based on an offset relative to the spatially coincident multiyear ice. The offset was computed by pooling all the observational data from SIMMS'92 to SIMMS'95 and computing the average offset in the surface albedo between first-year and multiyear sea ice. Details on the computation of the albedo offset are available elsewhere (Thomas 1996).

A distinction was made between pure multiyear ice and conglomerated multiyear ice. Pure multiyear ice albedo was used as the basis for the linear albedo offset for first-year ice since SIMMS multiyear albedo data were acquired primarily over pure multiyear ice forms. Albedo for pure multiyear ice was calculated based on the magnitude of its $\Delta\sigma^o$ in each of the unclassified change detection images. Due to the lack of a clear $\Delta\sigma^o$ threshold separating pure and young, compacted multiyear ice, $\Delta\sigma^o$ statistics for pure multiyear ice had to be extracted from polygons drawn around pure multiyear ice floes. Old, pure multiyear ice floes are easily detectable in SAR imagery by their rounded shape, the result of numerous collisions with surrounding ice. Statistics were collected from each polygon and summed up to calculate a mean pure multiyear ice $\Delta\sigma^o$ for each of the change detection images. First-year ice was assigned an albedo 4% greater than that for pure multiyear ice (Table 2).

A 3x3 averaging filter was applied to the classified albedo image to reduce the effects of coherent fading in the original data and emphasize large homogeneous albedo classes (Weydahl 1992). The albedo maps were then placed on a regional mosaic of ERS-1 imagery. Further details on these methods are available elsewhere (Thomas 1996).

**Table 2.** Albedo classes used in the classification of ERS-1 change detection imagery

| Albedo class<br>Albedo class | Multiyear ice $\Delta\sigma^o$ | Albedo class mean | Albedo Class | upper 90% CI<br>lower 90% CI |
|---|---|---|---|---|
| 1 | 0 to -1.4 dB | 0.741 | 0.773 | 0.720 |
| 2 | -1.5 to -3.2 dB | 0.693 | 0.718 | 0.684 |
| 3 | -3.3 to -4.7 dB | 0.645 | 0.682 | 0.653 |
| 4 | -4.8 to -6.5 dB | 0.597 | 0.651 | 0.617 |
| 5 | -6.6 to -8.0 dB | 0.549 | 0.615 | 0.586 |
| 6 | > -8.0 dB | 0.529 | 0.586 | 0.471 |

CI. Confidence Interval.

### 3.4.3.2
### *Results*

The regression plot for the optimal time series model with the 90% confidence intervals for mean predicted albedo (Fig. 15) shows that the width between the upper and lower confidence limits varies with the value of $\Delta\sigma^o$. The widest bands occur at the extremities of the linear relationship and become narrow at $\Delta\sigma^o$ values between -1.6 and -3.0 dB. Note that the shape of these confidence intervals is an artifact of the least squares estimator of the linear model.

The change detection approach resulted in a series of albedo maps showing the full evolution from winter to spring conditions. The comprehensive albedo map set is provided in Thomas (1996). We illustrate the nature of these maps by providing one colour composite for the May 13 to June 17 difference pair (Fig. 16). Seven composite albedo images showing the full evolution from winter to spring conditions are provided in Thomas (1996). Figure 16 encompasses the greatest $\Delta\sigma^o$ magnitude between repeat passes and shows the greatest spatial heterogeneity and diversity in derived albedo classes. Pure multiyear ice floes are distinguishable by their rounded shape and low estimated albedo. The low albedo values are associated with a large $\Delta\sigma^o$ due to microwave absorption in thick wet snow. Pure multiyear ice albedo falls into classes 6 and 7 (mean albedo 54.9% and 52.9%). The younger, conglomerated multiyear ice falls primarily into albedo classes 4 and 5 (mean albedo 64.5% and 59.7), while first-year ice is assigned a value of 63%, 4% greater than the mean albedo for pure multiyear ice.

Computation of albedo class histograms from the image maps further clarifies the spatial and temporal variability in the surface shortwave energy balance over the seasonal evolution from winter to advanced melt (Fig. 17). The mean surface albedo ranges from an initial value of 75.8% on May 29 to a minimum of 61.7% on June 14. The most rapid changes occur during the early melt period when moisture first appears in the snow cover (May 29 to June 6); a 9% drop in overall albedo occurs between these two dates. The temporal evolution of the first-year and multiyear sea ice albedo classes also illustrates that the spatial and temporal heterogeneity precludes effective prescription of single-value albedo values for these surfaces (Fig. 17).

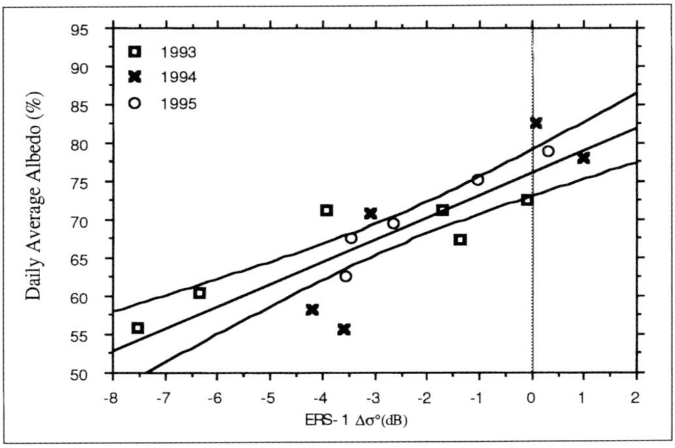


**Fig. 15.** Regression model of $\Delta\sigma^o$ and albedo for multiyear ice from winter to the end of melt onset; 90% confidence limits for mean predicted albedo are shown

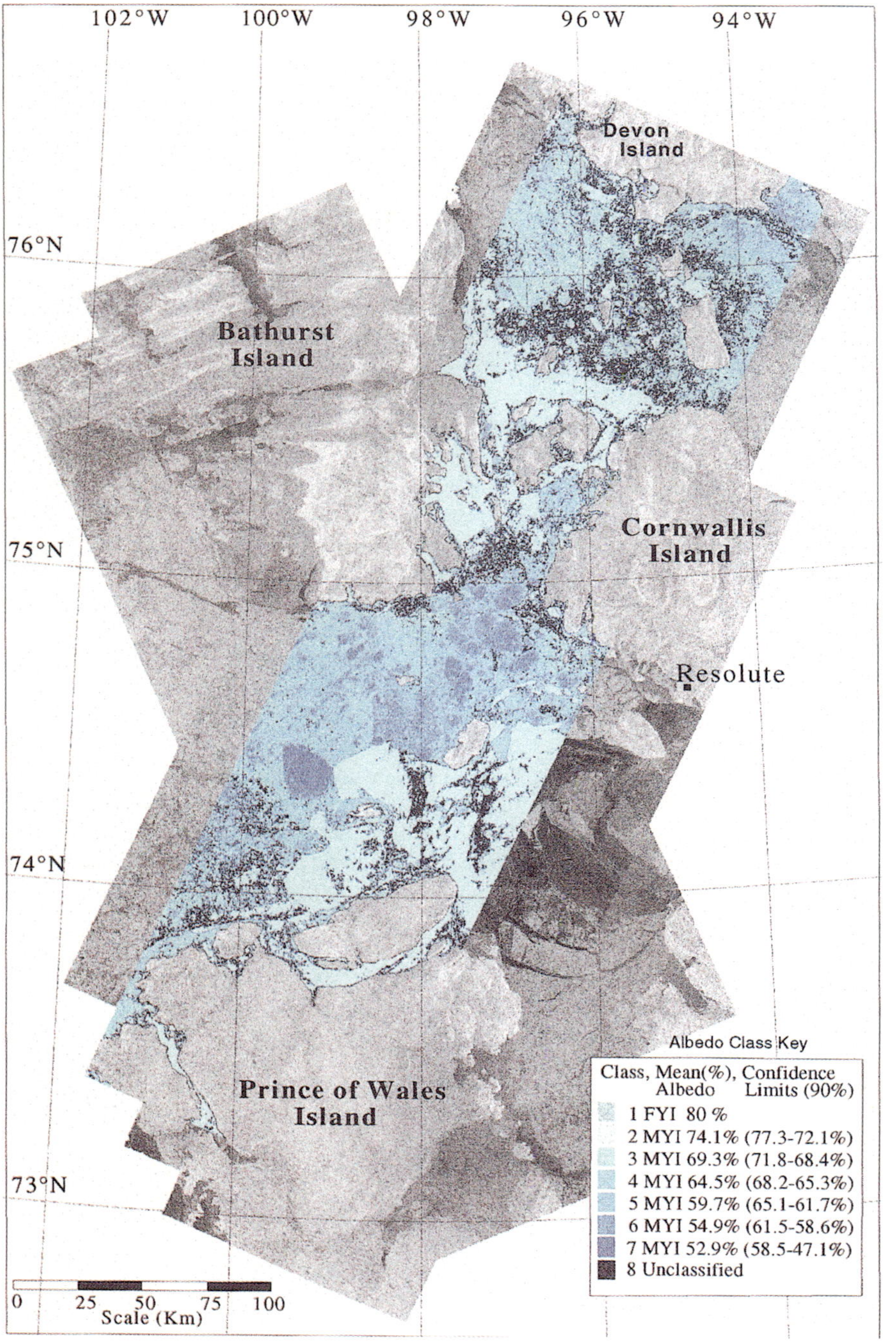


**Fig. 16.** Thematic image of sea ice albedo for June 17, 1995. The classes of albedo are calculated from changes in $\sigma^0$ between May 13 and June 17, 1995. (Adapted from Thomas 1996)

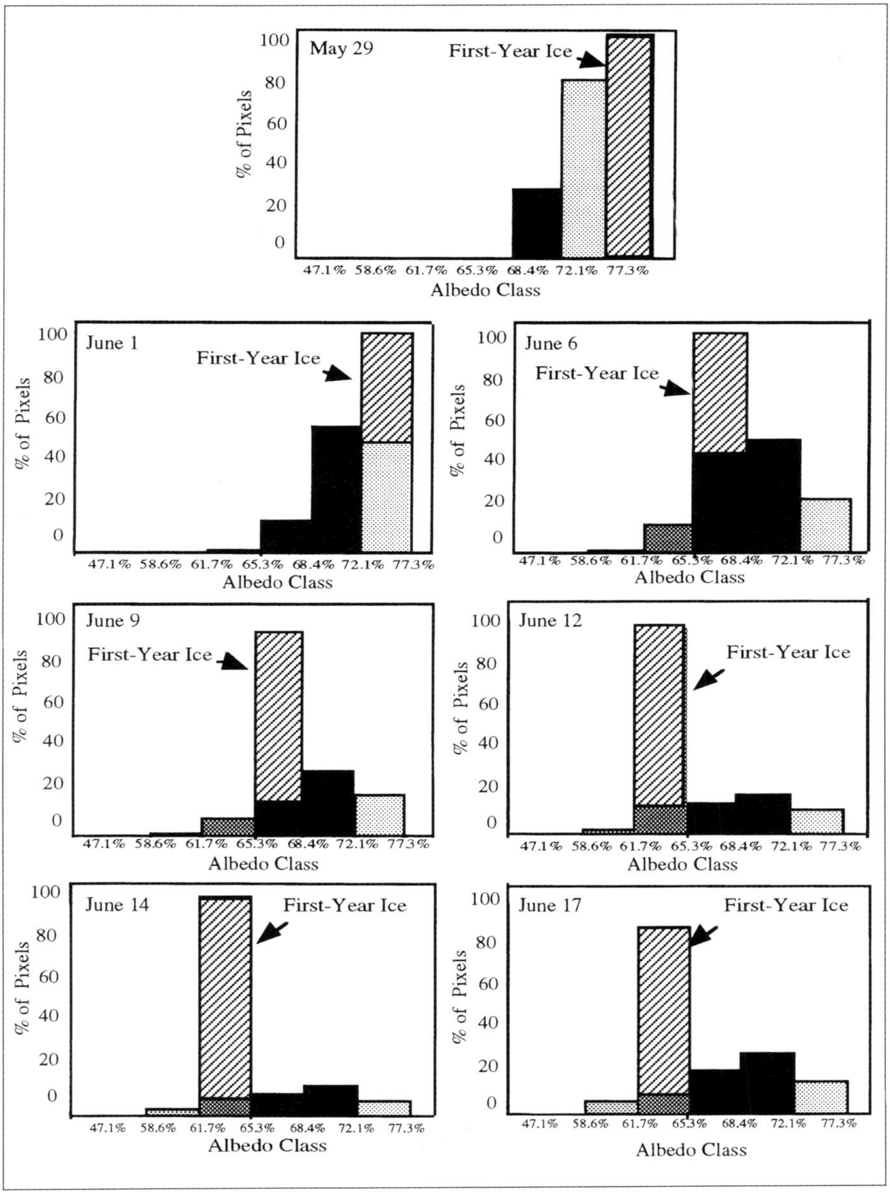


**Fig. 17.** Frequency distribution of first-year and multiyear ice albedo classes for the thematic albedo images. (Adapted from Thomas 1996)

## 3.5
## Conclusions and Future Directions

In this chapter we have introduced the notion that microwave energy interacts within the marine cryosphere as a function of the surface energy balance. Our long-term goal in this work is to develop a better understanding of how processes operate within the ocean-sea ice-atmosphere interface. To achieve this end we require spatially and temporally contiguous observations of both the energy balance and the physical characteristics of this interface. As these tools evolve we intend to use them to develop an understanding of the linkages, feedback mechanisms and potential anthropogenic effects on the dynamic and thermodynamic nature of the marine cryosphere.

In this chapter we partitioned the problem of estimating surface energy balance variables from the temporal evolution of the microwave scattering coefficient $\sigma^o$ into its constituent parts. In Sect. 3.2 we explored the nature of the interrelationships amongst the geophysical, electrical and microwave scattering properties of the snow-covered sea ice. We showed that there are sound physical reasons to suggest that transfer functions exist which would allow the estimation of energy balance variables from $\sigma^o$ due to the fact that both respond to the seasonal dynamics of the physical and electrical properties of the marine cryosphere. In Sect. 3.3 we investigated the nature of these transfer functions by using exploratory univariate and multivariate statistical analyses. Our objective here was to determine which components of the energy balance explained a statistically significant amount of the observed seasonal variability in $\sigma^o$. We found that the shortwave flux and the conductive flux at the snow/ice interface consistently accounted for a significant amount of observed variation in the seasonal evolution of $\sigma^o$. In Sect. 3.4 we began to exploit our knowledge of the "energy balance-physical property-microwave scattering" relationship through the development of image products. These image products, although still experimental, represent a significant step forward in the development of energy flux fields at a scale (in space and time) appropriate for studying marine cryosphere processes.

In the near term the work described here will evolve on two interrelated fronts:

1. **Development of RADARSAT image products.** Canada's RADARSAT is now operational and offers significant advantages over the ERS and JERS SAR satellites currently in operation. The payload instrument on board RADARSAT is a C-band, HH polarized SAR. RADARSAT offers a variety of mission-selectable resolutions, image swath widths, and incidence angle parameters. Orbit and image modes have been specified to allow maximum national and global coverage. In ScanSAR mode, RADARSAT will provide daily coverage of polar regions above 75°N (Ramsay et al. 1993). Two ScanSAR modes are offered by RADARSAT. The first mode alternates between two wide swath beams to image a 300-km swath with a 50-m pixel spacing. The second mode uses a four beam coverage to image a 500-km swath at a 100-m pixel spacing. A variety of research questions will be addressed with RADARSAT data to evaluate how well the principles developed here apply to the unique beam modes, polarization differences and increased temporal resolution of RADARSAT.
2. **Forward and inverse electromagnetic interaction modelling.** The Office of Naval Research (ONR) is nearing the end of a 5-year Accelerated Research Initiative (ARI) which is directed towards the development of forward and inverse modelling of various electromagnetic frequencies with snow-covered sea ice. The goals of this project

are (1) to improve upon existing forward scattering models, (2) to develop the mathematical framework required for the development of inverse scattering models, and (3) to explore the mathematical nature of transfer functions required to exploit the linkages amongst various electromagnetic frequency interactions within the marine cryosphere. Development of the forward scattering models (1) is being done at visible, thermal infrared and microwavelengths (over typical space-borne active and passive frequencies). These models are being used to improve our understanding of the primary scattering mechanisms at these frequencies and thereby the linkages which exist between the geophysics, dielectrics and electromagnetic interactions. The inverse scattering model development (2) is premised on the notion that, if we can precisely measure the scattered field, then the potential exists to invert through a physical model to solve for a single geophysical variable which gives rise to the scattering. This work holds promise as a means of geophysical inversion for variables such as ice thickness, snow thickness, snow grain size, brine volume, etc. The exploration of the nature of transfer functions (3) is, in effect, an extension of what has been presented in this chapter. The objective here is to create a model which would have as an input the scattered electromagnetic field at frequency $\lambda_1$, from which would be predicted the interactions at a different frequency $\lambda_2$ for a constant suite of geophysical and electrical characteristics of the snow-covered sea ice. An example of the application of one such model would be to observe the scattered field at 5.3 GHz ($\lambda_1$) and predict the interactions at climatological short wavelengths ($\lambda_2$).

These developments, included with the diversity of approaches represented in the other chapters in this volume, illustrate a broad, international effort to maximize the use of synthetic aperture radar remote sensing for studies of high-latitude processes. Collectively these research advances have resulted in a strong theoretical base upon which we are now developing specific-use image products, process model inputs and transfer functions amongst various frequencies of the electromagnetic spectrum. These incremental developments are in turn being used by Earth system scientists in development of a better understanding how the marine cryosphere interacts within an Earth system, which we now know is susceptible to anthropogenic effects.

**Acknowledgments.** This work was supported by an NSERC grant, an Office of Naval Research Grant, and a NASA grant for ASF data, each to D. Barber. Additional support was provided by the Polar Continental Shelf Project, Canadian Ice Services, and the Canada Centre for Remote Sensing. Numerous participants of the SIMMS field programs contributed their expertise: Ellsworth LeDrew, Roger De Abreu, Kevin Misurak, and Chris Derksen. Thanks also to the staff at the Alaska SAR Facility for help with acquisition and processing of ERS-1 data.

## References

Agnew TA, Silis A (1994) Spring season climate variability at resolute bay, NWT, Canadian climate centre, Rep No 94-1, Atmospheric Environment Service, Downsview, Ontario

Barber DG, LeDrew EF (1994) On the links between microwave and solar wavelength interactions within a seasonally dynamic snow covered sea ice volume. Arctic 47(3):298–309

Barber DG, Thomas A (1997) The influence of cloud on the radiation balance, physical properties and microwave scattering of first year and multiyear sea ice. IEEE Trans Geosci and Remote Sens (in press)
Barber, DG, Flett DG, De Abreu, RA, LeDrew EF (1992) Spatial and temporal variations in sea ice geophysical properties and microwave remote sensing observations: the SIMS'90 experiment. Arctic 45, no 3: 233–251
Barber DG, Shokr M, Soulis E, Fernandes R, Flett D, LeDrew EF (1993) A comparison of second order texture classifiers for SAR sea ice discrimination. Photogrammetr Eng Remote Sensing 59, no 9: 1397–1408
Barber DG, Papakyriakou TN, LeDrew EF (1994) On the relationship between energy fluxes, dielectric properties, and microwave scattering over snow covered first year sea ice during the spring transition period. J Geophys Res 99, no C11: 22401–22411
Barber DG, Papakyriakou TN, LeDrew EF, Shokr ME (1995) An examination of the relation between the spring period evolution of the scattering coefficient $\sigma^o$ and radiative fluxes over landfast sea ice. Int J Remote Sens 16, no 17: 3343–3363
Colbeck SC (1982) An overview of seasonal snow metamorphism. Rev Geophys and Space Phys 53, no 6: 4495–4500
Drinkwater MR (1989) LIMEX'87 ice surface characteristics: implications for C-band SAR backscatter signatures. IEEE Trans Geosci and Remote Sens 27, no 5: 501–513
Drinkwater MR, Crocker GB (1988) Modelling changes in the dielectric and scattering properties of young snow-covered sea ice at Ghz frequencies. J Glaciol 34, no 118: 274–231
Dunbar MJ (1981) Physical causes and biological significance of polynyas and other open water in sea ice. In: Stirling I, Cleator H (eds) Polynyas in the Canadian Arctic. Environment Canada, Canadian Wildlife Service, Occasional Paper No 45, pp 29–43
Hollinger JP, Troy BE, Ramseier RO, Asmus KW, Harman MF, Luther CA (1984) Microwave emission from high arctic sea ice during freeze-up. J Geophys Res 89, no C5: 8104–8122
Holt B, Digby SA (1985) Processes and imagery of first year sea ice during the melt season. J Geophys Res 90, no C3: 5045–5062
Kwok R, Cunningham GF (1994) Backscatter characteristics of the winter ice cover in the Beaufort Sea. J Geophys Res 99, no C4: 7787–7802
LeDrew EF, Barber DG (1994) The SIMMS program: a study of change and variability within the marine cryosphere. Arctic 47, no 3: 256–264
Livingstone CE, Drinkwater MR (1991) Springtime C-band SAR backscatter signatures of labrador sea marginal ice: measurements versus modeling predictions. IEEE Trans Geosci and Remote Sens 29, no 1: 29–41
Livingstone CE, Onstott RG, Arsenault LD, Gray AL, Singh KP (1987) Microwave sea-ice signatures near the onset of melt. IEEE Trans Geosci Remote Sens 25, no 2: 174–187
Martin S (1979) A field study of brine drainage and oil entrapment in first year sea ice. J Glaciol 22, no 88: 473–502
Maykut GA (1978) Energy exchange over young sea ice in the central arctic. J Geophys Res 83, no C7: 3646–3658
Miller DH (1981) Energy at the surface of the earth: an introduction to the energetics of ecosystems. Academic Press, New York
Oke TR (1987) Boundary layer climates. Methuen, London
Olmsted C (1993) Alaska SAR Facility Scientific SAR user's guide. Alaska SAR Facility Tech Rep ASF-SD-003

Onstott RG (1992) SAR and scatterometer signatures of sea ice. Geophys Monogr 68, 5

Papakyriakou TN (1997) The characterization of the surface energy balance over snow-covered sea ice: an examination of the relationships among, energy partitioning, surface properties, and climate. PhD Thesis, Geography Department, University of Waterloo, Waterloo, Ontario

Prinsenberg S (1988) Ice-cover and ice-ridge contributions to the freshwater contents of Hudson Bay and Foxe Basin. Arctic 41, no 1: 6–11

Ramsay B, Hirose T, Manore M, Falkingham J, Gale R, Barber D, Shokr M, Danielowicz B, Gorman B, Livingstone C (1993) Potential of RADARSAT for sea ice applications. Can J Remote Sens 19, no 4: 352–362

Sokal RR, Rohlf FJ (1981) Biometry. WH Freeman, New York

Stirling I (1980) The biological importance of polynyas in the Canadian Arctic. Arctic 33: 303–315

Thomas A (1996) Mapping the seasonal evolution of sea ice albedo using ERS-1 SAR signatures of multiyear ice. MA Thesis, Geography Department, University of Manitoba, Winnipeg, Manitoba

Ulaby FT, Moore RK, Fung AK (1986) Microwave remote sensing: active and passive, vol. 3 Addison-Wesley, Massachusetts

Vant MR, Ramseieer RO, Makios V (1978) The complex dielectric constant of sea ice at frequencies in the range 0.1 to 40 GHz. J Appl Phys 49, no 3: 1264–1280

Wakatsuchi M, Kawamura T (1987) Formation processes of brine drainage channels in sea ice, J Geophys Res 92, no C7: 7195–7197

Weydahl DJ (1992) Temporal change detection in ERS-1 SAR images. Proc Int Geosci Remote Sens Symp IGARRS' 92, May 26–29, Houston, Texas, pp 1346–1348

Winebrenner DP, Nelson ED, Colony R, West RD (1994) Observation of melt onset on multiyear arctic sea ice using the ERS-1 SAR. J Geophysi Res 99, no C11: 22425–22441

CHAPTER 4

# Extraction of Intermediate Scale Sea Ice Deformation Parameters from SAR Ice Motion Products

S. Li, Z. Cheng, and W.F. Weeks

## Contents



## 4.1 Introduction

In sea ice studies, there appear to be natural scale divisions (Overland *et al.* 1995), and the regional scale (10–50 km) is crucial because the processes on this scale interact with both the climate scale (100–300 km) sea ice processes and floe scale (1 km) processes. Understanding the intermediate regional scale sea ice deformation can help bridge the gap between the sea ice behavior on the floe scale, which is in situ measurable, and the one on the climate scale, which is used as a direct input to climate models. At the intermediate scale, the sea ice deformation behavior is observable from remote sensing data. In fact, the current generation of Earth-viewing radar satellites provides regular observations of the behavior of large areas of the Arctic pack ice. Only the extremely large image data stream poses difficulties for the rapid and efficient extraction of useful geophysical information, and thus requires the development of a variety of automated procedures. This is why the present chapter concentrates on the quantitative characterization of intermediate scale sea ice deformation.

Analysis of SAR Data of the Polar Oceans
Edited by C. Tsatsoulis and R. Kwok

Leads, pressure ridges, and near-shore shear zones form as a result of sea ice deformation. Leads are of particular geophysical interest because more than half of the turbulent energy loss from the ocean to the atmosphere in the central Arctic during the winter is believed to occur from the surfaces of recently formed leads (Badgley 1966; Andreas 1980; Maykut 1982; Makshtas 1991). The heat loss from leads results in the formation and thickening of the thin, new ice types in these features. In turn, the salt rejection and brine drainage from this new ice helps to maintain the arctic ocean halocline and also drives secondary circulation patterns in the upper ocean (Morison *et al.* 1992).

Ridges play a key role in determining the aerodynamic and hydrodynamic coupling between the ice, the atmosphere, and the ocean. The ridging process is not only a deformational energy sink, but in some regions ridges account for 20% or more of the total ice volume (Koerner 1973). In addition, the ridging process also results in brine expulsion simply because of the effect of warm temperatures in the resulting keels. Finally, both lead formation and ridging are important factors contributing to the overall rheological response of sea ice to the imposed external forcings of winds and currents. Clearly, a better understanding of leads and ridges will be useful in the analysis of the complex interactions between the ocean, the sea ice, and the atmosphere in the Arctic – interactions that are believed to play a significant role in the climate riddle.

In sea ice deformation studies, the direct quantity that needs to be measured is the relative motion of a number of regions or floes in the ice pack. Historically, sea ice deformation in the Arctic has been difficult to study because of the remoteness, the harsh environment, and the expense of carrying out such observations using surface-based techniques. The first successful regional field campaign was the Arctic Ice Dynamics Joint Experiment (AIDJEX), which utilized three manned camps separated by distances of roughly 100 km as well as several data buoys to characterize ice motions over a roughly circular region with a diameter of 500 km (Thorndike and Colony 1977; Pritchard 1980). AIDJEX also included a limited study of smaller-scale ice floe motions near one of the main camps (Hibler et al. 1973, 1974). One of the more important long-term results of AIDJEX has been that it was instrumental in starting the activities that ultimately led to establishing the Arctic Buoy Program, a program that has continued to this day and that has contributed significantly to our understanding of the movement of arctic pack ice (Colony and Thorndike 1984, 1985; Thorndike 1986).

Although data collected from buoy arrays is adequate for basin-scale sea ice models based on isotropic ice rheology, even the buoy program, in which typically 20 buoys are deployed over the Arctic Ocean at a given time, does not provide the detailed specification of the motion field that is desirable for studies of regional deformation processes. Overland et al. (1992) have also argued that this type of approach does not appropriately take into account the pertinent scale effects. Some attempts have been made to apply an anisotropic constitutive model to sea ice (Coon et al. 1993). For regional models, including mesoscale models, an examination of the state variables necessary to adequately describe the variation of sea ice (e.g., the principal directions and orientation of fracture lines) has not been performed. Overland proposed to examine these processes on a 1–20 km grid that could be used to predict regional ice responses on 100–500 km scales. For such types of study, the Advanced Very High Resolution Radiometer (AVHRR) aboard the NOAA satellite can be tremendously useful (Overland et al. 1992). However, on the other hand, persistent cloud cover during summer and the long dark period during winter in the Arctic may also significantly restrict the use of AVHRR in such studies.

An optimal approach would be to couple the results of the buoy program, which can, for a limited number of points, provide nearly continuous motion measurements in the temporal sense, with a satellite remote sensing program that can provide less frequent (every few days) but nearly continuous spatial observations. The remote sensing requirements for such a program would appear deceptively simple; the system must produce imagery that is of a sufficiently high resolution to allow one to follow identifiable ice features for a number of repeat periods. Unfortunately, this also means that the system must function under all-weather conditions so that data can be obtained whenever the satellite is over the region of interest. These requirements eliminate visual- and near-infrared-range sensors (limited by darkness and clouds), thermal infrared sensors (limited by clouds), and passive microwave sensors (limited by inadequate resolution). Until the launch of the two radar satellites ERS-1 (1991, ESA – the European Space Agency) and JERS-1 (1992, NASDA – the Japanese Space Agency), satellite-borne remote sensing systems have not held up to their portion of this task. However, the difficulty has now changed from a "no-suitable-sensor" problem to a problem of how to rapidly extract the necessary geophysical information from the extensive SAR coverage that is being obtained. This latter problem has become even more acute since the launch of the Canadian Radarsat in 1995. The wide-swath observation mode of Radarsat results in image strips that are 500 km wide (in contrast to present synthetic aperture radar satellites that provide image strips with widths of 100 km). Considering that ERS-1 and JERS-1 plus ERS-2, which is a follow-up of ERS-1, are still operating during the Radarsat time period, the motivation for the work reported in the present chapter becomes clear.

Our main objective is to develop a method for screening large volumes of synthetic aperture radar (SAR) products and extract significant ice deformation events without having to look at either the SAR images or at the SAR-derived ice motion products on which no significant ice deformation occurred. In the following sections, the background information on the satellite SAR derived ice motion products and the relationship between ice motion and ice deformation are given first. Then the details of the actual algorithms used for derivation of ice deformation are presented, the resulting ice deformation products are described and their accuracy discussed. Finally, as illustrations, two examples are given to demonstrate extraction of intermediate-scale sea ice deformation parameters using the algorithm.

In one example, the temporal variations in ice deformation are obtained for a 6-month period during the winter of 1991–1992, for a circular sampling area with a radius of 200 km centered at 81°N and 170°W in the Chukchi Sea portion of the Arctic Ocean. In the other example, the ice deformation as a function of time and latitude in the Beaufort Sea is extracted during a period between January 23 and March 15, 1994, as part of the SIMI (Sea Ice Mechanics Initiative) project.

## 4.2 Background

The following brief description provides some necessary background information about the already available ice motion data and the mathematical relationships between the ice motion and the ice deformation.

### 4.2.1
### Ice Motion Products

The overall operating system at the Alaska SAR Facility (ASF) includes a Geophysical Processor System (GPS) designed by the Jet Propulsion Laboratory (JPL) of the California Institute of Technology (Kwok et al. 1990; McConnell et al. 1991; Kwok and Pang 1992). This system produces a description of the movement field as specified by the motions of identifiable ice features in terms of their initial and final geographical positions and the displacements of individual ice motion vectors between two sequential SAR images. The system is able to derive correct ice motion vectors even if ice floes experience both translation and rotation.

By convention, the initial image is referred to as the "source image" and the final image as the "target image." In order to carry out this procedure, a special polar stereographic map projection (SSM/I) is used to generate a unique 2-D Cartesian system onto which all the ice motion vectors are mapped. The basic units of 5x5 km square grid cells are obtained from a uniformly spaced grid on the source image. Ice motion is then tracked by a best-match of the ice features in each initial source grid cell with the positions of the same features in the target image. The initial and final positions of the center of each grid cell provide a pair of tie points that define a specific ice motion vector. The ice motion vector can also be expressed in terms of the components of displacement in two perpendicular directions in the Cartesian coordinates defined by the SSM/I projection (Fig. 1).

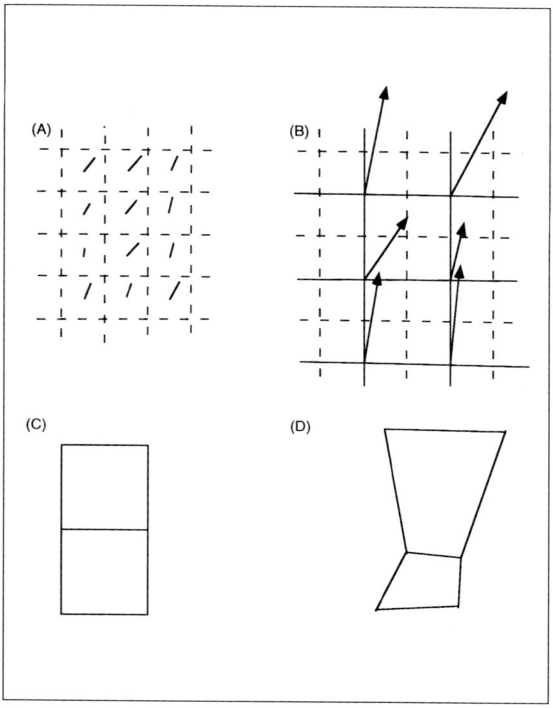


**Fig. 1.** **A** Representation of the GPS computer display showing a scaled array of ice motion vectors based on the tracking of common sea ice features observed on sequential SAR images. The dashed lines are added to represent the boundaries of the 5x5 km cells in the sampling grid. The vectors are scaled so that their magnitude is less than the grid spacing. **B** An enlargement of **A** showing the use of a modified grid such that the nodes are coincident with the initial vector locations. The *dashed lines* represent the original boundaries of the sampling grid (as in **A**). Here the vector lengths are scaled correctly relative to the grid. **C** The boundaries of the two modified grid elements in the initial image as represented by the tails of the six vectors in **B**. **D** The boundaries of the same grid elements in the final image as specified by the heads of the vectors. The upper cell has undergone divergence ($E_I > 0$) and the lower cell has converged ($E_I < 0$), while both cells have undergone changes in shape ($E_{II} > 0$)

Although the centers of the cells are uniformly spaced on the initial source image, they become somewhat irregularly positioned on the target image because of the differential motion of different portions of the pack ice. The results are displayed on the Geophysical Processor System (GPS) (Kwok and Cunningham 1993) as an array of vectors that are scaled such that their magnitude is less than the grid spacing. In the schematic representation of this process, shown in Figure 1A, the ice is drifting toward the NNE. Here the dashed lines, which are not part of the computer display, represent the boundaries of the $5 \times 5$ km grid cells on the source image. Because this fixed size of the grid cells was adopted in the production of the ASF ice motion products, and our purpose for developing the algorithm discussed here is to avoid or at least minimize the use of the SAR images in deformation studies, we presently only consider sea ice deformation problems addressable at a $5 \times 5$ km grid resolution which is suitable for the study of intermediate-scale sea ice deformation.

## 4.2.2 Derivation of Ice Deformation

Ice deformation can be defined in terms of ice motion by two invariants (Thorndike *et al.* 1975; Rothrock 1986; Fily and Rothrock 1990):

$$E_I = \frac{\partial u}{\partial x} + \frac{\partial v}{\partial y} \tag{1}$$

$$E_I = \left( \left( \frac{\partial u}{\partial x} + \frac{\partial v}{\partial y} \right)^2 + \left( \frac{\partial u}{\partial y} + \frac{\partial v}{\partial x} \right)^2 \right)^{\frac{1}{2}} \tag{2}$$

Here the first invariant quantifies the divergence or convergence of the ice field with a positive value indicating a divergent motion field, and a negative value indicating a convergent field. The second invariant quantifies the magnitude of the shearing. They depict different aspects of deformation and are invariant with translation and rotation.

In a grid configuration, differential operations can be approximated by division, the quantities $\partial x$ and $\partial y$ can be represented by the size of the grid cell in the $x$ and $y$ directions, while $u$ and $v$ can be regarded as the displacement, assuming a unit interval between the times of acquisition of the image pairs. Calculations of $\partial u$ and $\partial v$ then involve comparisons of the displacements at the two ends of the $x$ or $y$ side.

By making a slight rearrangement in the grid configuration by changing the boundary of each grid cell, the geometric meanings of the ice deformation parameters become clearer. In the original configuration, the cell boundaries consist of bisects between the center of the cell and those of the four neighboring cells (Fig. 1A). In the modified configuration, the boundaries of cells are defined by four neighboring tie points indicated directly by the ASF ice motion products (Fig. 1B). Figure 1B is an enlarged version of Fig. 1A on which the initial cell boundaries are indicated by dashed lines and the modified cell boundaries by continuous lines. In the modified representation (Fig. 1B), the motion vectors are drawn to scale and, therefore, are commonly longer than 5 km (typical ice drifts in this portion of the polar basin are 2–3 km/day). In the modified representation, ice deformation can be regarded simply as the change in the size and the shape of the involved cell. This is also shown in Fig. 1, where Fig. 1C represents the initial $5 \times 5$ km grid specified by the six initial positions of the six vectors shown in Fig. 1B while Fig. 1D shows the deformed grid specified by the final positions.

This slightly different, yet equivalent, view of deformation also clarifies the evaluation of the accuracy of deformation calculations in two ways: first, the difficulty in assessing the accuracy of the absolute ice motion field can be bypassed, and, second, only the accuracy in feature matching between the image pairs and the geometric fidelity of the SAR images, in terms of the relative positions of different objects within the images, are of importance.

## 4.3 Algorithm Description

To extract ice deformation parameters from the GPS ice motion products, a special algorithm has been designed that calculates the change of the size and the shape of each grid cell from the positions of its four corner points as given by the ice motion product. The algorithm consists of five major components, each of which carries out a specific function: a tie point locator, an edge point trimmer, a gap filler, a deformation generator, and an extension tracer (Fig. 2). In the following, these procedures are discussed briefly. For more detailed descriptions, readers are referred to Li et al. (1995).

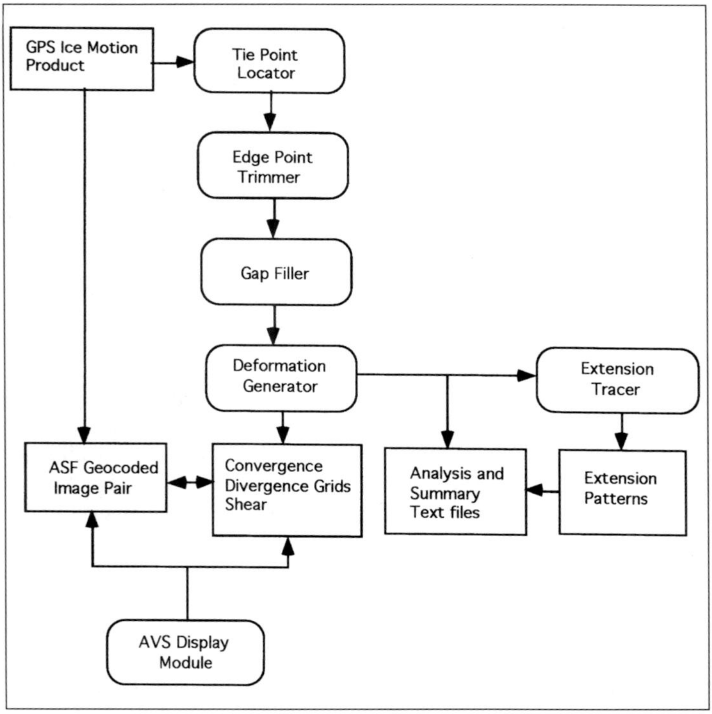


**Fig. 2.** A flow chart of the ice deformation algorithm

### 4.3.1 The Tie Point Locator

An ASF ice motion product covers the overlapped area of two images in a selected image pair whose times of acquisition are usually separated by periods of multiples of 3 days. The product provides the geographic locations of both ends of an ice motion vector. The two ends of each vector form a tie. Although the ice motion file itself is not a grid file, a grid file can be produced from it by using a tie point locator that focuses only on the geographic location of the starting point of a tie. This is the case because the starting positions of the ice motion vectors are evenly spaced on the SSM/I projection and therefore can be directly converted into grid locations. The actual procedure involves two steps: (1) converting the geographic location (i.e., the latitude and longitude of each tie point) into Cartesian coordinates, and (2) deriving grid coordinates, i.e., row and column positions from the Cartesian coordinates. Conversion between the geographic coordinates and the SSM/I coordinates is well defined (Bonbright 1984). The conversion of the Cartesian coordinates $X$, $Y$ into grid coordinates $i, j$ is made by the following formulas:

$$i = \frac{(Y - Y_{\min})}{\Delta Y}, \tag{3}$$

$$j = \frac{(X - X_{\min})}{\Delta X}, \tag{4}$$

where $X_{\min}$ and $Y_{\min}$ are the minimum Cartesian coordinate values found in the particular ice motion file and $\Delta X$ and $\Delta Y$ are the grid sizes in the $X$ and $Y$ dimensions. For

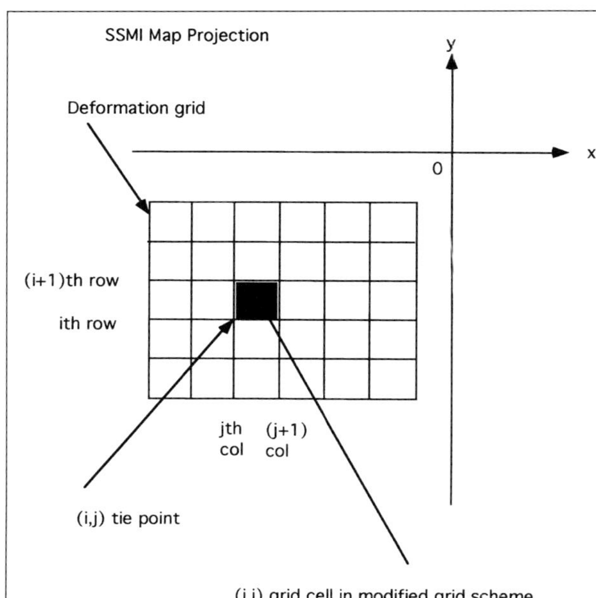


**Fig. 3.** Definition of a modified grid cell located at the $i$th row and $j$th column

the ASF ice motion products, $\Delta X$ and $\Delta Y$ are 5 km. Once the indices for the row and column positions of a tie point are found, the location of the tie point within our problem-specific 2-D matrix is set. All the other ice motion information, including the positions of both the starting and end points, the displacements in the *X* and *Y* directions, and the rotation angle of the neighborhood surrounding the tie point, are copied onto the relevant 2-D matrices using *i, j* as position indexes (Fig. 3).

### 4.3.2
### The Edge Point Trimmer

Defective points (or flyers) in the ice motion products can be a major source of errors in our procedure. Through inspection, we have found that the majority of points that exhibit large ice motion errors occur near the edges of the SAR images. This error is caused by the sharp boundary at the image edge. Before the ice motion product is generated, the images in the matching pair are geocoded and mapped onto a larger frame based on a standard polar stereographic (SSM/I) map projection. As a result, the area outside the original image is filled with zeroes as a background, thereby forming a sharp, but unreal, boundary. When such sharp boundaries exist in the matching cells on both images, the matching algorithm produces an erroneous best correlation based on the relative position of the sharp, imaginary boundaries, as the algorithm is unable to distinguish true image data from background. Thus, an erroneous ice motion vector is formed, resulting in a belt of unreasonably large deformation values near the edge of the matched area. Obviously, this type of error should be avoided.

Therefore, a special "edge point trimming" algorithm has been developed to eliminate these defective points. The procedure is conservative in that all of the edge points are eliminated to insure the reliable extraction of the ice deformation parameters. This procedure is performed after the row and column positions of all the gridded ice motion vectors are determined. Even the simplest approach, that of eliminating the end data points in each row and column, has been found to significantly improve the results by removing the majority of the bad vectors. However, this method commonly eliminates too many edge points, in that the end points for each row or each column are not necessarily the edge points of the matched area. Quite often, tie points occurring outside the end points were eliminated during the process of ice motion generation because of poor matching (Kwok and Cunningham 1993).

To eliminate all of the edge points while keeping all the data points inside the overlapped area, a more sophisticated, several-step procedure has been devised. It first searches through the first and the last rows and columns of the grid to find the vertices of a polygon that defines the overlapped area. At maximum, there can be a total of eight vertices, with two for each side of the rectangular grid. Then, a side is created by interpolating between the adjacent vertices of the resulting polygon. Next, the points on that particular side are marked by "2's" with all the other points initialized as "0's." Once all the points on the sides of the polygon are marked down to the grid by assigning them to nearest cells, a filling process is used to change the values at all the interior data cells into "1's." The resulting pattern forms a mask which is further trimmed removing all the edge points marked by "2's," and then complemented with the results of the previously mentioned simple elimination procedure to create a compound mask through a logic "OR" operation. Finally the compound mask is applied to the data panel to filter out all the true edge points (Fig. 4).

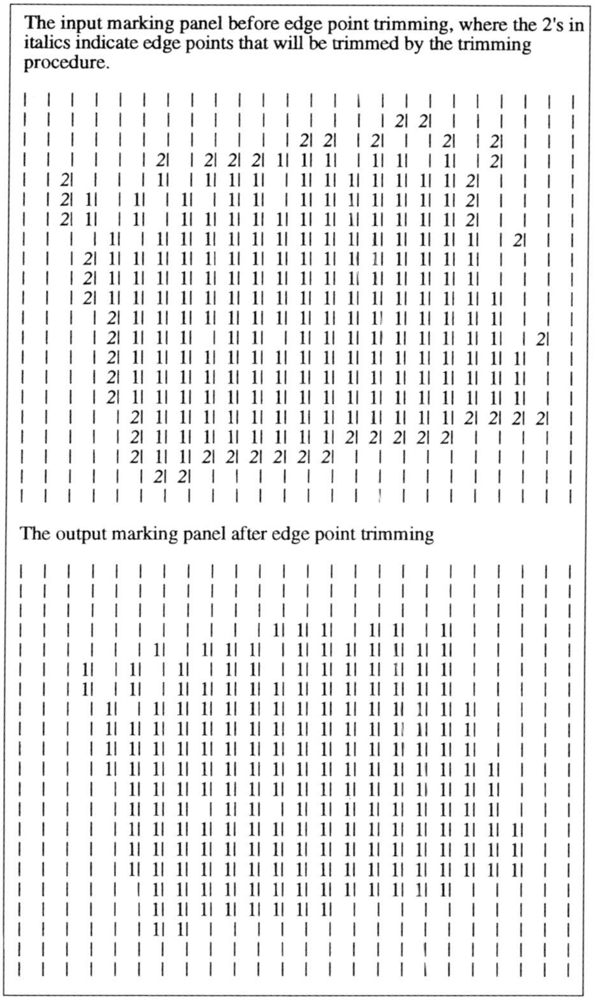


**Fig. 4.** The input and output marking panels of the edge point trimmer from the ASF ice motion product 1131010 which is generated from image 4535401 (Julian day 285, 1991) and image 4620401 (Julian day 288, 1991) for an area centered at 81°N, 172°42'W

## 4.3.3 The Gap Filler

Next, a 2-D template characterizing the data availability is produced. On this template, all the tie points where ice motion data are available are flagged with the integer value of 1. In addition, all the grid points representing background points as well as the edge points which have already been deleted are marked with "0," as are all the interior gaps. It is commonly found that wherever the ice undergoes very large, complex deformations, ice motion data can be missing. The gap filler estimates the ending positions of the missing ice motion vectors.

Initially, the linear interpolation is made along the rows and then along the columns for gap filling. The interpolation results are then further smoothed until the value at

```
| | | | | | | | | | | | | | | | | | | | | | | |
| | | | | | | | | | | | | | | | | | | | | | | |
| | | | | | | | | | | | | | | | | | | | | | | |
| | | | | | | | | | | 1| 1| 1| 2| 1| 1| 2| 1| | | | | |
| | | | | | 1| 2| 1| 1| 1| 2| 1| 1| 1| 1| 1| 1| 1| | | | | |
| | | 1| 2| 1| 2| 1| 2| 1| 1| 2| 1| 1| 1| 1| 1| 1| 1| | | | | |
| | | 1| 2| 1| 2| 1| 1| 1| 1| 1| 1| 1| 1| 1| 1| 1| 1| | | | | |
| | | | 1| 2| 1| 1| 1| 1| 1| 1| 1| 1| 1| 1| 1| 1| 1| 1| | | | |
| | | | 1| 1| 1| 1| 1| 1| 1| 1| 1| 1| 1| 1| 1| 1| 1| 1| | | | |
| | | | 1| 1| 1| 1| 1| 1| 1| 1| 1| 1| 1| 1| 1| 1| 1| 1| | | | |
| | | | 1| 1| 1| 1| 1| 1| 1| 1| 1| 1| 1| 1| 1| 1| 1| 1| 1| | | |
| | | | | 1| 1| 1| 1| 1| 1| 1| 1| 1| 1| 1| 1| 1| 1| 1| 1| | | |
| | | | | 1| 1| 1| 2| 1| 1| 2| 1| 1| 1| 1| 1| 1| 1| 1| 1| | | |
| | | | | 1| 1| 1| 1| 1| 1| 1| 1| 1| 1| 1| 1| 1| 1| 1| 1| 1| | |
| | | | | 1| 1| 1| 1| 1| 1| 1| 1| 1| 1| 1| 1| 1| 1| 1| 1| 1| | |
| | | | | 1| 1| 1| 1| 1| 1| 1| 1| 1| 1| 1| 1| 1| 1| 1| 1| 1| | |
| | | | | | 1| 1| 1| 1| 1| 1| 1| 1| 1| 1| 1| 1| 1| | | | | |
| | | | | | 1| 1| 1| 1| 1| 1| 1| 1| | | | | | | | | | |
| | | | | | 1| 1| | | | | | | | | | | | | | | | |
| | | | | | | | | | | | | | | | | | | | | | | |
| | | | | | | | | | | | | | | | | | | | | | | |
```

**Fig. 5.** The output of the gap filler for the same ice motion product as in Fig. 4. Here the "2's" represent the locations where missing data has been filled by the linear interpolation procedure of the gap filler

each gap point equals the average of its available neighbor points. Once these iterations are completed, all the gap positions are marked by "2," indicating that a less reliable ice deformation result is expected for the grid cell involved (Fig. 5).

### 4.3.4 The Deformation Generator

Next, the ice deformation parameters are calculated on a grid-by-grid basis as follows. First the size of the grid cell is calculated. Then the divergence and convergence values ($E_I$) are calculated as differences of grid cell sizes on the source and target images. Next the shear $E_{II}$ is calculated using Eq. (2) from the change of shape of the grid cells. This latter relation involves comparing the average rates of change of the lengths of the cell sides in two perpendicular directions, and calculating the rate of change of the tangential value of the angle made by two bisects of the cell. The $E_I$ values of the grid cells are then recorded on a 2-D matrix with the positive values indicating divergence and the negative ones indicating convergence. The non-negative $E_{II}$ values are recorded in a separate 2-D matrix.

### 4.3.5 The Extension Tracer

The purpose of the extension tracer is to identify possible contiguous areas that show substantial deformation. First, the thresholds of ±4% for divergence and convergence, and 0.04 for shear are selected, and any grid deformation values beyond these thresholds are regarded as significant. (The reasoning behind the selection of these particular values will become clear later in the section on accuracy.) Using these thresholds, the procedure searches through the $E_I$ and $E_{II}$ matrices, locating any cell with deformation values larger than the selected thresholds and keeping all such cells in a list. Once this search is completed, a sort is performed to reorganize the cells into a priority list according to decreasing deformation values. With the list established, a new marking template is created with all the initial values being zeros. Next, a recursive pro-

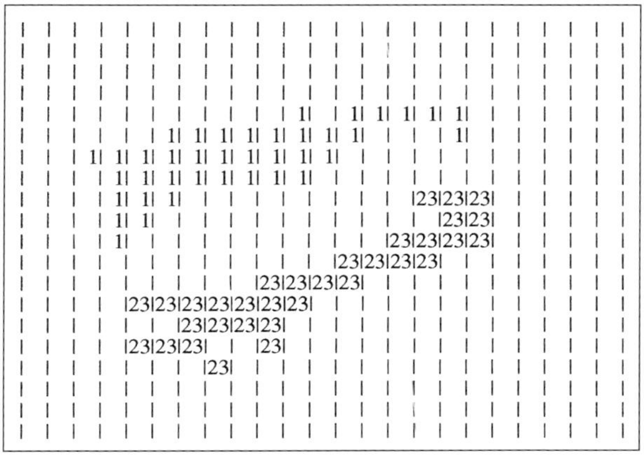

**Fig. 6.** The output of the deformation tracer for shear derived from the same ice motion product as in Figs. 4 and 5. The areas marked by "1's" and "23's" are two major new shear zones composed of combinations of complicated converence and divergence elements

cedure is applied which selects the first element in the priority list, i.e., the element with the largest deformation values, and flags its position on the marking template with a priority value. (In this initial case, the priority value would be "1" because the element is first on the priority list.) Then, its eight neighbors are checked to determine whether any of them show deformations larger than the threshold values. If any neighbor does, it is also included and marked with the same priority value, and neighbors of this newly-found element are further checked recursively. This search continues until no additional neighbors can be included. Now all the connected deformation elements will have the same priority values. Once an element is marked with one particular non-zero priority value, it is immune from subsequent searches. As soon as a search involving a particular priority number ends, the procedure goes down the priority list to the next available unmarked element. The entire search ends when there is no element left in the priority list.

To avoid including an isolated cell whose deformation value might be a result of random error, the procedure also eliminates patterns which contain single cells. The resulting pattern is shown in Fig. 6.

The same procedure is also used to find continuous areas without significant deformation. In such cases, the procedure searches the mutual connected cells which are not marked by the previous tracing process. Once tracing is done, a single value of divergence or convergence is calculated for the deformed or undeformed feature under tracing.

In addition to these five procedures which form modules of an integrated single program, a computer module using the AVS (Advanced Visualization System) has been developed to provide a link between the ice deformation and the original SAR images. This module has proven to be particularly useful in facilitating the detailed inspection of features exhibiting severe ice deformation.

## 4.4 Output

The basic output of the initial four component algorithms consists of two grid files; an additional ASCII file lists the deformation values calculated from each individual ice motion file. Since users often handle many ice motion files in a single batch process, the algorithm also produces a summary file that records the most critical deformation results. This facilitates the fast screening and extraction of significant ice deformation events from a series of ice motion files.

The grid files are in byte format consisting of integer values ranging from 0 to 250. In the $E_I$ file, the number 125 is taken as a neutral value, with 250 representing divergence of 31.25% or more and zero representing convergence of 31.25% or more. The larger the deviation from neutral, the larger the deformation. The other basic grid file is the shear file, in which zero represents non-shearing and 250 indicates an amount of 0.625 or larger in shear. The grid files can be easily displayed by most image processing and graphics software packages.

The ASCII file is a line-oriented text file that provides the deformation values for all the individual grid cells traceable by the algorithm. Also, some intermediate results are printed out for the user's information. The summary file is also a text file. It lists the number of ice motion records, the number of rows and columns of tie points, and the total and average divergence and convergence derived from each input ice motion file.

## 4.5 Accuracy in Estimation of Deformation Parameters

According to Fily and Rothrock (1990), the error sources in deformation calculations made on a grid basis consist of (1) mislocation of the tie points, and (2) nonlinearity of the boundaries of the deformed grids. The mislocation of tie points can be minor or severe depending on the situation. In this section we will discuss the impact of these errors in calculation of the net divergence, which is the difference of divergence and convergence, and for the separately aggregated divergence and convergence values for a set of grid cells.

The magnitude of errors caused by minor mislocation of tie points is our primary concern because of the pervasiveness in errors of this type. Fortunately, the magnitude of this type of errors can be quantified. According to a previous study (Li et al. 1995), it is safe to state that the random error in the corner locations of a deformed grid caused by minor mislocation due to the limitation of the resolution of images used in the motion tracking is within ±50 m. In the extreme case, for a cell with sides of 5 km, the size error caused by this random mismatch error can have a maximum value of ±4% if the random errors of the four corners of the cell happen to act additively. For a large number of cells, the standard deviation of the errors in deformation caused by minor mismatch of tie points is 0.82%. For features such as leads and large ice floes that are composed of $m$ cells, this type of error in deformation is further reduced by a factor of $\sqrt{m}$.

Separate aggregation of the divergence and convergence values for a set of cells introduces errors if there are errors in the individual cells. Those errors cannot cancel each other if the deformation is small. Suppose there is a uniform deformed area with a divergence value of $E_I$ and minor mislocations in the tie points which produce normally distributed errors in the values for individual cells. In such cases the separate aggregation of divergence and convergence values will result in an error in the averaged value which can be determined by the formula

$$\Delta\overline{E}_I = \frac{1}{\sqrt{2\pi\sigma}} \int_0^{\infty} (E_I - \overline{E}_I) e^{-(E_I - \overline{E}_I)^2/(2\sigma^2)} dE_I \qquad (5)$$

where $\sigma$ is the standard deviation of errors, $E_I$ the deformation for individual cells and $\overline{E_I}$ the average of the deformation. The error is a function of the averaged deformation (Fig. 7) with the error caused by separation decreasing as the deformation increases. For a nondeformed field, separation of divergence and convergence will result in an error of about $0.4\sigma$. According to the analysis discussed earlier, minor mislocation due to the limitation of the image resolution can cause a standard deviation of 0.82% in divergence and convergence calculations. Using the result of Eq. (5) shown in Fig. 7, it is estimated that separation will cause a 0.33% error in divergence and convergence estimation for a nondeformed field. According to Fily and Rothrock (1990), such an estimation will be too large for studies of deformation in pack ice when the motion field is sampled at time periods of 3 days or less (as is the case with ERS-1 images).

However, using the criterion of 4% for true deformation of individual cells, we may regard all the cells with $E_I$ and $E_{II}$ values lower than this criterion as the ones without significant deformation. Those cells are then aggregated to larger ice floes using the tracer procedure without actually inspecting the images, and a single divergence or convergence value is then calculated for each floe by simple algebraic summation of the individual divergence and convergence values of the cells within the floe.

Since such an aggregation will reduce the error of the averaged divergence/convergence value by a factor of $\sqrt{m}$, in a nondeformed field with 100 cells, the error caused by separation is about 0.03%. For a deformed field, separation result in errors smaller than 0.33% (Fig. 7). For example, assuming leads with an averaged divergence of 2.5%, which is equivalent to three times the standard deviation about the mean caused by minor mislocations, the separation would cause an error of 0.003% in estimation of

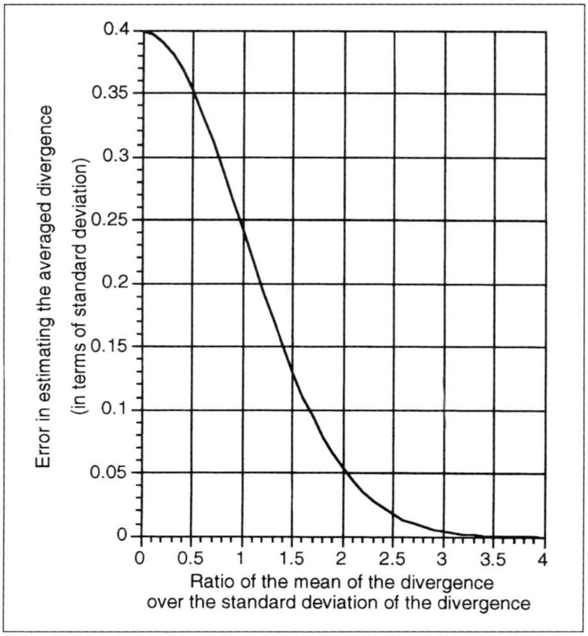


**Fig. 7.** Pattern of the errors in the separately aggregated divergence/convergence values as a function of the mean and standard deviation of the divergence/convergence.

divergence/convergence. Therefore, using open, closing, shear zones and ice floes identified in the tracer procedure as single entities, the error in the separate estimation of divergence/convergence can be significantly reduced. An accuracy of 0.1% in estimation of deformation for a field of 50 × 50 km or larger is achievable.

Severe mislocation caused by an erroneous match of tie points is another issue. Direct estimation of the impacts of this type of error is difficult. An evaluation of the ASF ice motion product (Stern et al. 1993) shows that 97–98% of the ice motion vectors are free of this type of error. The implication of this statement is that about 2–3% of the tie points are erroneous. However, four facts contribute to the minimization of the influence of this type of error on our analysis. (1) A strict quality assurance procedure is adopted at ASF. All of the questionable ice motion products, which are characterized by low values in an automatic rating of motion tracking, are subject to manual inspection by an ASF operator, and only those which pass the manual inspection are delivered to users. (2) The procedural use of the *edge point trimmer* eliminates the majority of bad tie points by trimming the edge of the ice motion products where most bad tie points are located. (3) Interior bad tie points with low values in the rating of motion tracking are treated as gaps, and their locations on the target image are recalculated in the *gap filler*. This tends to reduce the impact of location errors on the results of our analysis. (4) Aggregation of divergence and convergence values in leads and large ice floes through the *tracer* procedure further eliminates errors in deformation calculations when there are bad tie points in the interior of the concerned features. Furthermore, the random nature of this type of error tends to result in nonsignificant changes in the aggregated statistics.

Nonlinearity of the boundaries of the deformed grid cells can also be a source of error. It can have significant impact on the accuracy of the deformation calculation of the individual cells with severe shear. However, Fily and Rothrock (1990) suggest that the treatment of individual features, such as a lead or a large ice floe, as single entities also reduces errors of this type, because aggregation reduces subdivisions, and as a result, reduces the possibility of errors. Therefore, separate derivation of divergence and convergence values of the image area using the feature-aggregated deformation values instead of values from the individual grids is adopted in the following analysis.

Some extreme cases of severe mislocation or nonlinearity may cause large errors in aggregation of deformation values. However, those cases are also characterized by large deformation values. Our method does not eliminate the necessity of inspecting cases with large aggregated deformation values. However, it helps the investigator identity those events where further inspection is warranted.

## 4.6
## Examples and Discussion

The automatic algorithm described here makes possible the quick screening of a large number of ASF ice motion data files for regions of significant ice deformation. It also makes possible a quick extraction of intermediate-scale sea ice deformation data. Two examples are given here to demonstrate the use of this algorithm.

In the first example, 140 GPS ice motion data files, covering a period between October 10, 1991, and March 30, 1992, were collected from a circular region with a radius of 200 km, centered at 81° N, 170° W in the Chukchi Sea sector of the Arctic Ocean.

Although three of the ice motion data files from the month of October 1991 were based on a time interval of 6 days, the remaining 137 files were produced by determining the positions of identical ice features separated by a time interval of 3 days. A scaling factor of 0.5 has been applied to the results obtained from the three "6-day" files. Although this 6-month time series has an appreciable data gap in November and December 1991, and a minor gap in February 1992, certain basic features can still be observed from the time series and the distribution of the deformation values (Fig. 8).

First, although the majority of the 3-day scene aggregated convergence and divergence values are smaller than 1%, there are a few large events which exhibit deformation values well above 3%. Those events occur sporadically and occasionally may involve errors in ice motion tracking for vectors lying near the edge of the available motion field. For example, the extreme divergence value (26%) in our data set occurred between March 8 and 11. An examination of the associated image pair revealed that this unusually large value was an artifact caused by the fact that the input data bypassed the *edge trimmer* because the overlapping ice motion field on the two images was too small to survive the trimming procedure. This is an example of what we believe to be an important rule: when automated procedures result in unusually large (or small) values of a parameter, one should check the associated image pairs to verify that the values are real.

The mean value of the overall divergence between the ERS-1 SAR 3-day revisit time interval is 0.90% with a standard deviation 1.43%, while the mean value for convergence is 0.75% with a standard deviation of 1.04%. A simple linear regression calcula-

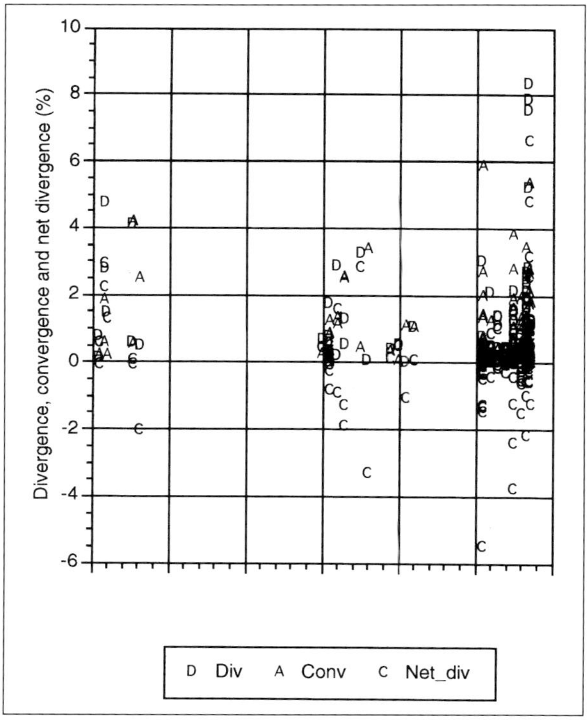


**Fig. 8.** Divergence and convergence in the central Arctic sample area as a function of time

tion shows that convergence and divergence are relatively uncorrelated ($r^2 = 0.076$, indicating that the regression line only explains 7.6% of the total variance).

The distribution of the intensity of the deformation can also readily be investigated (Fig. 9). Here it is obvious that more than 85% of the deformation values are less than 2%. When the accumulative curves of the aggregated convergence and divergence values are compared for the case of small deformations, both match each other quite well, indicating a fairly balanced alternation between the opening and closing of leads. However, for deformation values beyond 0.8%, the differences between two curves become significant, indicating that the pack ice is more susceptible to tension than to compression. This is reasonable in that the region under study contains a large amount of thick multiyear ice that is generally believed to be relatively undeformable in compression, yet fairly weak in tension.

The causes of the ice deformation, especially the cause of major ice divergence events, can also be studied. When ice divergence is compared with the average ice motion field for each particular time interval, it is found that the largest divergence values occur either during or just after periods of maximum ice motion (Fig. 10). For example, one of the largest observed ice divergence events occurred between March 26 and 29, 1992. The divergence and convergence values during this 3-day period are 1.56 and 0.91 respectively when averaged from 18 different motion files. Since ice divergence and convergence between March 27 and 30, 1992, are 0.82 and 0.29, the major deformation must have occurred on or about March 26. Prior to this date, the surface pressure dropped from 1012.7 mb at 0:00 GMT on March 24 to 999.4 at 12:00 GMT on March 25. Surface

**Fig. 9.** Accumulative percentages of divergence (*thick line*) and convergence (*thin line*)

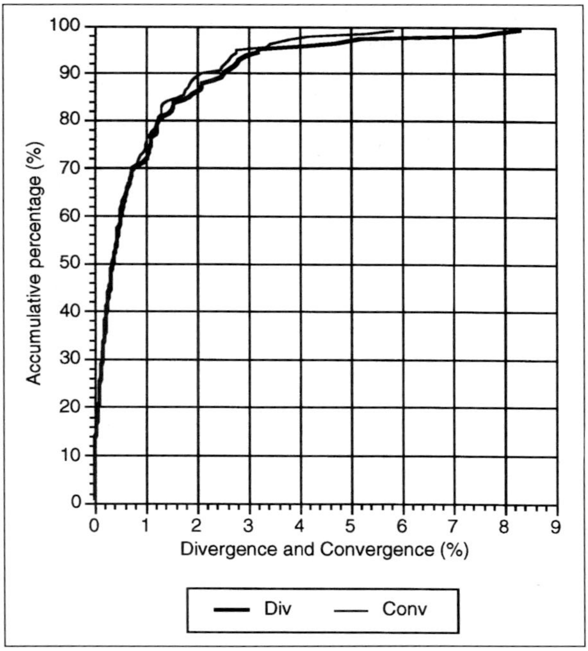

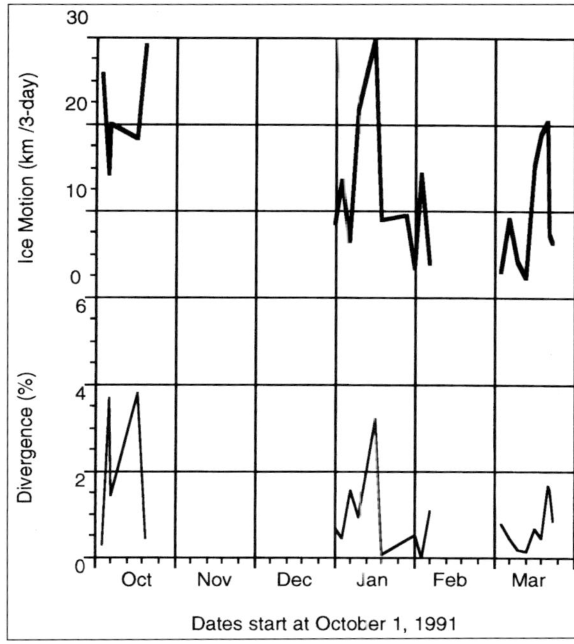


**Fig. 10.** The time series of ice motion and ice divergence in the central Arctic sampling area

wind speeds also dropped from around 10 m/s on March 24 to 2 m/s on March 26, and the wind direction changed from south to east. Consequently, the average ice drift dropped from 20 km between March 25 and 28 to 8.3 km between March 26 and 29. These facts suggest that a major storm passed through this area and that a large ice deformation event occurred following this storm when the wind decreased.

The algorithm also makes it easy to identify individual opening, closing, and shearing events on the SAR images and to readily obtain deformation measurements of mesoscale deformation events. For example, ice motion product 1182010 (Fig. 11) shows two new deformation features occurring between October 12 and 15, 1991. On the source image 5021401, the ice concentration is essentially 10/10s. However, on the target image 4947401, there are two major leads, both of which are correctly identified. Also, the areas of new openings can be estimated by the program; the upper lead has a total area of 41 km$^2$ while the middle lead has an area of 62.6 km$^2$, values that are printed out in the associated text file. Note that the lead at the lower right corner of the target image is not shown on the ice deformation product because the tie points which would have delineated this particular feature have been deleted by the edge point trimmer.

In the second example, 186 GPS ice motion data files were collected between January 23 and March 15, 1994, for the SIMI (Sea Ice Mechanics Initiative) project. During that time period, the SIMI ice camp, while drifting within 74.3–75.1°N and 153.4–156.3°W, was either inside or very close to the 3-day repeat swaths covered in the ERS-1 Second Sea Ice Phase. Therefore, we have detailed field data to compare with satellite data. We collected not only the GPS ice motion data files at the ice camp, but also data from areas further to the north and south of the camp along the satellite ascending paths. The algo-

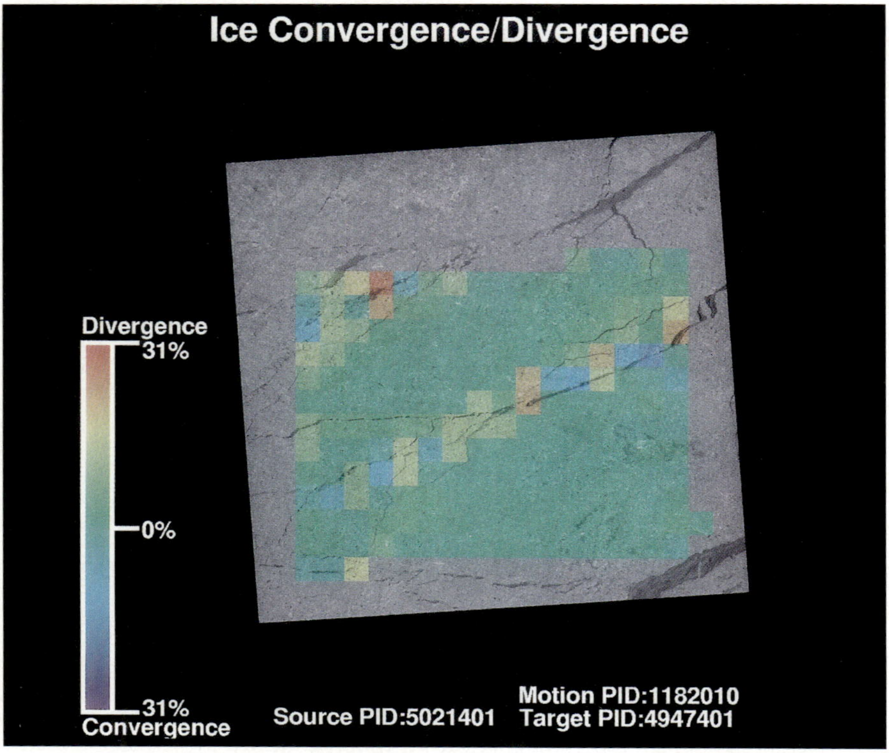


**Fig. 11.** New openings as shown on a pair of SAR images acquired on Julian days 285 and 288, 1991 and grid output from the ASF ice motion product 1182010 which is centered at 81°N, 172°20'W

rithm was then utilized to extract sea ice net divergence (Fig. 12) and shear (Fig.13) as a function of time and latitude along satellite swaths which ran from 71.5°N, 150.6°W to 78.1°N, 162.6°W diagonally. Wind velocity data (Fig. 14) were also extracted from Unidata for this period.

The results show several interesting phenomena. Between February 5 and 10, there was a major wind reversal event during which the wind direction reversed from southwest to northeast, while the wind speed dropped from more than 10 m/s to 3–4 m/s and then returned to more than 10 m/s. Because the southern part of the swath is very close to near-coastal open water and the sea ice there was thin and less consolidated, sea ice in this region easily changed direction under wind stress and yielded to compression when it was pushed against the coast by the northeast wind. As a result, this part of sea ice experienced significant convergence. By contrast, the northern part was covered by multiyear pack ice, which was thicker with a larger inertia and therefore tended to maintain the original moving direction. With the event of wind reversal, the area in the middle experienced divergence and shear due to the discrepancy between directions of motion in the areas to the north and to the south of the camp. The camp site happened

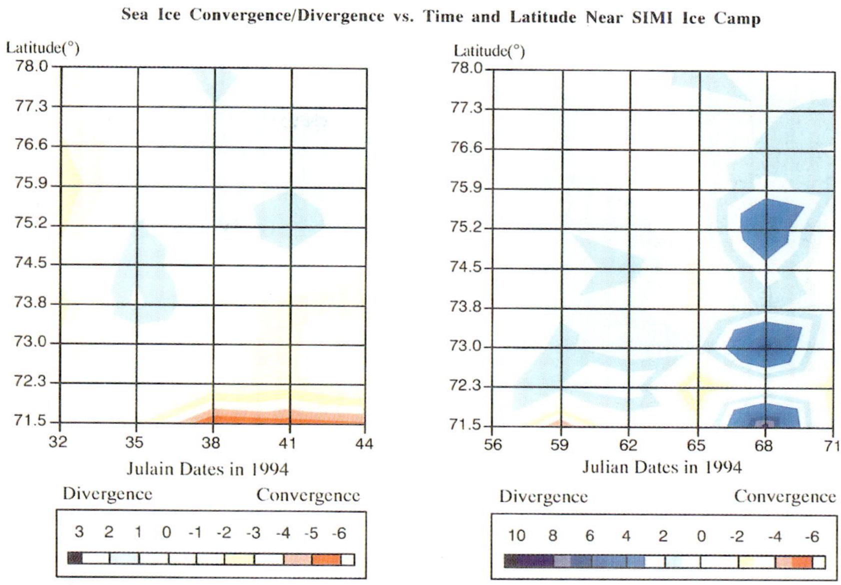


**Fig. 12.** Sea ice convergence/divergence as a function of time and latitude around the SIMI (Sea Ice Mechanics Initiative) camp site between January and March, 1994

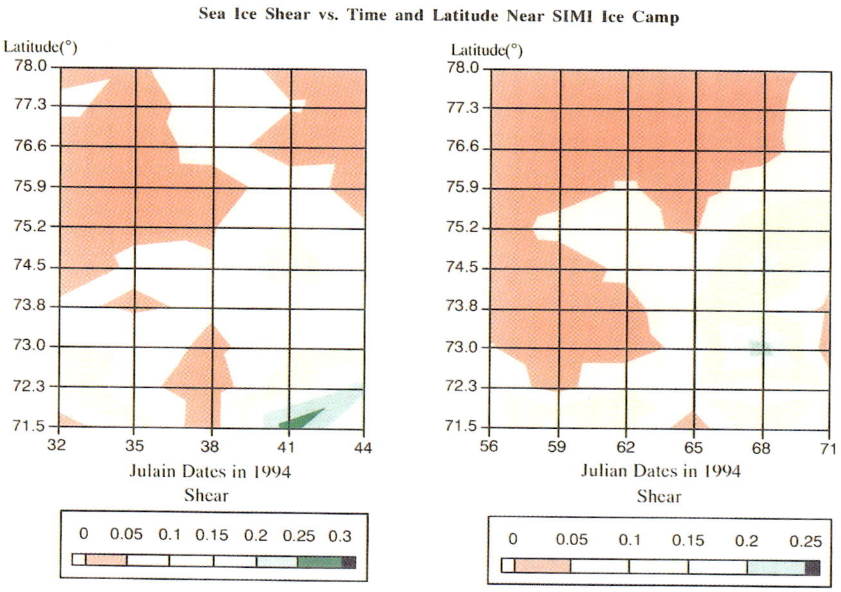


**Fig. 13.** Sea ice shear as a function of time and latitude around the SIMI camp site between January and March, 1994

**Fig. 14.** Wind velocity around the SIMI camp site between February and March, 1994. *u-vel* is the eastward velocity and *v-vel*

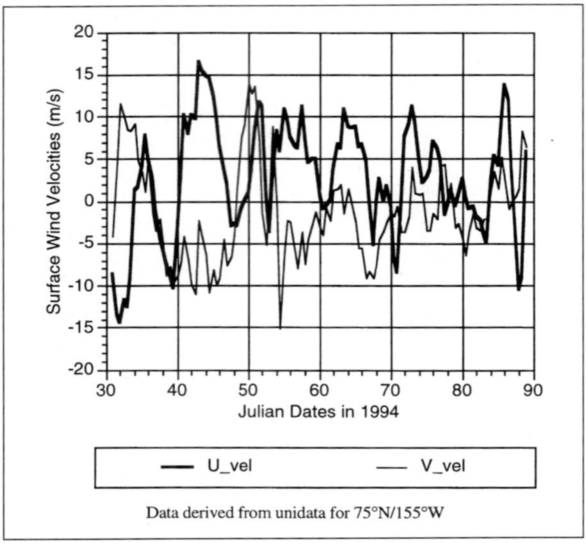


to be at this transitional location and recorded the divergence and shear by buoy measurements.

Another significant event occurred between March 9 and 12, 1994. Before that time period, there was an interval of about 10 days during which persistent west winds prevailed with an average speed around 8–10 m/s. According to Overland (1994, pers. discussion), fractures had already developed under stress but did not show simply because they did not present themselves as conspicuous features such as leads. However, a change of wind direction by 90° from the original west to north gave the fractures an opportunity to open, possibly with significant shear. This is exactly what happened during the time period between March 9 and 12 (marked on the $x$-axis as day 68). Because the same mechanism applies to all pack ice, the entire pack experiences significant divergence and shear with the northern part undergoing less significant deformation simply because of the greater strength of the thicker and larger multi-year ice there.

## 4.7 Summary

An understanding of ice deformation, especially of sea ice divergence, which produces leads resulting in thin ice and open water, and sea ice convergence which produces pressure ridges, is critical to many aspects of sea ice research. This chapter describes one way in which ASF ice motion products can be used to extract quantitative descriptors of such ice deformation processes. A special, fully automatic computer program has been designed, tested, and applied for this purpose. The program consists of five components: a tie point locator, an edge point trimmer, a gap filler, a deformation generator, and an extension tracer. The accuracy of the program is ±1% for a single grid cell and ±0.1% for a motion field with a size greater than 50 × 50 km.

By using this program, a large number of the ASF ice motion products that are produced by the GPS can be easily analyzed without the investigator looking at either the original images or the ice motion fields. The results can then be compiled as statistics for intermediate-scale analysis. Also, since the majority of the ice motion products do not exhibit substantial deformation, they can easily be excluded from further investigation, allowing one to focus on the relatively small number of ice motion fields that do show appreciable deformation.

**Acknowledgments:** This work was supported by NASA grant NAGW-2716 to S. Li and W.F. Weeks. We would like to thank the ASF data processing group for providing the ERS-1 SAR images and the ASF ice motion products, Harry Stern at the Applied Physics Laboratory of the University of Washington for allowing us to use the ice motion display tool that he developed, and Rick Guritz for technical assistance with the ASF Interactive Image Analysis System (IIAS).

## References

Andreas EL (1980) Estimation of the heat and mass fluxes over arctic leads. Monthly Weather Rev 108: 2057–2063.

Badgley FI (1966) Heat budget at the surface of the Arctic Ocean. Proc Symp Arctic Heat Budget and Atmospheric Circulation, RM-5233-NSF, The RAND Corporation, Santa Monica, California, pp 267–278

Bonbright DI (1984) PODS SSM/I functional requirements (version 1.0). Jet Propulsion Laboratory Document 715–63, Pasadena, California

Colony R, Thorndike AS (1984) An estimate of the mean field of arctic sea ice motion. J Geophys Res 89, no C8: 10623–10629

Colony R, Thorndike AS (1985) Sea ice motion as a drunkard's walk. J Geophys Res 90, no C1: 965–974

Coon MD, Echert DC, Knoke TS (1993) Pack ice anisotropic constitutive model. Proc IAHR Ice Symp 1992, Banff, Alberta, International Association for Hydraulic Research, Delft, Netherlands, pp 1188-1205

Fily, M, Rothrock, DA (1990) Opening and closing of sea ice leads: digital measurements from synthetic aperture radar. J Geophys Res 95, no C1: 789–796

Hibler III WD , Weeks WF, Ackley SF, Kovacs A, Campbell WJ (1973) Mesoscale strain measurements on the Beaufort Sea pack ice (AIDJEX 1971). J Glaciol 12, no 65: 178–206

Hibler III WD , Weeks WF, Kovacs A, Ackley SF (1974) Differential sea ice drift I: spatial and temporal variations in sea ice deformation. J Glaciol 13, no 69: 437–455

Koerner RM (1973) The mass balance of sea ice of the Arctic Ocean. J Glaciol 12, no 65: 173–185

Kwok R, Pang A (1992) Performance of the ice motion tracker at the Alaska SAR Facility. Proc Int Geosci Remote Sens Symp IGARSS '92, May 26–29, Houston, Texas, pp 588–590

Kwok R, Cunningham G (1993) Geophysical processor system data user's handbook (version 2.0). Jet Propulsion Laboratory Document 715–63, Pasadena, California

Kwok R, Curlander JC, McConnell R, Pang SS (1990) An ice-motion tracking system at the Alaska SAR Facility. IEEE J Ocean Eng 15, no 1: 44–54

Li S, George T (1991) ERS-1 data products review meeting, Jet Propulsion Laboratory Document, Dec 17, 1991, Pasadena, California

Li S, Cheng Z, Weeks WF (1995) A grid-based algorithm for the extraction of intermediate-scale sea-ice deformation descriptors from SAR ice motion products. Int J Remote Sen 16, no 17: 3267–3286.

Makshtas AP (1991) The heat budget of Arctic ice in the winter, Int Glaciol Soc, Cambridge, England

Maykut GA (1982) Large scale heat exchange and ice production in the Central Arctic. J Geophys Res 87, no C10: 7971–7984

McConnell R, Kwok R, Curlander JC, Kober W, Pang SS (1991) Y–S correlation and dynamic time warping: two methods for tracking ice floes in SAR images. IEEE Trans Geosci Remote Sens 29, no 6: 1004–1012

Morison JH, McPhee MG, Curtin TB, Paulson CA (1992) The oceanography of winter leads. J Geophys Res 97, no C7: 11199–11218

Overland JE, Walter BA, Davidson KL (1992) Sea ice deformation in the Beaufort Sea. Preprints of the 3rd Conf Polar Meteorology and Oceanography, American Meteorological Society, pp 64–67

Overland JE, Walter BA, Curtin TB, Turet P (1995) Hierachy and sea ice mechanics: a case study from the Beaufort Sea. J Geophys Res 100, no C3: 4559–4571

Pritchard RS (ed) (1980) Sea ice processes and models. Proc Arctic Ice Dynamics Joint Experiment (AIDJEX). University of Washington Press, Seattle, Washington

Rothrock DA (1986) Ice thickness distribution–measurement and theory. In: Untersteiner N (ed) The geophysics of sea ice. Series B: Physics, vol 146, Plenum Press, New York, pp 551–575

Stern H, Kwok R, Li S, Milkovich M (1993) Ice motion validation team report. In: Kwok R, Cunningham G (eds) Alaska SAR Facility geophysical processor system data user's handbook (version 2.0). Jet Propulsion Laboratory Document D-9526, Pasadena, California, pp 101-108

Thorndike AS (1986) Kinematics of sea ice. In: Untersteiner N (ed) The geophysics of sea ice. Series B: Physics, vol 146, Plenum Press, New York, pp 489–550

Thorndike AS, Colony R (1977) Estimating the deformation of sea ice. Proc. 4th Int Conf on Port and Ocean Engineering Under Arctic Conditions POAC'77, vol 1, Memorial University of Newfoundland, St. John's, Newfoundland, pp 506–517

Thorndike AS, Colony R (1980) Large-scale ice motion in the Beaufort Sea during AIDJEX, April 1975–April 1976. In: Pritchard RS (ed) Sea ice processes and models: Proc Arctic Ice Dynamics Joint Experiment (AIDJEX), University of Washington Press, Seattle, Washington, pp 249–260

Thorndike AS, Rothrock DA, Maykut GA, Colony R (1975) The thickness distribution of sea ice. J Geophys Res 80, no 33: 4501–4513.

# CHAPTER 5

# Fusion of Satellite SAR with Passive Microwave Data for Sea Ice Remote Sensing

S.G. Beaven and S.P. Gogineni

## Contents



## 5.1 Introduction

In this chapter we discuss an approach for fusing satellite synthetic aperture radar (SAR) with passive microwave data for improved ice type concentration estimates. The algorithm is demonstrated on satellite image data obtained after the onset of freeze-up in the central Arctic. During freeze-up new ice growth results in significant amounts of young first-year and thin ice. These are as important as multiyear ice and open water in modulating the heat flux between the ocean and atmosphere in the central Arctic

Ice type concentration estimates based on the NASA Team (NT) algorithm are inaccurate during the melt and freeze-up seasons because of confusion between multiyear and first-year ice. Ambiguities between ice types are caused by dynamic changes in the passive microwave signatures of sea ice during these seasons and are related to the use of tie points to define pure ice type signatures.

We combined data from the European Remote Sensing Satellite (ERS-1) SAR and the Defense Meteorological Satellite Program (DMSP) Special Sensor Microwave/Imager (SSM/I) to improve estimates of ice type concentration after the onset of freeze-up. We

Analysis of SAR Data of the Polar Oceans
Edited by C. Tsatsoulis and R. Kwok

developed a technique for incorporating information from the ERS-1 SAR into a passive microwave algorithm to improve ice type concentration estimates. We used ERS-1 SAR data to estimate multiyear ice fraction through a dynamic thresholding technique. Using the SAR-derived multiyear ice concentration in a modified version of the NT algorithm we obtained an improved first-year ice concentration estimate. This technique demonstrates that the use of both active and passive satellite data can remove or reduce the ambiguities that occur in estimating ice type concentration.

## 5.2
## Data Fusion Background

### 5.2.1
### Data Fusion for Sea Ice Remote Sensing

It has been speculated that a combination of active and passive microwave measurements will be important in sea ice remote sensing (Cavalieri et al. 1990). A number of multisensor studies of the microwave properties of sea ice have been carried out over the last three decades with the primary goal of determining backscatter and emission characteristics of sea ice. Current techniques for exploitation of satellite microwave data use only single sensor information. The use of single-sensor data results in ambiguities in the estimations of geophysical parameters. The fusion of multisensor data has been discussed as a means for removing these ambiguities. Current spaceborne SAR systems used by the remote sensing community (ERS-1 and RADARSAT) are capable of resolution of a few tens of meters, whereas the current series of spaceborne passive microwave imagers (SSM/I) have spatial resolution on the order of tens of kilometers, depending on the wavelength. This limits techniques for combining these data sets to those that can overcome or exploit this disparity in resolution.

Combining active and passive imagery has been previously suggested as an approach to improving estimates of key sea ice geophysical parameters such as thickness, mass distribution and extent (Burns et al. 1987; Livingstone, et al. 1987; Drinkwater et al. 1991; Collins 1992; Grenfell et al. 1992). However, until recently multisensor fusion approaches have not been applied to this problem. Below we discuss the three basic architectures for multisensor fusion, followed by a hybrid approach that shows promise for fusion of active and passive microwave data.

### 5.2.2
### Data Fusion Architectures

The purpose of fusing multisensor data is to obtain better estimates of geophysical parameters or new information that could not be obtained with any single sensor. Three basic architectures are utilized in multisensor data fusion processes. These are characterized by the level at which the data are combined or fused – data level, feature level and decision level (Waltz and Llinas 1990) – and are defined for an identity declaration process requiring fusion of multisensor data. Although multisensor fusion problems may not necessarily fall into the identity declaration category, any fusion process is typically characterized by one of these three categories, or a hybrid of these. The architectures are shown in Fig. 1.

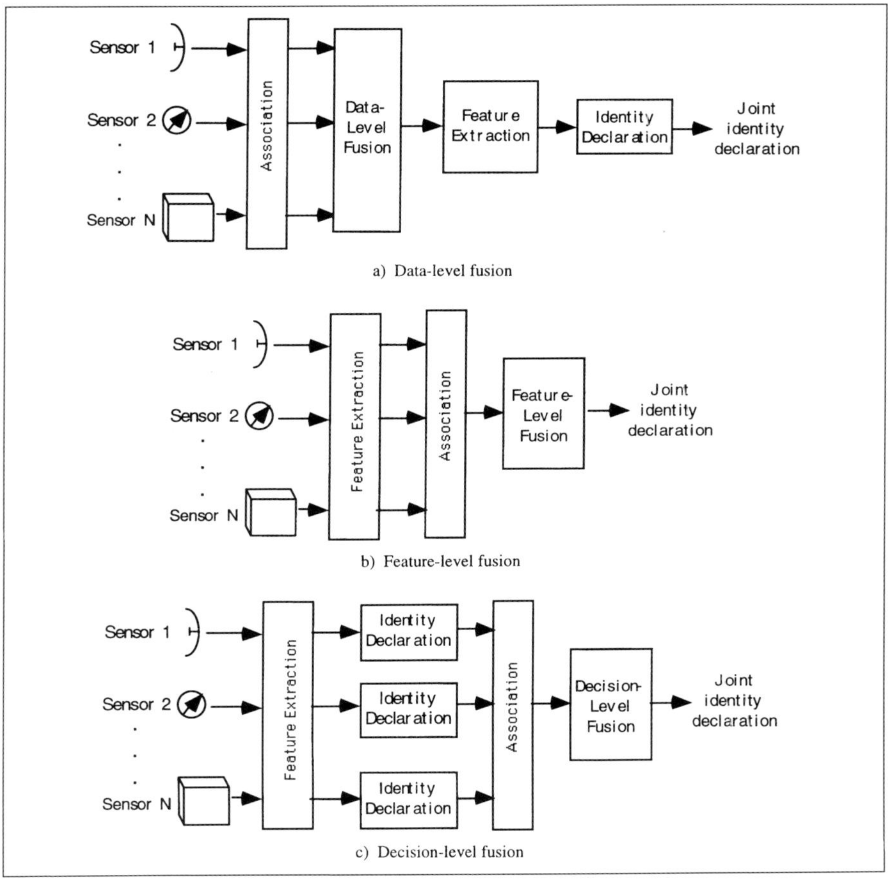


**Fig. 1** a-c. Data fusion architectures for identity declaration from multiple sensors: **a** data-level fusion, **b** feature-level fusion, and **c** decision-level fusion.

Data-level fusion occurs in techniques in which the raw data or pixels from different sensors are fused prior to any feature extraction or association operations. Usually, limited preprocessing is required for data from each source. Therefore, most of the extensive processing follows the fusion of the data. This architecture is limited to sensors that are identical or very similar. If the feature extraction process is applied to the data before they are combined, but individual identity declaration is not performed separately on each sensor's data, then the process is considered a feature-level fusion process. Decision-level fusion occurs when the data are merged after products are derived from each of the sensors and the results are combined and a joint declaration of identity is made. This architecture differs from the feature-level fusion because identity declaration is obtained on each sensor individually before data are combined for a joint identity declaration. Decision-level fusion is used when quite different sensors are to be employed.

In addition to the basic architectures outline above, it is possible to develop a technique that is actually a hybrid of two of these. Data fusion is thought to be most beneficial if it is performed "as close to the sensor as possible," although the level at which fusion is performed depends on the nature of the problem and the characteristics of the sensors to be used or developed.

## 5.3 Satellite Passive Microwave Algorithms

Satellite microwave radiometers have been used for estimating the fractional coverage of sea ice nearly continuously since the 1970s (Carsey 1982; Gloersen et al. 1992). The NT algorithm and the bootstrap algorithm have been used to determine sea ice concentration and type in the Arctic (Cavalieri et al. 1984; Comiso 1986; Gloersen and Cavalieri 1986; Gloersen et al. 1992). The NT algorithm was originally developed to determine sea ice concentration from daily averaged Scanning Multichannel Microwave Radiometer (SMMR) microwave brightness temperatures over the Arctic and has been adapted to DMSP SSM/I data. It is based on the differences in the emissivity characteristics of multiyear ice, first-year ice and open water at 37 GHz and 19 GHz and the difference between vertically and horizontally polarized emissivity at 19 GHz. It was developed primarily for use during winter months.

The mean difference between the estimates of total ice concentration during fall (late September through November) obtained using the NT algorithm and from the Landsat imagery is less than 1%, with a standard deviation of ±7.4% (Steffen and Schweiger 1991). The results for fall are indicative of the performance of the algorithm in winter because of similar ice conditions. Steffen and Schweiger (1991) have also shown that the mean difference between the NT and Landsat-based estimates of total ice concentration is 2.4% during spring, with a standard deviation of ±5.9%. These results were obtained using hemispheric tie points; they have shown that some improvements can be made through the use of local tie points.

During the summer melt period, however, ice concentration estimates from the NT algorithm differ from Landsat estimates by as much as 46%, with a mean difference of 11% (Steffen and Schweiger 1991). In addition, the NT algorithm underestimates multiyear ice concentration during fall and through the early winter. When air temperatures are around 0° C, it overestimates first-year ice concentration and underestimates multiyear ice concentration (Carsey 1985; Cavalieri et al. 1990). This effect continues into the onset of freeze-up.

Using SMMR data, Rothrock and Thomas (1992) have shown that multiyear ice concentration in the central Arctic just after freeze-up is inconsistent with the total ice concentration at the end of the summer melt. The NT algorithm yields a multiyear ice concentration of about 0.4 in the central Arctic, whereas the total ice concentration at the end of summer was about 0.7. Thomas (1993) has shown that the multiyear concentration estimates from the NT algorithm reach a minimum during September, and this minimum is not believed to be caused by a real decrease in multiyear ice concentration, but by changes in the signature of multiyear ice. The underestimation of multiyear ice is attributed to the use of constant pure ice type signatures in the NT algorithm. He has shown that by accounting for this variability through the use of daily mean signatures, estimates of multiyear concentration may be improved. This appears to be associated with the use of constants as tie points for defining each of the surface types: multiyear ice, first-year ice, and open water (Thomas 1993).

A detailed summary of the performance of the algorithm and the validation of derived ice classification for SSM/I data are given in the literature (Cavalieri et al. 1991; Steffen and Schweiger 1991). The 25-km ice concentration algorithm is also described in detail in the SSM/I brightness temperature grid user's guide (NSIDC 1992). Since the algorithm is explained in several of the references cited above, it is not repeated here.

## 5.4 Fusion Approach for Sea Ice Classification

When sensors differ widely in resolution it is impossible to merge the data at the pixel level. In this case the most likely level to merge the data is at the decision level (Waltz and Llinas 1990; Collins 1992). One possible approach that may allow for fusing data from sensors with different resolutions is to enhance or improve the resolution of the lower-resolution sensor and combine the data at the data level. When the resolution scales differ greatly, as in the case of satellite SAR and SSM/I data, an alternative is to develop a fusion technique that uses features or products derived from the high resolution sensor data with the data from the low resolution sensor. This solution may be considered a multilevel, or hybrid fusion approach, since the SAR data are used at the information level and passive microwave data are used at the data level (Fig. 2). This hybrid-level fusion approach has been applied to the fusion of active and passive microwave data for ice type concentration. This approach removes the ambiguity between multiyear and first-year ice that is present in passive microwave data during the freeze-up season. This relies on the fact that SAR-derived multiyear ice concentration can reliably be obtained from SAR under cold conditions (Kwok et al. 1992).

The current state of our algorithm for combining satellite active and passive data consists of two steps. First, the SAR data are used to obtain an estimate of multiyear ice concentration. This estimate is currently obtained through the use of a local-area thresholding method. Second, this estimate is used to constrain the inversion of multichannel passive microwave data for ice type concentration.

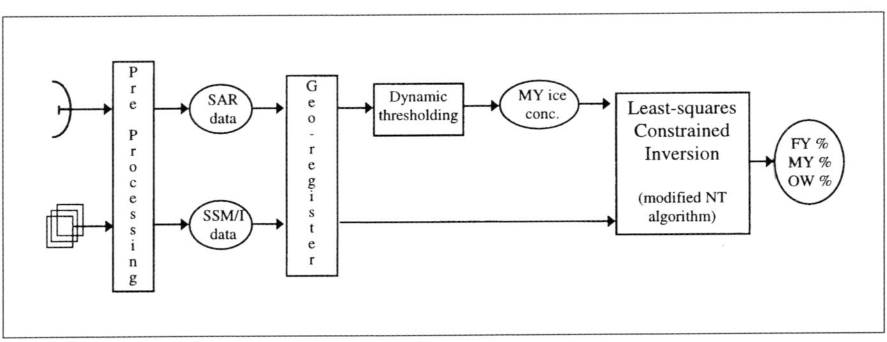


**Fig. 2.** Hybrid approach for fusion of ERS-1 SAR and SSM/I data for sea ice type concentration. *MY* multiyear ice, *FY* first-year ice, *OW* open water. [Copyright 1996 IEEE (Beaven et al. 1996)]

### 5.4.1
### Determination of Multiyear Ice Concentration from ERS-1 SAR

Multiyear ice concentration can be derived from spaceborne SAR systems by exploiting the higher backscatter from multiyear ice than from first-year ice and calm open water (Kwok et al. 1992; Haverkamp et al. 1993, 1995). We used the dynamic local thresholding (DLT) method described in detail by Haverkamp et al. (1995), because it appears to have some advantages for ice classification during seasonal transitions. This method does not depend on absolute image intensity levels and predetermined thresholds, which can be problematic during these periods. However, since it was used as a first stage of a knowledge-based algorithm for determining ice types, it requires addition-

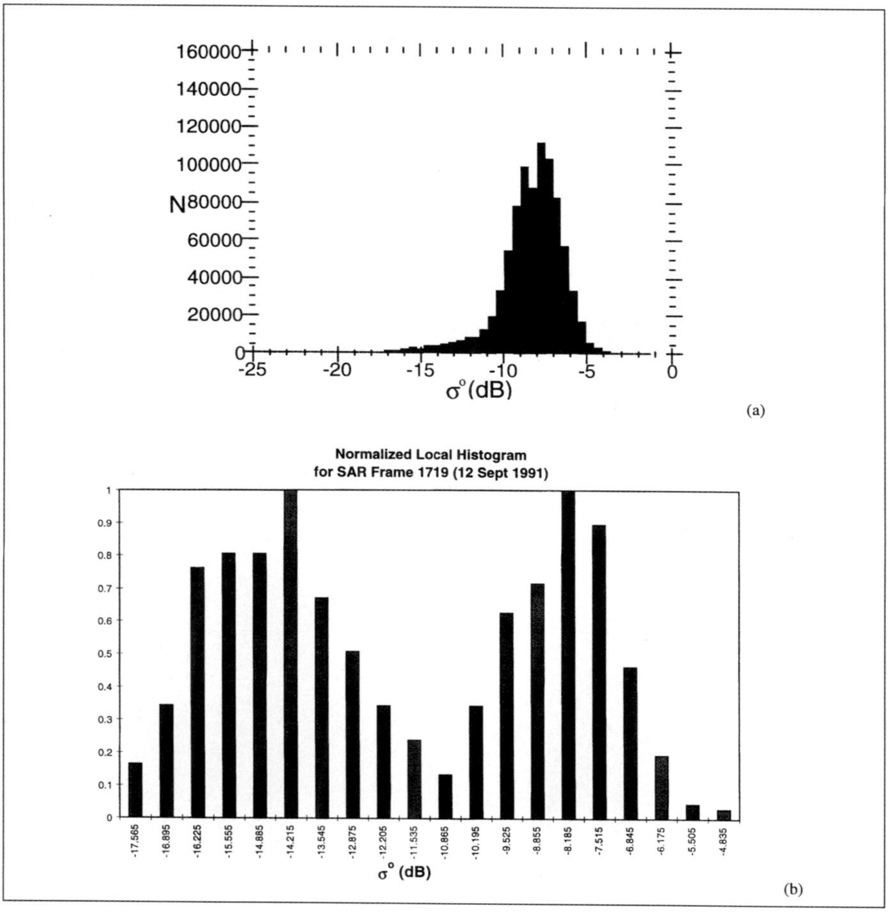


**Fig. 3 a,b.** Comparison of local and global histograms from ERS-1 SAR (frame 1719, September 12, 1991). **a** Global histogram shows no distinct bimodality, whereas **b** local histogram shows distinct bimodal distribution. This demonstrates that distributions over local areas may be used to separate classes of targets, although the global distribution does not. [Copyright 1996 IEEE (Beaven et al. 1996)]

al effort to assign the result of local thresholding to each particular ice type. The technique, as used in our multisensor fusion process, is described below.

Local area thresholding is based on computation of histograms over small, overlapping regions of an image that are assumed to be small enough to contain only two distinct classes based on image intensity. The threshold is established on the basis of estimates of the distributions of the two classes in each subarea. Thresholds based on local area statistics are used since local changes in mean backscatter often occur within SAR image frames. These local changes in backscatter cause fluctuations in image intensity, which may cause problems in global thresholding techniques. We used subregions that were $32 \times 32$ pixels, with 50% overlap, from low-resolution ERS-1 SAR imagery. Histograms computed from entire ERS-1 SAR image frames do not appear to be bimodal, as shown in Fig. 3a, for example. However, subregions of this same image yield distributions that appear to be bimodal, as shown in Fig. 3b. The local area histogram shows a bimodal distribution over the subregion, which is not apparent from the global histogram. This demonstrates that although no clear threshold may be obtained from a global distribution of SAR data, local distributions may show distinct classes. Dynamic thresholding is based on this premise and is described in detail in Haverkamp et al. (1995).

We used the DLT algorithm to separate each SAR image frame into three classes. These classes are not defined by the method, and must be assigned to ice type through further processing or through inspection by an expert. To simplify the study of multisensor fusion for ice classification we identified which category corresponds to multiyear ice through visual identification of multiyear floes within the class. An automated method could require use of shape or other information as described in Haverkamp et al. (1995). The multiyear ice concentration is then estimated over areas defined by SSM/I grid boundaries. These boundaries were determined by the colocation methods discussed below.

## 5.4.2 Active/Passive Image Registration

The use of multisensor data first requires registration or mapping of the data to a common coordinate system. When the two sensors have drastically different resolution it is best to map the low-resolution data to the same coordinates as the high-resolution data. We remapped the SSM/I brightness temperature grids to the ERS-1 SAR image frame coordinates. The process consists of computing the latitude/longitude pairs for each pixel in the SAR image, then determining the corresponding pixel location within the SSM/I grids for the Arctic. These are stored in look-up tables for mapping the SSM/I pixel locations to the SAR image frame. A summary of both steps is given below.

The geographic locations of the corner points of each SAR image frame are supplied by the ERS-1 Processing and Archive Facilities (PAFs) in the header file for each ERS-1 SAR image. We use these corner points to derive the latitude and longitude coordinates for each pixel within the image using the algorithm given by Li (1994). This algorithm uses the fact that ERS-1 SAR imagery covers a small spatial area, eliminating the need for using the more general formulation given by Curlander and McDonough (1992). The transformation from ERS-1 SAR image coordinates to geographical coordinates assumes that the boundaries of the image follow the great circle on the Earth. This assumption results in a maximum error of about 10 m for SAR frames of 100 × 100 km. Image coordinates along each side of the image frame are first computed by interpolation from the end points using a bilinear interpolation in Cartesian coordinates. The corner points must first be converted to three-dimensional Cartesian coordinates, using the local Earth radius. This radius is computed based on an elliptical model of the Earth and the center latitude and longitude coordinates of the image. Once the corner points are converted to Cartesian coordinates, the coordinates along any line are computed using the bilinear interpolation,

$$X' = X_1 \frac{d_2}{d_1 + d_2} + X_2 \frac{d_1}{d_1 + d_2} \tag{1}$$

for each of the coordinates, where $X'$ is any point between the end points $X_1$ and $X_2$ and the distance from $X'$ to the end points is given by $d_1$ and $d_2$. Equations for the other two coordinates, $Y'$ and $Z'$, are similar to Eq. (1). In practice, the interpolation is first performed along two opposite sides of the image. Then each row or column uses the interpolated edges as end points. The final step involves transforming from Cartesian coordinates into spherical coordinates to obtain the latitude and longitude for each pixel within the SAR image.

The SSM/I data, obtained from the US National Snow and Ice Data Center (NSIDC), are mapped to polar stereographic coordinates and the brightness temperature data are stored on a rectangular grid. Latitude and longitude coordinates for each pixel of the SSM/I data were obtained with a conversion program supplied by NSIDC. The reported accuracy of these location programs is less than 1 km, which is approximately the same accuracy as the SSM/I sensor data record geographic location values (NSIDC 1992). The conversion from latitude/longitude values to image pixel locations was performed in two steps. First, we transformed the latitude/longitude into polar stereographic coordinates for the polar regions. Then we used these to find the row and column values for the particular pixel in which the latitude/longitude point was located. The algorithm is described in detail in the NSIDC report (1992).

Using the two-step process for obtaining the SSM/I grid locations for each SAR image pixel we generated SSM/I-gridded brightness temperature images that were mapped to the coordinates of each SAR image frame used in this study. We then used these pixel values to find the brightness temperature that corresponded to each ERS-1 SAR pixel and a resampled SSM/I brightness temperature image was obtained. The brightness temperatures from these resampled images were then used in the fusion algorithm with the corresponding SAR-derived multiyear ice fraction estimate.

### 5.4.3
### Relationship to Freeze-Up Detected with SAR

We used the coregistered ERS-1 SAR and SSM/I imagery to investigate the relationship between passive microwave-based estimates of ice concentration and radar backscatter signature. We used the NT algorithm to compute sea ice type concentrations on the resampled SSM/I image data using hemispheric tie points. The mean values over each SAR frame area are plotted in Fig. 4 to demonstrate the changes in relative concentrations of first-year and multiyear ice estimated by the NT algorithm during this time frame. The mean ice type concentrations for SSM/I pixels are located within the SAR frame are listed in Table 1. The estimated changes in first-year and multiyear ice con-

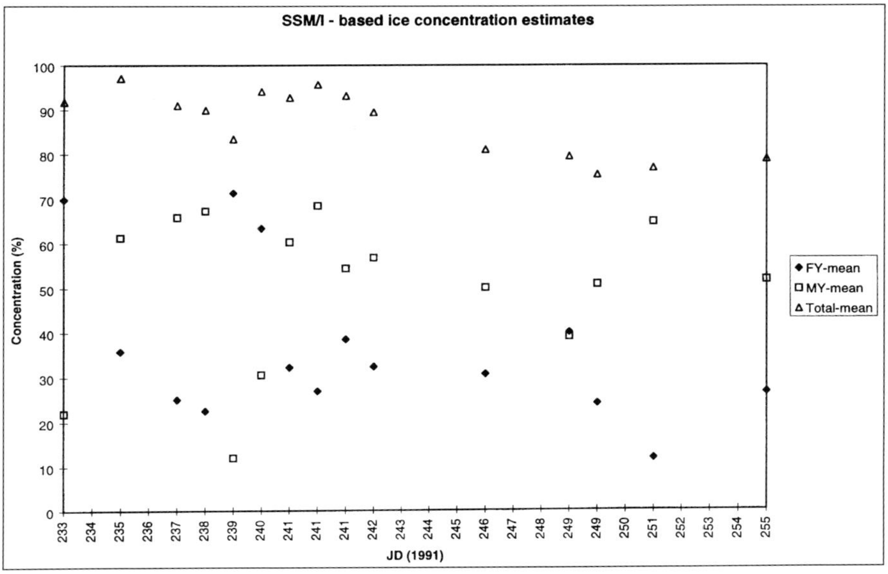


**Fig. 4.** Mean values of ice type concentration, derived from the NASA Team (NT) algorithm. The averages are computed for areas enclosed by the SAR images used later in the fusion process

**Table 1.** SSM/I-based ice concentrations

| Date and corresponding SAR frame numbers | | Ice concentration | | |
|---|---|---|---|---|
| Julian day | SAR frame | MY ice (%) | FY ice (%) | Total ice (%) |
| 233 | 1827 | 22.0 | 69.8 | 91.8 |
| 235 | 1863 | 61.3 | 35.8 | 97.1 |
| 237 | 1791 | 65.8 | 25.1 | 91.0 |
| 238 | 1809 | 67.3 | 22.6 | 89.9 |
| 239 | 1701 | 12.1 | 71.3 | 83.4 |
| 240 | 1791 | 30.7 | 63.4 | 94.1 |
| 241 | 1755 | 60.4 | 32.3 | 92.7 |
| 241 | 1791 | 68.5 | 27.0 | 95.6 |
| 241 | 1809 | 54.5 | 38.6 | 93.1 |
| 242 | 1701 | 56.9 | 32.5 | 89.4 |
| 246 | 1719 | 50.2 | 30.9 | 81.0 |
| 249 | 1719 | 39.3 | 40.2 | 79.5 |
| 249 | 1827 | 51.0 | 24.4 | 75.4 |
| 251 | 1701 | 65.0 | 11.9 | 76.9 |
| 255 | 1719 | 52.0 | 26.9 | 78.9 |

MY multiyear ice; FY first-year ice

centration are the result of misclassification of some multiyear ice as first-year ice, as explained below.

Under summer melt conditions the frequency dependence of emissivity of multiyear ice becomes similar to that of first-year ice (Cavalieri et al. 1984). Field measurements of emissivity of multiyear ice obtained by Grenfell (1992) during the coordinated Eastern Arctic Research Experiment (CEAREX'88) demonstrate the change in spectral response of emissivity as the temperature drops during the early stages of the freeze-up season. The spectral response of multiyear ice is relatively flat. The emissivity difference between 19 GHz and 37 GHz becomes more pronounced as the freeze-up progresses. This results in the gradient ratio for multiyear ice moving from being similar to first-year ice to cold multiyear ice. These field measurements showed that variations in emissivity from multiyear ice during this season can vary anywhere between the first-year and multiyear tie points of the NT algorithm as the freeze-up progresses.

Apparent changes in concentration from multiyear to first-year ice occur at approximately the same time as observed decreases in radar backscatter from multiyear ice. High values of SSM/I-derived first-year ice concentration correspond to drops in ERS-1 SAR average $\sigma^o$ as shown in Fig. 5. This was shown to be caused by refreezing of moisture in the upper layers of the sea ice volume (Beaven and Gogineni 1994; Beaven 1995). The changes in the radar backscatter signatures were caused by the effects of moisture on or near the ice surface as the temperature fluctuates about freezing. At temperatures around 0 °C the of sea ice increases dramatically, decreasing penetration depth. The decrease in penetration depth causes a reduction in backscatter and a corresponding increase in emissivity. The passive microwave data appear to be more sensitive to this

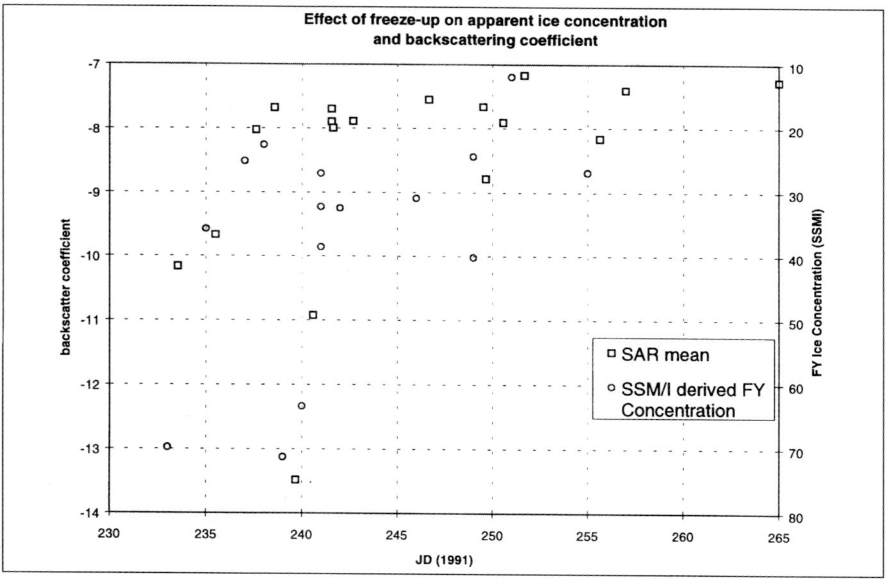


**Fig. 5.** Comparison of mean $\sigma^0$ values computed from the ERS-1 SAR images and the first-year ice concentration obtained with the NT algorithm from SSM/I data

moisture than the SAR data. Fluctuations of 1–2 dB in backscatter roughly correspond to SSM/I-derived concentration fluctuations between 10 and 40%. SSM/I appears to be more sensitive because of the higher frequencies (19, 37 GHz), relative to ERS-1 (5.3 GHz). Since the backscatter fluctuations after freeze-up are small it may be feasible to estimate the multiyear ice fraction with SAR to augment the passive microwave algorithms for ice classification during the early freeze-up period.

### 5.4.4 Modification of the NASA Team Algorithm for Multisensor Ice Classification

To fuse the SSM/I and ERS-1 SAR data using the hybrid algorithm outlined above, the passive microwave algorithm for ice concentration estimates is modified to use SAR-derived information. The multiyear ice concentration from SAR data constrains the NT algorithm. If the multiyear ice concentration ($C_M$) is known, the two equations in the NT solution can be solved for the first-year ice concentration ($C_F$). These solutions can be expressed in the following form,

$$\begin{bmatrix} a_1 \\ a_2 \end{bmatrix} \cdot C_F = \begin{bmatrix} g_1 \\ g_2 \end{bmatrix}, \qquad (2)$$

where *a1, a2, g1* and *g2* are functions of the following: (1) the NASA Team algorithm tie points, (2) the measured brightness temperatures from the SSM/I, and (3) the multiyear concentration derived from the ERS-1 SAR. Detailed definitions of these functions are given in Beaven et al. (1996). The multiyear ice concentration is used to obtain the overconstrained system of equations given by Eq. (2). The first-year ice concentration is then obtained by solving this using the least-squares solution (Twomey 1977). This results in an updated first-year ice concentration from the modified NT algorithm. Therefore, the SAR and SSM/I sensor data are fused by using the SAR-derived multiyear ice concentration to constrain the multispectral inversion process.

## 5.5 Active/Passive Multisensor Fusion Results

To demonstrate the algorithm we used the SAR images obtained from August 30 through September 22, 1991, in which the DLT algorithm worked well for separating the multiyear ice from the remainder (presumably young or thin first-year ice and open water). These SAR images were obtained from a region and in a time frame in which field measurements of radar backscatter were obtained during the International Arctic Ocean Expedition (IAOE) from aboard the US Coast Guard icebreaker *Polar Star* (Beaven and Gogineni 1994). Two examples of the DLT algorithm applied to the SAR imagery are shown in Figs. 6 and 7. The IAOE cruise track passed almost exactly through the center of image frame 1719 obtained on September 6 (Julian day 249) and September 12 (Julian day 255). The *Polar Star* passed through this region during Julian day 242 through Julian day 243, and based on observation of large features in these SAR frames there is less than a 25-km movement of large features from September 6 through September 12. Although the use of ship-based observations may not reflect the ice characteristics across the entire SAR frame we will use ice observations recorded aboard

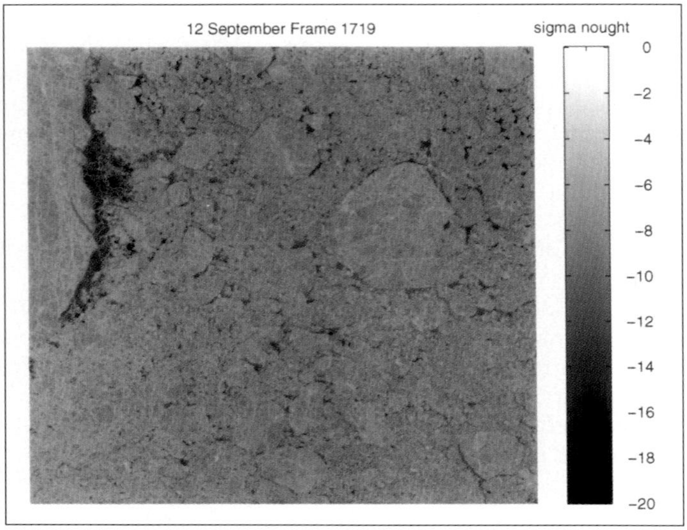


**Fig. 6.** ERS-1 SAR frame number 1719 from September 12, 1991 (Julian day 255). (Copyright 1991 ESA, reprinted from Beaven et al. 1996)

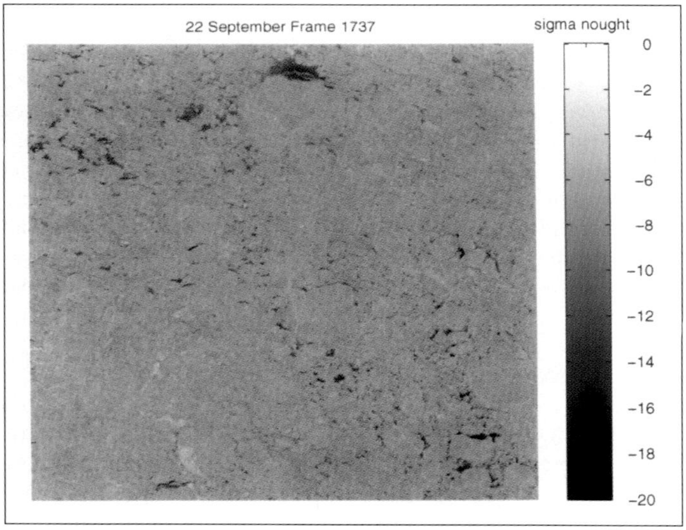


**Fig. 7.** ERS-1 SAR frame number 1737 from September 22, 1991(Julian day 265). (Copyright 1991 ESA, reprinted from Beaven et al. 1996)

the *Polar Star* for semiquantitative assessment of the combined active and passive estimates of ice concentration. We also used SAR images from August 30 (Julian day 242, frame 1701) and September 22 (Julian day 265, frame 1737) to test the hybrid fusion technique. The image obtained from September 22 has about 30% overlap with frame 1719 from September 12, and the image from August 30 is just south and west of these frames, within 50 km of the IAOE cruise track.

These images were thresholded using the DLT algorithm and an example of the resulting classified image is shown in Fig. 8. Of the three classes separated by the thresholding technique we determined which type is most likely multiyear ice (shown in red)

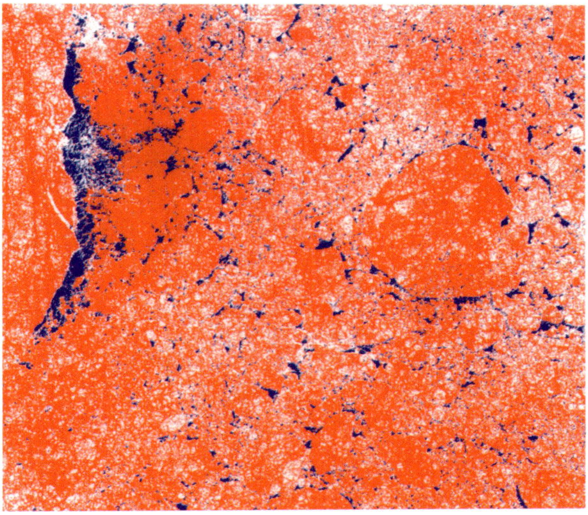


**Fig. 8.** Result of dynamic local-area thresholding to determine multiyear ice concentration from ERS-1 SAR frame 1719, September 12, 1991 (Julian day 255). [Copyright 1996 IEEE (Beaven et al. 1996)]

by examining the shapes of individual floes. The multiyear ice should also be the dominant ice type for this region and time frame, based on ship observations. Further improvement in SAR classification may be obtained through the use of feature extraction and/or heuristic techniques (Haverkamp et al. 1995). Our purpose here, however, is to demonstrate that the fusion of information from these two sensors can improve derived sea ice concentrations.

Assuming that multiyear ice is the dominant ice type in each image, we computed its concentration from these classified SAR images for each SSM/I pixel region within the SAR frame. We used these estimates in our hybrid fusion approach to constrain the NT algorithm to determine a revised estimate of the first-year ice concentration for each of these images. The ice concentration results averaged over each entire image frame are shown in Table 2, along with results from using the NT algorithm without SAR input. The hybrid fusion algorithm results in an estimate of first-year ice concentration that is reduced by between 3 and 21%. The largest differences are for the frame areas obtained on September 6 and 12. The ice concentration results from the NT algorithm for the region covering the ERS-1 SAR frames were computed and the mean values are

**Table 2.** Results of fusion of ERS-1 SAR and SSM/I data for sea ice concentration[a]

| | | NT Results | | Hybrid Fusion Results | |
|---|---|---|---|---|---|
| Julian day | SAR Frame | FY ice conc. (%) | MY ice conc. (%) | FY ice conc. (%) | MY ice conc. (%) |
| 242 | 1701 | 32.5 | 56.9 | 23.0 | 65.7 |
| 249 | 1719 | 40.2 | 39.3 | 18.7 | 62.3 |
| 255 | 1719 | 26.9 | 52.0 | 14.5 | 63.7 |
| 265 | 1737 | 11.3 | 62.5 | 8.0 | 65.5 |

a Ship-based estimates: MY: 70% FY+new: 14%

shown in Table 1 and Fig. 4. These results are obtained over the region of the SSM/I grid in which the SAR image was used for the multiyear ice concentration estimates in the fusion process. Results are shown in Figs. 9 and 10, with the fused ice type percentages overlaid on each of the individual SAR images.

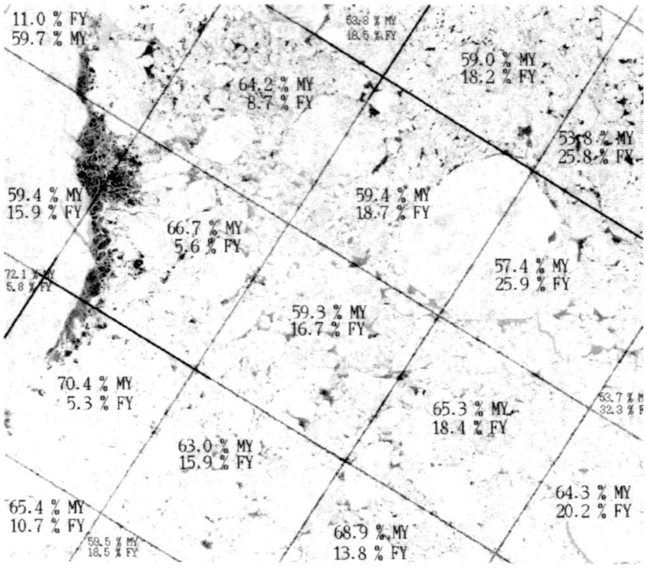


**Fig. 9.** Fused estimates of first-year and multiyear ice concentration from fusion of ERS-1 SAR and SSM/I data for the region corresponding to the SAR image frame 1719 from September 12, 1991(Julian day 255). [Copyright 1996 IEEE (Beaven et al. 1996)]

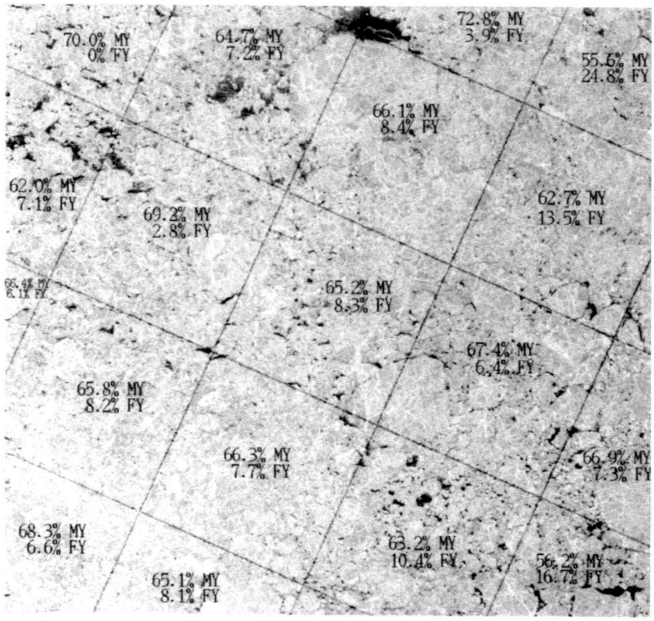


**Fig. 10.** Fused estimates of first-year and multiyear ice concentration from fusion of ERS-1 SAR and SSM/I data for the region corresponding to the SAR image frame 1737 from September 22, 1991 (Julian day 265). [Copyright 1996 IEEE (Beaven et al. 1996)]

For the region observed in SAR frame 1719, the ice observations logged during the IAOE field experiment aboard the *Polar Star* were used to compute average multiyear and average first-year/new ice concentrations for this region. The ice observations were obtained in this region over a period of 3 days from Julian day 243 to Julian day 245. The average ship-based estimate for multiyear ice concentration for this region was 70% and the first-year plus new ice concentration was about 14%, for a total ice concentration of approximately 84%. The ice observations for the young ice categories were approximately 10%. The hybrid fusion estimates for the frame 1719 region yield multiyear ice concentrations of 62 and 64% and first-year ice concentrations of 14.5 and 18.7%. Using the NT algorithm, the multiyear and first-year ice concentration ranges were 39–52% and 27–40%, respectively. The hybrid fusion estimates are much closer to the ship-based estimates from this region than the ice concentrations obtained using the NT algorithm. The multiyear ice concentration results are also consistent with the 60 to 70% multiyear ice concentration obtained by using new sets of pure ice type signatures in the NT algorithm (Thomas and Rothrock 1993). Our results demonstrate that fusion of SAR and passive microwave data can improve estimates of ice type concentration in the fall season.

## 5.6 Discussion and Future Direction

During the initial portion of the freeze-up season, any accurate first-year ice concentration should include thin and young first-year ice in the 10–50 cm range of thickness. Published measurements suggest that the passive microwave signature of thin ice approaches the first-year ice tie point in the NT algorithm as it grows to a thickness of approximately 10 cm (Wensnahan et al. 1993). Therefore, the first-year ice concentration estimates obtained using our hybrid fusion technique include new first-year ice during this time of the year. The mean signatures of these subclasses of first-year ice are described below. This is followed by discussion of the effect of the error in first-year ice concentration on derived heat flux estimates.

Widespread initiation of sea ice growth occurs during the freeze-up season in the Arctic. Thin sea ice, in the form of frazil, dark nilas, and light nilas, was observed in concentrations of up to 5% during the IAOE experiment. The passive microwave signature of new sea ice experiences rapid changes in the initial growth phase. Based on laboratory measurements of thin ice grown at the US Army Cold Regions Research Laboratory (CRREL), these changes in emissivity may be caused by salinity enhancement of the ice surface (Wensnahan et al. 1993). The evolution of passive microwave signatures of thin ice appears to be rapid in the initial growth phase and approaches the first-year ice signature as ice thickness reaches approximately 10 cm. Beyond the initial stages of growth, the passive microwave signatures of young first-year ice gradually evolve towards the first-year ice signature (Eppler et al. 1992). Passive microwave signatures from SSM/I are typically described in terms of the polarization ratio (PR) and gradient ratio (GR), which are defined as

$$\mathrm{PR} = \frac{T_B(19,v) - T_B(19,h)}{T_B(19,v) + T_B(19,h)}. \tag{3}$$

$$\mathrm{GR} = \frac{T_B(37,v) - T_B(19,v)}{T_B(37,v) + T_B(19,v)}. \tag{4}$$

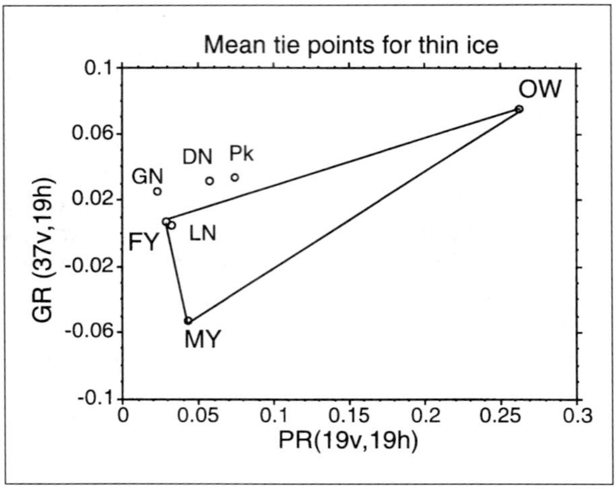


**Fig. 11.** Mean values of PR and GR for young ice types tabulated in Eppler et al. (1992). The thin ice signature evolves rapidly towards the first-year ice signature as ice growth exceeds 10 cm. *OW* open water, *FY* first-year, *MY* multiyear, *Pk* consolidated pancake ice (3 cm), *DN* dark nilas, *GN* gray nilas, *LN* light nilas. [Copyright 1996 IEEE (Beaven et al. 1996)]

The signatures of several types of thin ice are shown in Fig. 11, along with the tie points for first-year ice, multiyear ice and open water. These signatures were computed from Table 4-1 of Eppler et al. (1992), which is a compilation of mean values for a number of measurements reported in the literature for the emissivity of different types of sea ice. Here, we observe the signature of light nilas approaches the first-year ice signature. Light nilas is typically the thickest of the thin ice types shown here and may be 10–15 cm thick. Therefore, at approximately 10 cm thickness, new ice appears very similar to first-year ice in terms of the NT algorithm tie points. Furthermore, the thinner ice types including dark nilas, which is typically around 5 cm thick, also have signatures that are between those of first-year ice and open water. Based on these signatures, the fused estimates of first-year ice concentration that we have computed should include some contribution from thin ice, which significantly affects heat transfer in the Arctic (Maykut 1978). Misclassification of multiyear ice as first-year ice may result in significant errors in the resulting heat flux estimates, particularly if first-year ice consists of significant amounts of thin ice as shown in Beaven et al. (1996).

## 5.7 Conclusions

The use of active and passive microwave data for determining sea ice geophysical parameters offers the potential to resolve ambiguities present in single sensor data. Here we have shown the improvement in estimation of the relative concentrations of first-year and multiyear ice and open water through the use of ERS-1 SAR data to constrain the inversion of multichannel SSM/I data.

The concentrations of first-year and multiyear ice derived from multispectral satellite radiometer data alone, are not reliable during the early freeze-up season in the Arctic. This is caused by fluctuations in the ice signatures, which result in confusion between first-year and multiyear ice. To reduce this problem, our multisensor fusion approach capitalizes on the fact that C-band backscatter from Arctic multiyear ice is relatively invariant under cold conditions, allowing reliable estimates of multiyear ice concentration from SAR (Kwok et al. 1992; Kwok and Cunningham 1994). Improved estimates of ice type concentration are obtained from a combination of satellite active and passive microwave data. First, we derived the multiyear ice concentration from ERS-1 SAR using a local-area threshold method. We then used the information derived from SAR as a constraint in a modification of an existing satellite passive microwave algorithm. The estimates of first-year ice concentration obtained from the fused result for the region in which we had ship observations were 18.7 and 14.5% for Julian day 249 and 255, respectively. Ship-based observations of ice type were obtained during Julian days 240–242 from the same region as the satellite measurements. These observations showed that the first-year ice coverage was approximately 10–15% and the multiyear ice coverage was approximately 70% along the ship's track in this region. The estimated concentration from the satellite passive microwave data alone yields 27–40% first-year ice concentrations. These results demonstrate that an improved estimate of first-year ice concentration during the freeze-up season is obtained by fusing of active and passive satellite data. The estimate of multiyear ice concentration shows similar improvement.

Combining SAR and passive microwave data has been demonstrated here and shows great potential. Further research is needed to determine the usefulness of this method for different seasons and, potentially, to extend the technique to subdivide the first-year ice category into thick and thin first-year ice. RADARSAT now provides wide-swath data so that the coverage of the Arctic will be greatly increased. This increased coverage affords an opportunity for improvements in this approach since the SAR coverage will be more commensurate with the SSM/I coverage than that from the ERS-1 SAR.

## References


Beaven SG (1995) Sea ice radar backscatter modeling, measurements, and the fusion of active and passive microwave data. PhD Dissertation, The University of Kansas, Lawrence, Kansas (also: RSL Tech Rep 8243-4, March 1995)

Beaven SG, Gogineni SP (1994) Shipborne radar backscatter measurements from Arctic sea ice during the fall freeze-up. Remote Sens Rev 9, nos 1–2: 3–25

Beaven SG, Gogineni SP, F.D. Carsey FD (1996) Fusion of satellite active and passive microwave data for sea ice type concentration estimates. IEEE Trans Geosci Remote Sens 34, no 5: 1172-1183

Burns BA, Cavalieri DJ, Keller MR, Campbell WJ, Grenfell TC, Maykut GA, Gloersen P (1987) Multisensor comparison of ice concentration estimates in the marginal ice zone. J Geophys Res 92, no C7: 6843–6856

Carsey FD (1982) Arctic sea ice distribution at end of summer 1973–1976 from satellite microwave data. J Geophys Res 87, no C8: 5809–5835

Carsey FD (1985) Summer Arctic sea ice character from satellite microwave data. J Geophys Res 90, no C3: 5015–5034

Cavalieri DJ, Gloersen P, Campbell WJ (1984) Determination of sea ice parameters with the Nimbus 7 SMMR. J Geophys Res 89, no D4: 5355–5369

Cavalieri DJ, Burns BA, Onstott RG (1990) Investigation of the effects of summer melt on the calculation of sea ice concentration using active and passive microwave data. J Geophys Res 95, no C4: 5359–5369

Cavalieri DJ, Crawford JP, Drinkwater MR, Eppler DT, Farmer LD, Jentz RR, Wackerman CC (1991) Aircraft active and passive microwave validation of sea ice concentration from the Defense Meteorological Satellite Program Special Sensor Microwave Imager. J Geophys Res 96, no C12: 21989–22008

Collins MJ (1992) Information fusion in sea ice remote sensing. In: Carsey FD (ed) Microwave remote sensing of sea ice, Geophys Monogr 68, pp 431–441

Comiso JC (1986) Characteristics of Arctic winter sea ice from satellite multispectral microwave observations. J Geophys Res 91, no C1: 975–994

Curlander JC, McDonough RN (1992) Synthetic aperture radar: systems and signal processing. Wiley, New York

Drinkwater MR, Crawford JP, Cavalieri DJ (1991) Multifrequency, multipolarization SAR and radiometer sea ice classification. Proc Int Geosci Remote Sens Symp IGARSS '91, June 3–6, Espoo, Finland, pp 107–111

Eppler DT, Farmer LD, Lohanick AW, Anderson MR, Cavalieri DJ, Comiso J, Gloersen P, Garrity C, Grenfell TC, Hallikainen M, Maslanik JA, Mätzler C, Melloh RA, Rubenstein I, Swift CT (1992) Passive microwave signatures of sea ice. In: Carsey FD (ed) Microwave remote sensing of sea ice, Geophys Monogr 68, pp 47–71

Gloersen P, Cavalieri DJ (1986) Reduction of weather effects in the calculation of sea ice concentration from microwave radiances. J Geophys Res 91, no C3: 3913–3919

Gloersen P, Campbell WJ, Cavalieri DJ, Comiso JC, Parkinson CL, Zwally HJ (1992) Arctic and Antarctic sea ice, 1978–1987: satellite passive-microwave observations and analysis. NASA SP-511, Washington, DC

Grenfell TC (1992) Surface-based passive microwave studies of multiyear sea ice. J Geophys Res 97, no C3: 3485–3501

Grenfell TC, Lohanick AW (1985) Temporal variations of the microwave signatures of sea ice during the late spring and early summer near Mould Bay NWT. J Geophys Res 90, no C3: 5063–5074

Grenfell TC, Cavalieri DJ, Comiso JC, Drinkwater MR, Onstott RG, Rubenstein I, Steffen K, Winebrenner DP (1992) Considerations for microwave remote sensing of thin sea ice. In: Carsey FD (ed) Microwave remote sensing of sea ice, Geophys Monogr 68, pp 291–301

Haverkamp D, Soh LK, Tsatsoulis C (1993) A dynamic local thresholding technique for sea ice classification. Proc Int Geosci Remote Sens Symp IGARSS '93, August 18–21, Tokyo, Japan, pp 638–640

Haverkamp D, Soh LK, Tsatsoulis C (1995) A comprehensive, automated approach to determining sea ice thickness from SAR data. IEEE Trans Geosci Remote Sens 33, no 1: 46–57

Kwok R, Cunningham G (1994) Backscatter characteristics of the winter ice cover in the Beaufort Sea. J Geophys Res 99, no C4: 7787–7802

Kwok R, Rignot E, Holt B, Onstott RG (1992) Identification of sea ice type in spaceborne SAR data. J Geophys Res 97, no C2: 2391–2402

Li S (1994) Development and applications of an algorithm for a two-way transformation between the geographical coordinates and the ASF SAR image coordinates. Technical Memorandum, Geophysical Institute/UAF, Fairbanks, Alaska

Livingstone CE, Singh KP, Gray AL (1987) Seasonal and regional variations of active/passive microwave signatures of sea ice. IEEE Trans Geosci Remote Sens 25: 159–173

Maykut GA (1978) Energy exchange over young sea ice in the Central Arctic. J Geophys Res 83: 3646–3658

NSIDC Special Report (1992) DMSP SSM/I brightness temperature and sea ice concentration grids for Polar regions on CD ROM: user's guide, National Snow and Ice Data Center. Special Report-1, Cooperative Institute for Research in Environmental Sciences, University of Colorado, Boulder, Colorado

Rothrock DA, Thomas DR (1992) Ice modeling and data assimilation with the Kalman smoother. In: Carsey FD (ed) Microwave remote sensing of sea ice, Geophys Monogr 68, pp 405–418

Steffen K, Schweiger A (1991) NASA Team algorithm for sea ice concentration retrieval from Defense Meteorological Satellite Program Special Sensor Microwave Imager: comparison with Landsat satellite imagery. J Geophys Res 96, no C12, 21971–21987

Thomas DR (1993) Arctic sea ice signatures for passive microwave algorithms. J Geophys Res 98, no C6: 10037–10052

Thomas DR, Rothrock DA (1993) The Arctic Ocean ice balance: a Kalman smoother estimate. J Geophys Res 98, no C6, 10053–10067, 1993

Twomey S (1977) Introduction to the mathematics of inversion in remote sensing and indirect measurements. Elsevier Scientific , New York

Waltz E, Llinas J (1990) Multisensor data fusion. Artech House, Norwood, Massachusetts

Wensnahan MR, Grenfell TC, Winebrenner DP, Maykut GA (1993) Observations and theoretical studies of microwave emission from thin saline ice. J Geophys Res 98, no C5: 8531–8545

# Wavelet Analysis of SAR Images in the Marginal Ice Zone

A. K. Liu and C. Y. Peng

## Contents



## 6.1 Introduction

Ocean-ice interaction processes in the marginal ice zone (MIZ) by waves and mesoscale features, such as upwelling and eddies, have been studied by using ERS-1 SAR imagery and wave-ice interaction models (Liu and Peng 1992, 1993a, 1994; Liu et al. 1993, 1997). Satellite observations of mesoscale features and ice motion can play a crucial role in ocean-ice interaction study. Ocean surface waves from the open sea may penetrate into the MIZ and contribute to the breakup of floes and to other processes that modify the ice cover. Using SAR imagery, the spatial variability of the wave and current field can be estimated through spectral analysis and wave models. The wave-ice interaction models take into account the physical characteristics of the ice cover and the wave parameters to predict wave dispersion, attenuation, refraction, and possibly ice thickness (Liu et al. 1991, 1992; Liu 1995).

Mesoscale eddies are key features associated with the ice margin and are usually attached to the ice edge. The models of surface effect associated with grease ice, wave-current interaction, and atmospheric instability due to upwelling have been used to interpret the mesoscale features of eddies and fronts from ERS-1 SAR images (Liu and Peng 1993a; Liu et al. 1994a). Typical scales of these eddies are 20–40 km. Rotation was mainly cyclonic with a maximum speed of up to 40 cm/s. Eddies play important roles in the distribution of heat, mass, and momentum fluxes in polar regions and in the control of the ice edge and its locations. They may directly affect the loss of ice by moving ice into warmer water which enhances ice melt and thereby the lateral mixing of heat. The ice edge has been found to be highly productive for spring bloom and fishery feed-

Analysis of SAR Data of the Polar Oceans
Edited by C. Tsatsoulis and R. Kwok

ing. In the Bering Sea, fish abundance is found to be highly correlated with yearly ice extent. Fish select cold pools after ice retreat for survival as the food web descends toward the ocean bottom. The cold pools depend on the yearly ice extent.

High-resolution SAR images are very useful for tracking the ice edge motion. A sequence of seven SAR images of the MIZ in the Chukchi Sea from September 27 to October 18, 1991, with 3-day intervals, have been investigated for ice edge advance/retreat by Liu et al. (1994b). Simultaneous current measurements from the northeast Chukchi Sea, as well as Barrow wind records, are used to interpret the MIZ dynamics. On the basis of ice edge locations determined from the SAR images, the ice edge motion appears to be highly dynamic with hundreds of kilometers advance/retreat in 3 days. When the ice edges are relatively stationary, the formation of mesoscale eddy near the meandering ice edge is evident.

A new method for time-varying signal analysis, the wavelet transform, has been developed for remote sensing applications at NASA/GSFC (Goddard Space Flight Center) during the past 3 years and provides spectral decompositions via the scale concept. Basically, wavelet transforms are analogous to Fourier transforms, but are localized in both frequency and time (e.g. Combes et al. 1989). Recent investigations in physical oceanography (Ruskai et al. 1992) verify the efficiency and abilities of these transforms to analyze nonlinear dynamical ocean systems. Wavelet analysis of wind fluctuations over wave groups has been reported by Liu et al. (1995) and Peng et al. (1995).

The two-dimensional wavelet transform is a highly efficient band-pass filter, which can be used to separate various scale processes and show their relative phase/location information, as done, e.g., in SAR imagery by Liu and Peng (1993b) and Liu et al. (1997). The two-dimensional Gaussian wavelet (often referred to as Mexican-hat or Laplacian of Gaussian) transform of a SAR image for small-scale features can be used with a threshold as an edge detector (Canny 1986). The SAR scene can be wavelet transformed with various scales to separate various texture or features. The Laplacian of Gaussian (Mexican-hat) wavelet can also be used as a band-pass filter, and its first derivative as a threshold based on the histogram of the transformed results. In the MIZ study, the ice edge location and ice floe motion in each SAR image can be delineated by using a two-dimensional wavelet transform. The evolution of mesoscale features such as fronts, ice edge meanders, leads, ice floes, surface film, and eddies can be easily tracked by the wavelet analysis using multitemporal SAR images. The numerical simulation of a two-dimensional ocean-ice interaction model can be used as a guide for the search of ice floe, leads, and ice edge in the SAR images during the repeating cycles. This kind of SAR processing by wavelet transform as described here can be used interactively as needed and can provide a more cost-effective monitoring program that would keep track of changes in important elements such as eddies, open leads, ice edge, and ice floe in coastal polynyas and in the MIZ.

The ASF Geophysical Processor System (GPS) has been quite successful in tracking ice motion in the winter Arctic, but has some difficulties with summer ice, in polynyas, and in the MIZ. The wide-swath ScanSAR mode of RADARSAT will have the capability to image the Arctic sea ice cover every 3 days. The MIZ is a critical boundary condition for large-scale climate change study. The assimilation of RADARSAT SAR data with the ocean-ice interaction model can provide a more cost-effective monitoring program that would keep track of changes in important processes (such as ice motion) of the climate system. The advance or retreat of the ice edge, ice floe motion, and mesoscale processes in polynyas and in the MIZ can be tracked by the wavelet analysis using

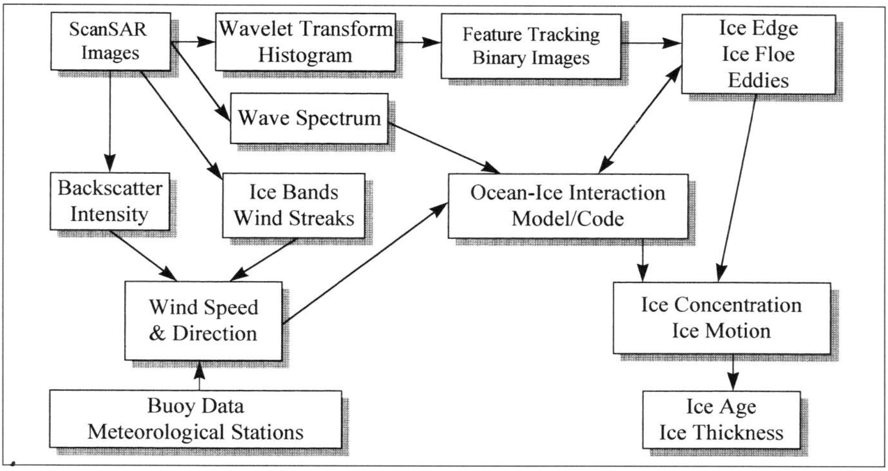


**Fig. 1.** Analysis of marginal ice zone (MIZ) dynamics using a two-dimensional wavelet transform technique for synthetic aperture radar (SAR) data and ocean-ice interaction model to track eddies, ice edge, and ice floe motion

RADARSAT SAR data with its superior repeat coverage. A flow chart for the analysis of MIZ dynamics is shown in Fig. 1 for reference.

In this paper, the two-dimensional Gaussian (Mexican-hat) wavelet transform is introduced briefly in the next section. The wavelet analysis for ice feature tracking, such as ice edge in the MIZ and polynya, and ice floes is then presented with some detailed discussion of algorithms and models. Finally, the application of satellite images for sea ice monitoring and the results of wavelet analysis are discussed in the last section.

## 6.2 Two-Dimensional Gaussian Wavelet

The wavelet transform, $W_s(a,b)$, of a function, $s(r)$, where r = $(x,y)$, is expressed in terms of the complex valued wavelet function, $w(r)$, as follows:

$$W_s(a,\underline{b}) = \frac{1}{\sqrt{a}} \int s(\underline{r}) w^* \left( \frac{\underline{r}-\underline{b}}{a} \right) dt \tag{1}$$

in which the wavelet function is dilated by a factor $a$, and shifted by $b$. The function $w(r)$ is the basic wavelet which must satisfy the admissibility condition, but is otherwise subject to choice within certain limits; the superscript * indicates complex conjugate. For data analysis, the mother wavelets frequently used are the Gaussian modulated sine and cosine wave packet (the Morlet wavelet), and the second derivative of a Gaussian (the Mexican hat). In this study, the analyzing wavelet is defined as the second derivative of the Gaussian function as follows:

$$w(x,y) = \frac{1}{a} \left( 2 - \frac{x^2+y^2}{a^2} \right) e^{\left( -\frac{x^2+y^2}{2a^2} \right)} \tag{2}$$

Since convolution is commutative with respect to differentiation, the resulting wavelet transform is the Laplacian of a Gaussian smoothed function. Thus, its zeroes correspond to the inflection points of the original function (Canny 1986). The contours of zero crossing indicate the edges in the pattern of the input function. The detailed procedure of wavelet analysis is discussed in the following section.

## 6.3 Wavelet Analysis for Ice Feature Tracking

The flow chart of wavelet analysis of SAR images for ice feature (polynya, ice floe, and open leads) is shown in Fig. 2 for reference. The SAR subscenes are first wavelet trans-

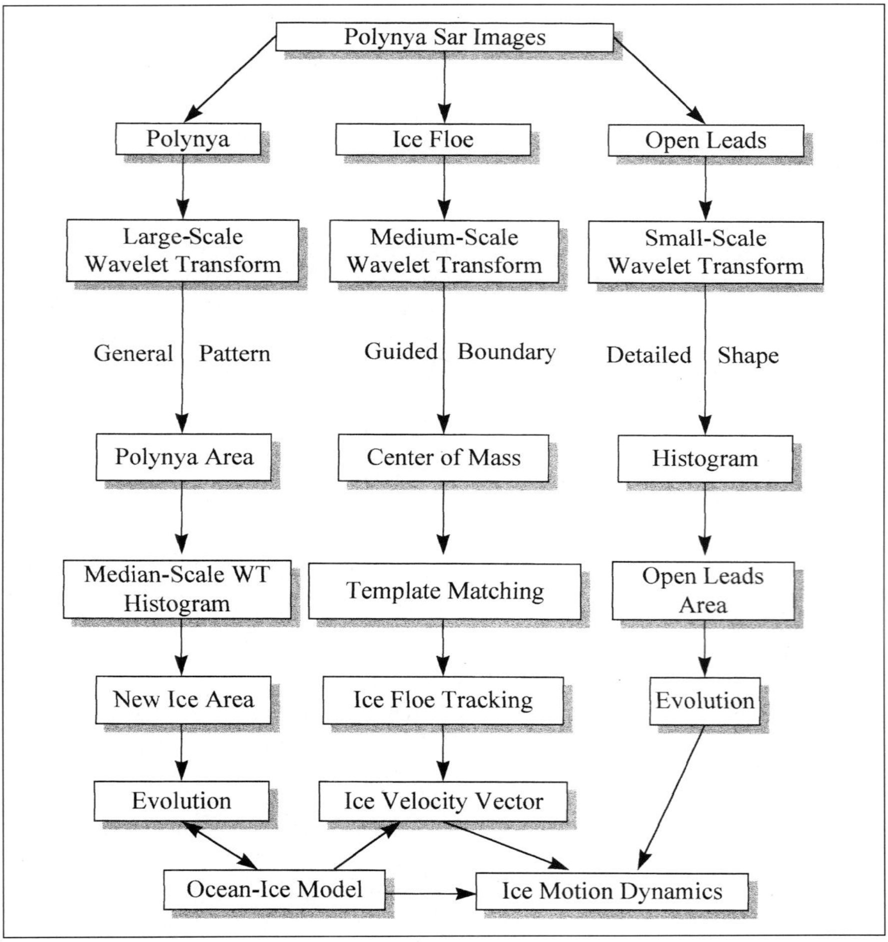


**Fig. 2.** Flow chart of satellite image analysis using a two-dimensional wavelet transform technique of different scales and thresholds to detect varying textures of mesoscale oceanic features

formed with various scales to separate various textures or features. The Mexican-hat wavelet is used as a band-pass filter to separate various scale processes and show their relative location information. The two-dimensional Mexican-hat wavelet transform of a SAR image for small-scale features is used with a threshold as an edge detector. A binary image can then be produced in order to reduce the data volume. By tracking the ice feature evolution from repeat satellite path and using the ocean-ice interaction model (Liu et al. 1993), an ice velocity field has been derived for the study of sea ice dynamics and processes.

## 6.3.1 Chukchi Sea Marginal Ice Zone

In the Chukchi Sea, a sequence of seven SAR images from September 27 and October 3 to October 18, 1991, with 3-day intervals were studied for ice edge advance/retreat. Each SAR image was 100 x 100 km in size and the pixel sizes were 100 m. The ice edge in each SAR image was delineated using two-dimensional Mexican-hat wavelet transform as an edge detector with a scale of 1.6 km, and the maximum intensity gradient lines above a threshold (70%) were extracted from the images. The scale of 1.6 km was chosen to resolve the curvature of the ice edge meanders. Thus, the mesoscale features such as ice edge and the boundaries of eddies could be effectively delineated by such wavelet transforms. When the ice edge was not all well defined, the intensity gradient lines was found to be piecemeal. The disconnected joints were then linked by straight lines.

The locations of the ice edge as determined from the SAR images are summarized in Fig. 3. The SAR coverage is in the rectangular box, and the locations of the four current meter moorings are indicated. Notice that the ice edge motion is highly dynamic, with hundreds of kilometers of advance/retreat in 3 days. When the ice edges are relatively

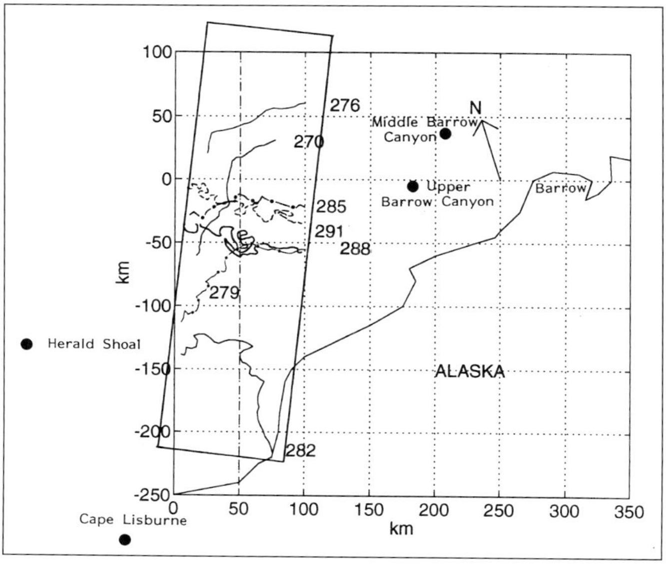


**Fig. 3.** Summary of ice edge locations with SAR coverage area by using a sequence of seven ERS-1 SAR images of the MIZ in the Chukchi Sea from September 27 to October 18, 1991, with 3-day intervals current moorings are indicated by *solid circles*

stationary (among day 285, day 288, and day 291), the formation of a mesoscale eddy near the meandering ice edge is clearly evident, as shown in Fig. 4. Figure 4 shows the SAR image of the MIZ in the Chukchi Sea from October 15, 1991 (day 288 in Fig. 3), with ice edge and eddy formation delineated by the wavelet transform and presented in white curves.

The current measurements from the northeast Chukchi Sea (Cape Lisburne, Herald Shoal, Upper Barrow Canyon, and Middle Barrow Canyon) as well as the wind record at Barrow are used to interpret the dynamics of the MIZ during the period of the satellite imagery. The Barrow wind record was obtained from measurements at 3-h intervals. The Alaska coastal current may provide a relatively uniform advection. However, the dominant driving force for ice edge advance/retreat is clearly the shifting wind.

Figure 5a shows an ERS-1 SAR image of the MIZ in the Chukchi Sea consisting of open water, brash ice, and large floes. Figure 5b shows the Mexican-hat wavelet transform with a large scale to separate the large ice floes from the open water and brash ice as a binary image. Notice that in Fig. 5b the large dark areas (open water) and small floes (brash ice) disappear after application of this band-pass filter. Thus, different textures and size of features can also be extracted by using the wavelet transform with different scales. Another case of ice edge locations delineated by wavelet transform as an edge detector is summarized in Fig. 6 from ERS-1 SAR images collected near the St. Lawrence Island during February 1 to February 28, 1992, with a 3-day repeat cycle. Notice how the ice edges meander and advance/retreat quickly around the island.

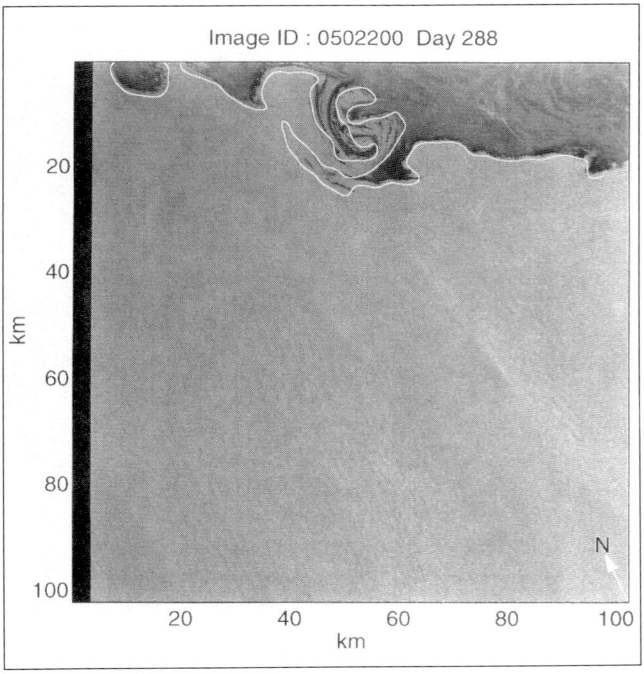


**Fig. 4.** ERS-1 SAR image of the MIZ in the Chukchi Sea from October 15, 1991, with ice edge and eddy formation delineated as the *white curve* by wavelet transform

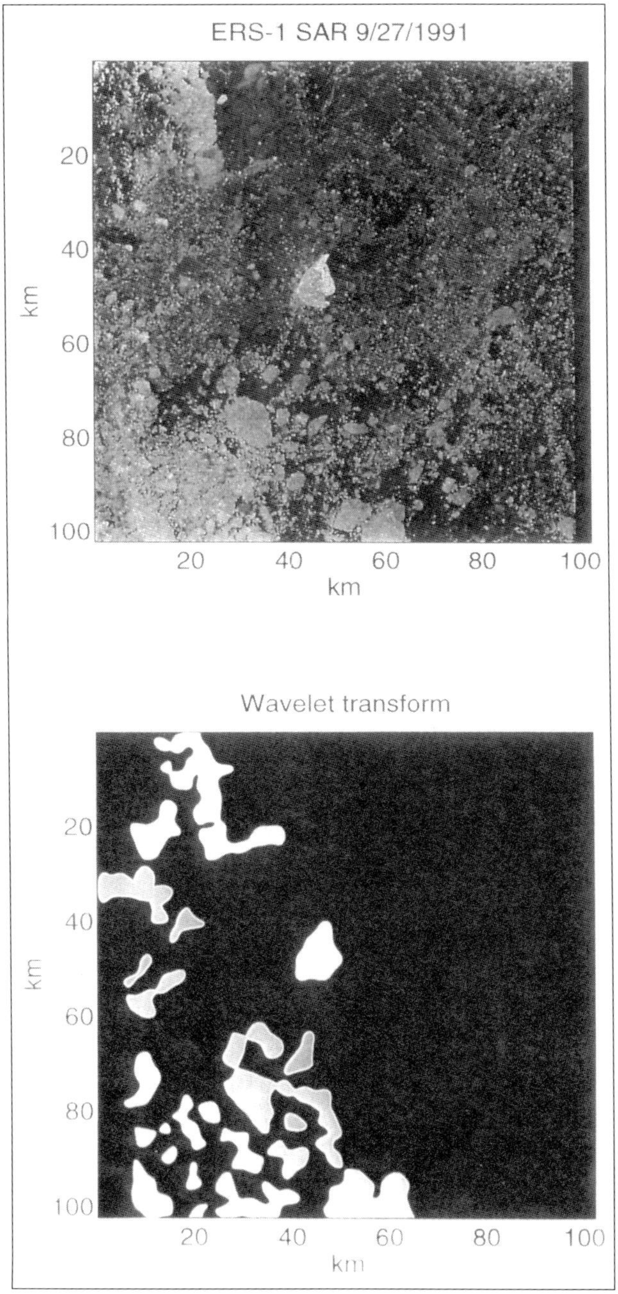


**Fig. 5.** a ERS-1 SAR image of the MIZ in the Chukchi Sea consisting of open water, brash ice, and large floes, and b the Mexican-hat wavelet transform with a large scale to separate the large ice floes from the open water and brash ice

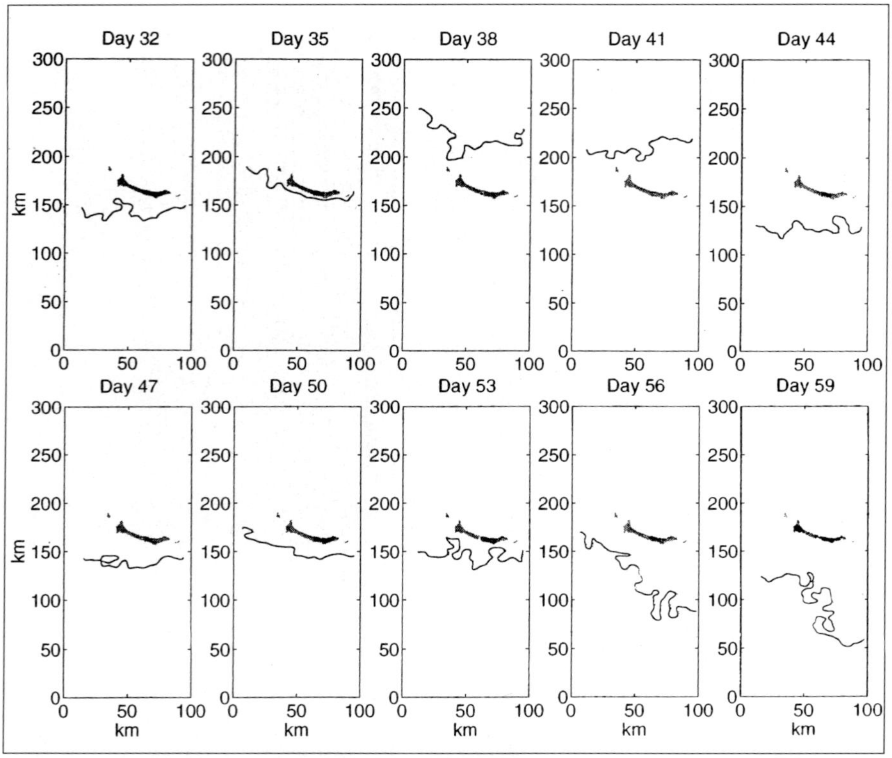


**Fig. 6.** Summary of the ice edge locations near the St. Lawrence Island from February 1 to February 28, 1992, with 3-day intervals

## 6.3.2 Bering Sea Polynya

In the Arctic, the coastal polynyas occur along coasts with strong steady off-shore winds. In the open water in polynya, the cold winds cause oceanic heat loss and generate waves and turbulence. Figure 7 shows a low-resolution SAR image (100 m resolution) from February 27, 1992, where the opening of a polynya to the south side of the St. Lawrence island is clearly visible. In this SAR image 1024 x 3072 pixels in size, the wavelet transform using the Mexican-hat wavelet with a scale $a = 128$ units of pixel spacing [see Eq. (1)] was computed. Each pixel is 100 x 100 m$^2$. With such a large wavelet scale, zero-crossings can hardly exist. In general, using a 50% threshold for the results of the wavelet transform with positive values yields an approximate ice edge. Note that because of the large scale of the wavelet, the edge effect of the Fourier transforms applied to carry out the wavelet transform may cover a significant portion of the domain so that it is impossible to extract any meaningful approximate boundary. To

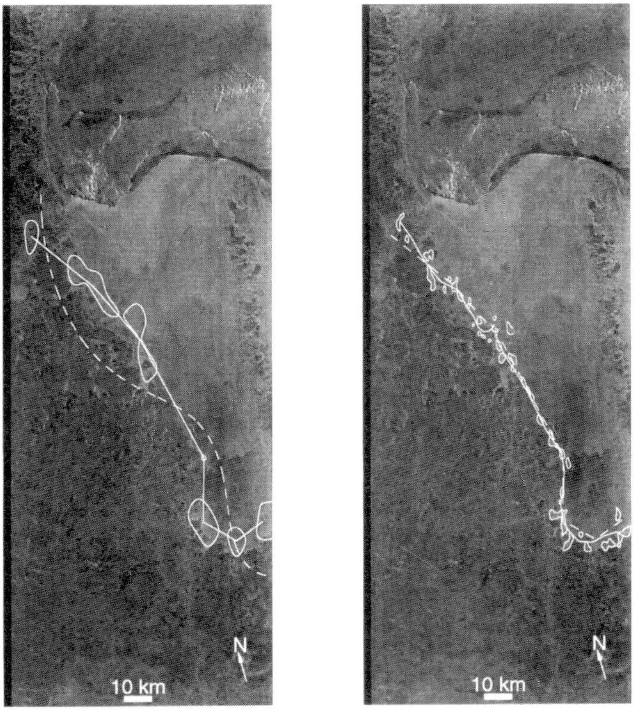


**Fig. 7.** **a,b**. Wavelet transform of SAR image from day 27, 1992, with **a** large *(dashed line)* and median scale *(solid contour)* and **b** small scale. The contours are connected by the sliding averages of the center of mass

reduce to a minimum the distortion due to the edge effect, the image is first extended in the north-south direction by reflection; next the extended image is extended in the east-west direction, again by reflection, to form a new image four times the original size. The wavelet transform is then computed for the new extended image, but only a quarter of the computed domain needs to be retained. Because of the continuity of the pixel values at all the boundaries of the computational domain, the edge effect inherent in the Fourier transform of a finite domain is significantly reduced so that an approximate boundary corresponding to the ice edge can be extracted, as shown by the dashed line in Fig. 7a.

In the SAR images of the MIZ and polynyas, the boundaries between open water and ice, or young gray ice and older brighter ice, are sometimes easily identified. In such circumstances, a single integral boundary may be detected through a single-scale wavelet transform by using the Mexican-hat wavelet as demonstrated in Fig. 5 and 6. The maximum gradient change in pixel intensity along a contour of zero-crossing determines the boundary location (Liu et al. 1994b). Because of forcing such as by winds, waves, and ocean currents, the ice edge in general consists of ice floes of multiple scales with a complex edge configuration. In such cases the use of a single-scale wavelet transform may not be applicable. Therefore, in these cases, it is difficult to delineate a representative ice edge by locating the zero crossing of the Mexican-hat wavelet transform as a single-scale band-pass filter. If one uses a large-scale wavelet transform to suppress the noisy small-scale features, a simple approximate boundary can be located, but it may deviate from the true ice edge considerably. On the other hand, if a small scale is chosen, the wavelet transform yields so many edge elements of zero-crossing that it is hard to sort out by threshold techniques those belonging to the true boundary.

To overcome these deficiencies of the edge segmentation with a single-scale wavelet transform, we use a more general algorithm to locate the ice edge in the SAR images by assembly of the edge elements from the results of a multiscale wavelet transform which include both large and small scales. In the algorithm, some of the techniques developed in the field of computer vision are used, such as the method of "proximity to an approximation" (Ballard and Brown 1982) to locate a more accurate boundary through use of a distance criterion (evaluation function) to obtain a new point for an approximate boundary. The distance operator measures the minimum distance of the new point to the approximate boundary. With the ice edge detected from the large-scale wavelet transform used as the approximate boundary, the edge elements from the small-scale wavelet transform are then used as boundary points for a more accurate boundary. Thus, the linking of these edge elements can serve as an approximate boundary for an even smaller-scale wavelet transform. Several scales can be used to delineate the ice edge as accurately as required.

Thus, in the case of the St. Lawrence Island polynya a wavelet transform with a median scale, $a = 16$ units of pixel spacing, is computed as a band-pass filter. The edge elements are not obtained directly from the contours of zero-crossing of wavelet transform. As mentioned above, because the smoothing effect of the Mexican-hat wavelet is reduced as its scale becomes smaller, the zero-crossing associated with noise proliferates in the whole image, and it becomes difficult to sort out the true boundary even by threshold. Here we obtain the edge elements by selecting a contour value of the wavelet transform equal to a fraction of its maximum, as shown in Fig. 7a indicated by white contours. A threshold value of 5% was used in this case. The result is the appearance of a number of small closed contours, most of them located in the neighborhood of the approximate ice edge given by the dashed line. Because of the fragmented nature of the marginal ice zone, a few of the closed contours may appear anywhere in the image. These closed contours can be effectively eliminated by the method of proximity mentioned above.

The center of mass of these closed contours is computed first. Then the distance from the point of the center of mass $X_c$ to the approximate boundary is calculated. If the distance is below a certain threshold value, the point $(x_c, y_c)$ is considered to be a new point on a refined boundary of the ice edge. A threshold value of 100 m was used in this case.

These new points still have to be linked to form an integral boundary corresponding to the large-scale ice edge. There are a number of methods for edge linking, such as the method of "divide and conquer" (Selfridge et al. 1979). Since the boundary points given by the median scale are in general sparse and well separated, the boundary can be delineated by successively connecting each center of mass to its nearest neighbor.

Finally, a wavelet transform with a scale of $a = 4$ units of pixel spacing is computed. Repeating the same procedure as above, we obtain a new boundary closer to the true ice edge by using the previous median scale results as the approximate boundary. The new boundary points obtained by the small-scale wavelet transform may not be regularly distributed near the approximate boundary, so that a running average of the coordinates of these points may be necessary to facilitate the line linking as shown in Fig. 7b.

Figure 8 summarizes the ice boundary results from February 21, 24, and 27, 1992, respectively. The results of the large-scale wavelet transform are indicated by dash-dotted lines, the median-scale by dashed lines, and the small-scale by solid lines. Note that because of the edge effect of the island, there are no results from the large-scale wavelet transform for day 21. Figure 9 shows the schematic diagram of the evolution of the young ice in the St. Lawrence Island polynya from day 21 to 24 to 27, 1995. Thus, by combined use of several scales of wavelet transform, we can detect (with a large scale), track (with a small scale for detailed shape), and enhance (with edge linking) the satellite images of the ice boundary.

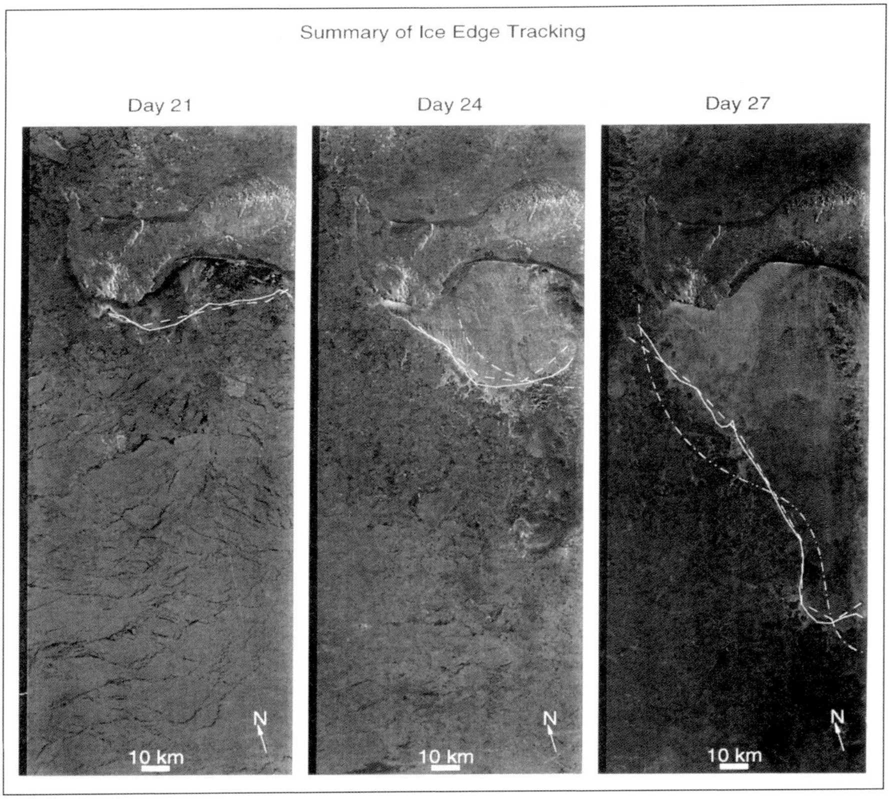


**Fig. 8.** a-c. Summary of the ice edge tracking results of wavelet transform of SAR images from **a** day 21, **b** day 24, and **c** day 27, 1992. *Dashed dotted lines* indicate results of large-scale wavelaet transform, *dashed lines* median-scale, and *solid lines* small-scale

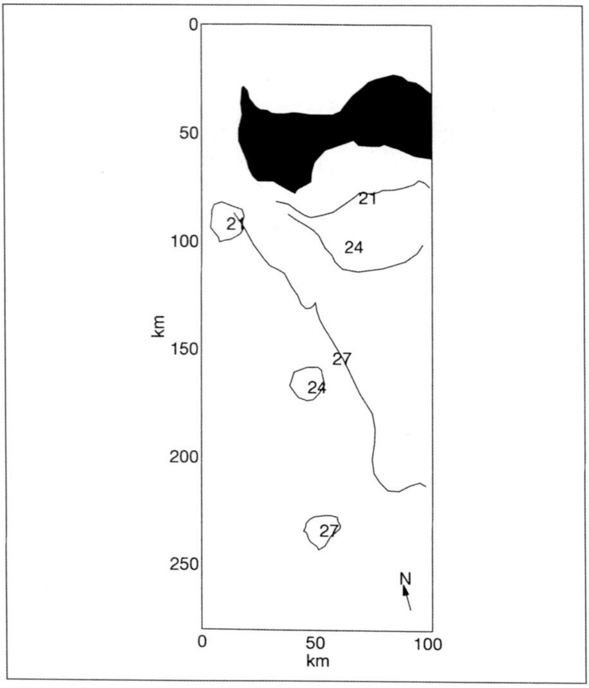


**Fig. 9.** Schematic diagram of the evolution of the St. Lawrence Island polynya from February 21 to 24 to 27, 1992. The *solid lines* are the edges of polynya and the *closed contours* are the matched ice floes for ice tracking

### 6.3.3 Tracking of Ice Floes

An algorithm for tracking ice floes in the interior of ocean basins has been developed by Kwok et al. (1994) by matching distributions of pixel intensity in a fixed-size window scanning a sequence of SAR images. It has been used by ASF at the University of Alaska to provide operational maps of ice flow vectors in the Bering and Chukchi Seas. In the MIZ and the polynyas, because the processes are so dynamic, ice motion tracking at 3-day intervals on a grid-to-grid basis may become very challenging. Although most ice floes in these areas may change beyond recognition in a 3-day interval, some large ice floes may still be identifiable and can be tracked. In this study, an algorithm for ice-floe tracking in these areas has been developed for demonstration. It is based on template matching of the ice floe shapes approximated by closed contours of ice floes obtained from the wavelet transform. Details of the procedures are described in the following case study.

The procedure for ice floe tracking is similar to the tracking of ice edge. We first compute the wavelet transform of SAR images of St. Lawrence Island on days 21, 24, and 27 with a scale $a = 16$, corresponding to a typical ice floe size in this area. Then, we plot the contours of the wavelet transform with a contour level setting to 5% of the maximum. With a finite contour level instead of a zero value, edge elements corresponding to noisy features are effectively filtered out, and closed contours are generated except near the edge of the image. Each closed contour corresponds to the approximate boundary of a large floe or an area with the same texture, as shown in Fig. 10a, b for

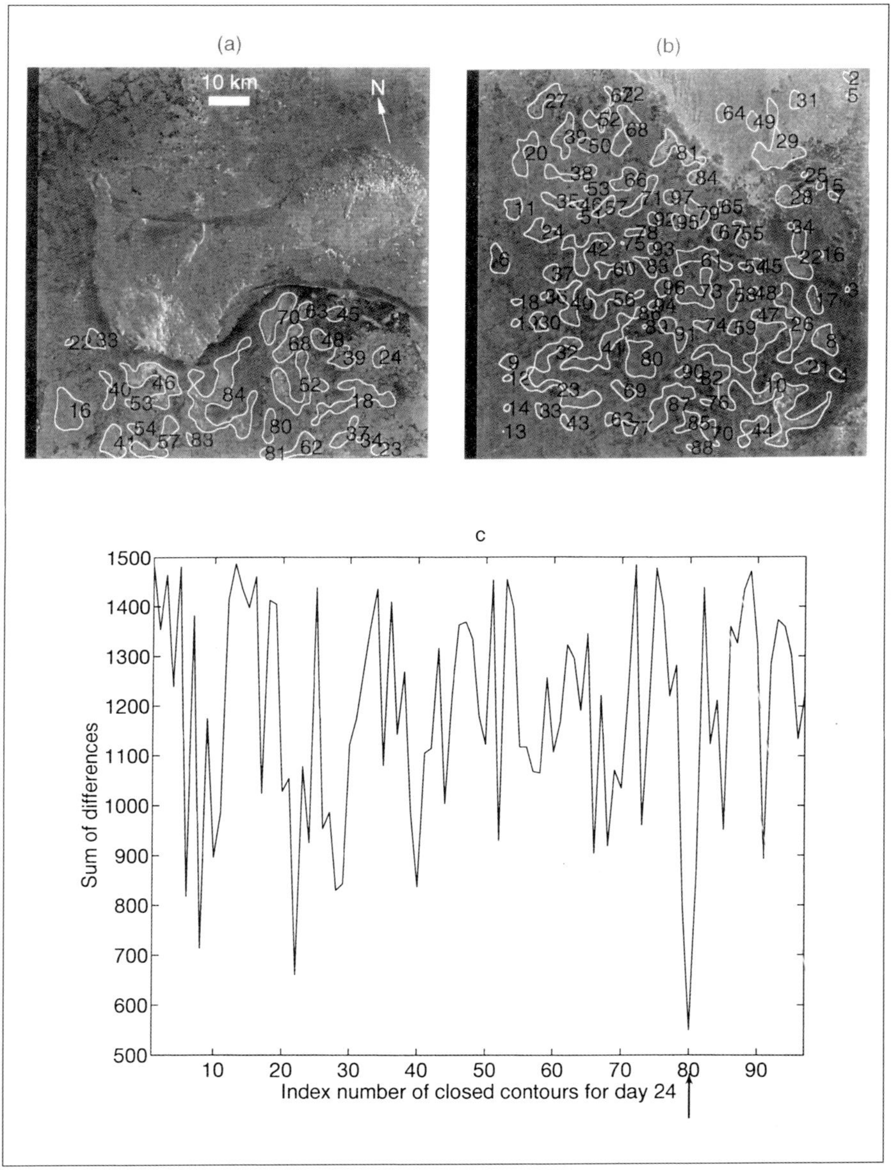


Fig. 10. a-c. Wavelet transform results with median scale for **a** day 21 and **b** day 24, 1992, with index for each closed contour, and **c** sum of difference to contour 16 on day 21 as a function of the index number of the closed pattern from day 24

days 21 and 24, 1992, near St. Lawrence Island. Then, each closed contour is indexed and framed in a rectangular window with its four sides just tangent to the four extreme locations of the closed contour respectively. For each window, a binary image is generated with those pixels within the boundary set to one and those outside the boundary but within the rectangle set to zero. We also select a closed contour at the starting date as the template of the ice floe. In this study, we selected a prominent contour, number 16 in Fig. 10a, as the initial ice floe.

We then perform template matching on the binary image of the template with each instance of binary image 3 days later (day 24, Fig. 10b). The two binary images are first superimposed at their respective center of mass. For template matching, a metric which measures the degree of mismatch between the template and an unknown pattern is defined as follows (Schalkoff 1992):

$$m = \sum_R |f - g| \tag{3}$$

where $f$ denotes the template pattern, $g$ is the unknown target pattern, and $R$ is the range of the template pattern, $f$, namely, the window enclosing the closed contour for the template. A small value of $m$ indicates two patterns are similar, while a large value of $m$ indicates they are different. Because the shapes of the ice floes are in the form of binary images, $f$ and $g$ are both in binary digits. Thus the metric in Eq. (3) can be further simplified by:

$$m = \sum_R XOR(f, g) \tag{4}$$

This follows simply from the fact that the method of exclusive-or, XOR, of two binary digits always gives 1 when they are of the same value, otherwise 0. Thus, we only need to perform a logical XOR operation on the coincident pair of pixel values for each position in the window of the template shape (in the form of a binary image). The summation of the result of the XOR operation at each pixel is used as a measure of the degree of match of the possible ice floe shapes. The correlation of the binary images which gives the minimum $m$ indicates where their shapes are most matched. Figure 10c shows the sum of the differences to contour number 16 on day 21 as a function of index number of the closed pattern from day 24. On day 24, the contour number 80 has the lowest sum of differences and will be selected as the matched ice floe. Once the shapes have been matched, the velocity vector can be easily estimated.

The matched ice floe is then used in turn as the ice floe template to be correlated with the ice floe shapes (in the form of binary image from the wavelet transform) at a later time by repeating the above procedures, so that ice floes can be tracked sequentially. Figure 11 shows the summary of our results of matched floe framed with templates for day 21, 24, and 27 for reference. Note that the shape of the floe is only approximated from the contouring wavelet transform of a median scale with a threshold above zero.

After the ice floe has been matched and tracked, a refined boundary can be obtained again by the method of proximity of approximation discussed above for the ice edge tracking. First, we assemble edge elements obtained by wavelet transform using a small scale ($a = 2$) within the neighborhood of the approximate boundary, and then we link the center of mass of these edge elements consecutively as shown in Fig. 12. The summary of floe tracking is also shown in Fig. 9. Notice that the detailed shape of

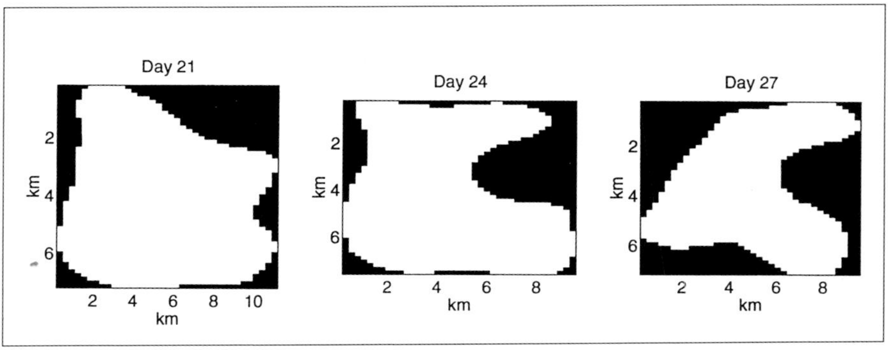


**Fig. 11.** Summary of the results of matched floe framed with templates for days 21, 24, and 27, 1992, near the polynya

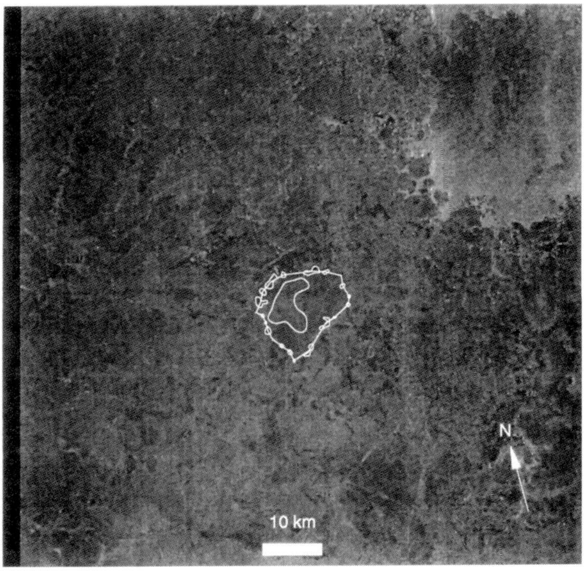


**Fig. 12.** Floe edge elements (small contours) obtained by wavelet transform using a small scale within the neighborhood of the approximate boundary obtained by wavelet transform with a median scale. The small contours are connected by the center. The small contours are connected by the center of mass to construct the detailed shape of the the ice floe.

the ice floe with sharp edges and corners is well delineated by this method. Based on these detailed floe shapes, both translation and rotation of the floe can be estimated. The floe speeds are approximately 0.32 m/s from day 21 to day 24, and 0.26 m/s from day 24 to day 27. The rotations of the floe due to the shear through the opening of the polynya are approximately 50° from day 21 to day 24 and 40° from day 24 to day 27. Within 6 days, the floe moved approximately 150 km and rotated by 90° due to the strong shear associated with the evolution of polynya as summarized in Fig. 9.

## 6.4 Discussion

In this paper, the two-dimensional Gaussian wavelet transform of SAR images has been used with transforms of various scales to separate texture or features. The use of wavelet transforms for tracking ice features in the SAR imagery has been demonstrated, especially in situations where the feature correlation techniques such as the ones used in the ASF Geophysical Processor System fail to yield reasonable results. The evolution of the St. Lawrence polynya in the Bering Sea was studied using ERS-1 SAR data with a 3-day repeat cycle. This wavelet analysis procedure is able to systematically extract geophysically relevant parameters from SAR imagery of polynyas. The procedure is robust and can contribute to the monitoring of polynyas with quantitative information on a large scale.

For the St. Lawrence polynya study, the evolution of the polynya can be easily tracked by wavelet analysis due to the large difference in backscatter between the water and the ice. The "proximity to an approximation" method has been developed in order to locate a more accurate boundary through the use of a distance criterion. With the ice edge detected from the large-scale wavelet transform used as the approximate boundary, the edge elements from the small wavelet transform are then used as boundary points for a more accurate boundary. Several scales can be used to delineate the ice edge as accurately as required. Therefore, wavelet analysis of SAR images may be useful for large-scale polynya monitoring. The procedure of ice floe tracking near the polynya is similar to tracking of the ice edge. In the case of St. Lawrence Island, first the wavelet transform of SAR images is computed with a median scale, corresponding to typical ice floe size in this area. Then the method of XOR is applied for template matching and the matched ice floes are used for tracking. The detailed shape of an ice floe can be constructed by assembling edge elements obtained by wavelet transform using a small scale.

Open leads (dark areas) can be tracked similarly by using the two-dimensional wavelet transform with a small scale as an edge detector to separate the leads areas and produce a binary image as shown in Fig. 2. In this case a threshold below zero will be used to delineate the dark leads area. In addition, a small scale of 2 units of pixel spacing will be used for sharp boundaries between water and ice. Directional wavelet transform, such as the Morlet wavelet, may be very useful for leads aligned in a certain direction. Further study is underway and will be reported separately in the future.

In this study, wavelet analysis was applied primarily for feature or pattern recognition, and therefore it was fairly insensitive to the absolute calibration of the SAR data. Note that the present method of template matching of ice floe shapes using a window of a binary image obtained after manipulating the results of the wavelet transform is very efficient computationally. This is mainly due to the fact that the only computation operations involved are bitwise logical operations and additions, while classical template matching involves the operations of addition, squaring, and multiplication. Furthermore, it is only necessary to match the template pattern to a limited number of target patterns generated by the results of the wavelet transform, not to every location in the images as with classical template matching. It should also be noted that, although template correlation was applied here only to find the translation of the target pattern with respect to the template pattern, it can be extended to find the rotation of the target pattern by incremental rotation of the target pattern and then matching the degree of agreement.

In this work the wavelet transform was applied to satellite SAR imagery and the feasibility of this technique for feature extraction was demonstrated. Especially when the signal-to-noise ratio is relatively low in the satellite imagery, this technique of extracting features first and then matching the features in sequential images is very efficient and promising. This kind of SAR processing by wavelet transform as described here can be used interactively as needed for operation and also applied to the mapping of mesoscale ocean features in the coastal zone (Liang et al. 1995; Liu et al. 1997). The wavelet transforms of satellite images can be used for near real-time "quick look" screening of satellite data (feature detection), data reduction (binary image), and image enhancement (edge linking). By combining ocean color SeaWiFS/OCTS, ERS-2/RADARSAT SAR, and infrared AVHRR images (wavelet transformed binary images) by a data fusion technique, mesoscale features of various physical processes such as oil spills, surface slicks, fronts, upwelling, eddies, ice floes, and ice edge meandering can be detected and tracked in the coastal area and in the MIZ. Reliable imaging of such mesoscale features by certain sensors and definition of the conditions under which imaging is possible will provide a better understanding of the processes.

**Acknowledgments.** The authors wish to thank Ron Kwok and Ben Holt of the Jet Propulsion Laboratory and Seelye Martin of the University of Washington for their valuable discussions and suggestions. The ERS-1 SAR images were collected by ESA satellite and processed by ASF. This work was supported by the National Aeronautics and Space Administration and the Office of Naval Research.

## References

Ballard DH, Brown CM (1982) Computer vision. Prentice-Hall, Englewood Cliffs, New Jersey

Canny J, (1986) A computational approach to edge detection. IEEE Trans Pattern Analysis and Machine Intelligence 8: 679–698

Combes JM, Grossmann A, Tchamitchian Ph (eds) (1989) Wavelet: time frequency methods and phase space. Proc Int Wavelet Conf, December 14-18, 1987, Marseille, France, Springer, Berlin Heidelberg New York

Kwok R, Rothrock DA, Cunningham GF, Stern H (1994) Determination of ice age: a proposed scheme for a RADARSAT sea ice geophysical processor system. Proc Int Geosci Remote Sens Symp IGARSS' 94, August 8–12, Pasadena, California, pp 1750–1752

Liang NK, Liu AK, Peng CY (1995) A preliminary study of SAR imagery on Taiwan coastal water. Acta Oceanogr Taiwanica 34: 17–28

Liu AK (1995) Study of waves in ice-covered ocean using SAR. In: Ikeda M, Dobson F (eds) Oceanographic application of remote sensing. CRC Press, New York, pp 397–404

Liu AK, Peng CY (1992) Waves and mesoscale features in the marginal ice zone. Proc 1st ERS-1 Symp, November 4–6, Cannes, France, pp 343–348

Liu AK, Peng CY (1993a) Synthetic aperture radar for ocean applications. Proc of Symp Remote Sensing in Environmental Research and Global Change. Environment '93, Hong Kong, pp 167–182

Liu AK, Peng CY (1993b) Ocean-ice interaction in the marginal ice zone. Proc 2nd ERS-1 Symp, Hamburg, Germany, October 11–14, 1993. ESA SP-361 (January 1994): 317–322

Liu AK, Peng CY (1994) Marginal ice zone study using SAR. In: Satellite remote sensing of oceanic environment. Jones ISF, Stewart RW, Sugimori Y (eds), Seibutsu Kenkyusha Co, Tokyo, Japan, pp 341–348
Liu AK, Holt B, Vachon PW (1991) Wave propagation in the marginal ice zone: model predictions and comparisons with buoy and SAR data. J Geophys Res 96: 4605–4621
Liu AK, Hakkinen S, Peng CY (1993) Wave effects on ocean-ice interaction in the marginal ice zone. J Geophys Res 98: 10025–10036
Liu AK, Peng CY, Chang YS (1997) Wavelet analysis of satellite images for coastal watch. IEEE J Ocean Eng 22: 9–17.
Liu AK, Peng CY, Schumacher JD (1994a) Wave-current interaction study in the Gulf of Alaska for detection of eddies by SAR. J Geophys Res 99: 10075–10085
Liu AK, Peng CY, Weingartner TJ (1994b) Ocean-ice interaction in the marginal ice zone using SAR. J Geophys Res 99: 22391–22400
Liu AK, Peng CY, Chapron B, Mollo-Christensen E, Huang NE (1995) Direction and magnitude of wind stress over wave groups observed during SWADE. Global Atmos Ocean Syst 3: 175–194
Liu AK, Vachon PW, Peng CY, Bhogal AS (1992) Wave attenuation in the marginal ice zone during LIMEX. J Atmosphere Ocean 30: 192–206
Peng CY, Liu AK, Chapron B, Mollo-Christensen E (1995) Wavelet analysis of sea surface flux and wave modulation by swell. Global Atmos Ocean Syst 3: 195–208
Ruskai MB, Beylkin G, Coifman R, Daubechies I, Mallat S, Meyer Y, Raphael L (1992) Wavelets and their applications. Jones and Bartlett, Boston, Massachusetts
Schalkoff RJ (1992) Pattern recognition: Statistical, structural, and neural approaches. Wiley, New York
Selfridge PG, Prewitt JMS, Dyer CR, Ranade S (1979) Segmentation algorithms for abdominal computerized tomography scans. Proc 3rd COMPSAC Conf, November, pp 571–577

CHAPTER 7

# Mapping the Progression of Melt Onset and Freeze-Up on Arctic Sea Ice Using SAR and Scatterometry

D.P. Winebrenner, D.G. Long, B. Holt

## Contents



## 7.1 Introduction

The annual changes of seasons, their timing, and their variations from place to place reflect and drive many processes of geophysical import, as well as a wide range of human activities. In remote parts of the world, including the Arctic, knowledge and understanding of annual and interannual cycles has been limited by a paucity of observational data. Satellite observations promise to illuminate many relationships and underpin new theoretical understanding. Seasonal transitions on Arctic sea ice are especially germane to studies of global climate and links between high latitude and global geophysical processes.

For example, Scharfen et al. (1987) have reported evidence, based largely on satellite observations, that the date when Arctic sea ice begins to melt is related to the areal extent of sea ice remaining in late summer, which significantly affects global albedo (Ebert and Curry 1993). It is widely thought that the length of the melt season, i.e., the number of days between spring melt onset and autumn freeze-up, strongly influences the sea ice mass balance budget (Maykut and Untersteiner 1971; Ebert and Curry 1993), which in turn influences global ocean circulation (Aagaard and Carmack 1994). Parkinson (1992), using satellite passive microwave data, has noted significant geographical and interannual variability in melt season length. Carsey (1985) observed that Arctic sea ice backscatter seemed to respond to the end of the melt season, i.e., the autumn freeze-up, and that the progression of this event northward could be traced in the record.

In all cases, however, the relevant satellite records are short, their analysis is to some degree uncertain because of limited understanding of the physics underlying the observations, and the necessary analysis was highly labor intensive and therefore slow and limited. To realize the promise in satellite remote sensing, we must learn how to

Analysis of SAR Data of the Polar Oceans
Edited by C. Tsatsoulis and R. Kwok

interpret the observations in a quantitative physical sense, and we must implement the interpretative operations on machines that can keep pace with the torrent of raw data that satellites provide.

The fine spatial resolution provided by synthetic aperture radar (SAR) can aid quantitative, physical interpretation of observations in cases where, as in the case of sea ice, the geophysical medium of interest varies greatly on horizontal scales of tens of meters to hundreds of kilometers. Spatial resolution allows one to focus attention on sea ice of a particular type (e.g., leads, thick first-year ice, multiyear ice, ridges, etc.) for which a theoretical understanding of signatures may be available. The use of fine spatial resolution avoids mixing signatures of various ice types into an average which cannot, in general, be uniquely unmixed, or which can be unmixed only by a laborious analysis using ancillary information.

There are, however, prices for to be paid the advantages of SAR. One of these has been a limitation in swath width and temporal coverage; various design and information throughput constraints limit the width of the SAR image strip and the interval between revisits of a given geophysical target. Thus what is gained in interpretability may be lost in lack of context over the largest relevant spatial scales or shortest time scales. There is, in addition, a danger of drowning in data while trying to extract a relatively small volume of information. This is the background against which recent work to utilize SAR to observe seasonal transitions on Arctic sea ice should be viewed.

A SAR image of sea ice is essentially a snapshot of the spatially varying backscattering cross section of the ice. Ground-based and airborne measurements (Onstott et al. 1987; Livingstone et al. 1987; Barber et al. 1992; Carlstrom and Ulander 1993; Beaven and Gogineni 1994) show that like-polarized backscattering cross sections of sea ice, at frequencies from 5 GHz upward, change at the beginning and again at the end of the melt season. In the case of autumn freeze-up, the measurements of Carlstrom and Ulander (1993) and Beaven and Gogineni (1994) provided fine temporal resolution of the 6- to 10-dB change, with continual observations of the same ice samples. Their simple but convincing modeling indicated the physical cause of the change to be an increase in volume scattering from air bubbles in the upper 10–20 cm of ice as dielectrically lossy liquid water froze into low-loss, relatively fresh ice. The timing of the freeze-up signature change followed within a week the onset of sustained air temperatures below freezing. In the case of melt onset, fine temporal resolution observations of individual ice floes were rare or uncalibrated, but the observations did indicate a decrease for multiyear ice of roughly 7–12 dB between periods before and after the beginning of melting. Temporal changes in first-year ice signatures are evidently smaller and more subtle (Barber et al. 1992).

Winebrenner et al. (1994) used temperature records from drifting buoys and nearly coincident SAR imagery from the first European Remote Sensing (ERS-1) satellite to show that sharp, 9-dB drops in multiyear ice backscattering cross sections (at 5.3 GHz, VV-polarization) coincided, to within 2–4 days, with rises in air temperature to 0 °C at locations in the Beaufort Sea between 72°N and 82°N during June 1992. Theoretical analysis by these authors showed that the cross-section drop is due solely to the appearance of liquid water in the snow overlying the multiyear ice. Essentially, the strong 5.3-GHz backscattering from the bubbly upper layer of multiyear ice dominates the SAR return while the snow is dry, but moist snow strongly attenuates illumination of, and scattering from, the bubbly layer. Thus the physical event corresponding to the melt onset signature change is clear. For freeze-up, Winebrenner et al. (1996) combined a wide area SAR study in the Beaufort and Chukchi Seas with a buoy temperature/SAR study in the Beaufort and concluded that temporally erratic but generally low summer backscattering cross sections stabilize at winter multiyear ice values (roughly -9 dB) within 7 days of continuously subfreezing air temperatures. Schwartz et al. (1994) observed spring and autumn transitions in the Beaufort and Chukchi Seas, but compared with analyzed temperature fields based primarily on shore-based data, and reached similar conclusions.

In an effort to reap the advantages of SAR while mitigating the costs, Winebrenner et al. (1994; 1996) proposed and demonstrated machine-automatable algorithms for retrospective estimation of melt onset and freeze-up dates within geographical cells using ERS-1 SAR imagery. This approach utilizes many low-resolution (100-m) SAR images but immediately reduces the information volume to time series of backscattering cross section histograms, one for each geographic cell. This avoids any need to monitor backscattering cross section histories for individual ice floes.

More or less simultaneously with these developments, several investigators found that characteristic seasonal backscattering changes could also be identified for several terrestrial targets in addition to sea ice. An illustrative, rather than exhaustive, list of examples includes the following. Smith et al. (1997) found that ERS-1 SAR backscatter observations of alpine glaciers show clearly the onset and progression of spring melt. On the Greenland ice sheet, Drinkwater and Long (1994) found that 14.6 GHz backscattering observed using the Seasat wind scatterometer not only delineated snow zones in winter but showed clearly the progression and extent of the melt season at all altitudes. Early et al. (1994) found evidence of temporal variation in snow facie boundaries between 1978 and 1991 by comparing scatterometer observations from ERS-1 with those from Seasat. Wismann et al. (1996) found continuous temporal variation from 1991 to 1995 in 5.3-GHz ERS-1 scatterometer observations of Greenland, and found valuable seasonal information in scatterometer observations of Siberian tundra, the Sahel, and tropical and boreal forests as well. Thus, the observations of seasonal transitions on sea ice may be seen within a wider geophysical and remote sensing context, and methods developed in connection with sea ice may prove useful in developing automated processing methods for other geophysical targets and variables.

In this chapter, we begin with a review of the methods and results of Winebrenner et al. (1994, 1996) for automatable mapping of seasonal transition dates and melt season length on Arctic sea ice. This review includes neither new data nor results, but we focus particularly on the problems and essential features of machine automation that contribute to a useful geophysical data product. In the following section, we present new results based on Seasat wind scatterometer observations and the resolution enhancement algorithm of Long et al. (1993) for the freeze-up period in the Beaufort and Chukchi Seas during 1978. We briefly compare these results with those obtained by Drinkwater et al. (1994) for seasonal transitions on Antarctic sea ice. The observations clearly show the observability of freeze-up and its temporal progression in 14.6-GHz observations, as well as the applicability of Long's resolution enhancement method to moving targets such as sea ice during a period of temporal change. Thus it appears that scatterometry data may usefully serve to set the context for, augment, or in some cases substitute for SAR observations even for targets such as sea ice that vary strongly on small horizontal scales. Insight gained in the development of methods for SAR and scatterometry appear to be mutually complementary. We conclude with a discussion of what we believe to be the most important issues to be addressed in future work, both to extend methods and to begin to address important geophysical questions using automated methods.

## 7.2
## Synopsis of Transition Date Mapping Using the ERS-1 SAR

The essential phenomena on which automatable mapping of melt onset and freeze-up relies are, respectively, the striking drop in multiyear ice 5.3-GHz backscattering cross sections in spring, and the restabilization of those cross sections at relatively high winter values in the autumn.

As summarized in the Introduction, and discussed in detail by Winebrenner et al. (1994), the springtime drop in cross sections coincides with the rise of local air temperatures to 0°C to within 2–4 days. This is shown graphically in Fig. 1 for a case with a particularly reliable temperature record. Figure 2 shows the visible manifestation of the phenomenon in ERS-1 SAR imagery. The physical cause of the change is the appearance of liquid water in the snowcover on the ice. Volumetric snow moisture of as little

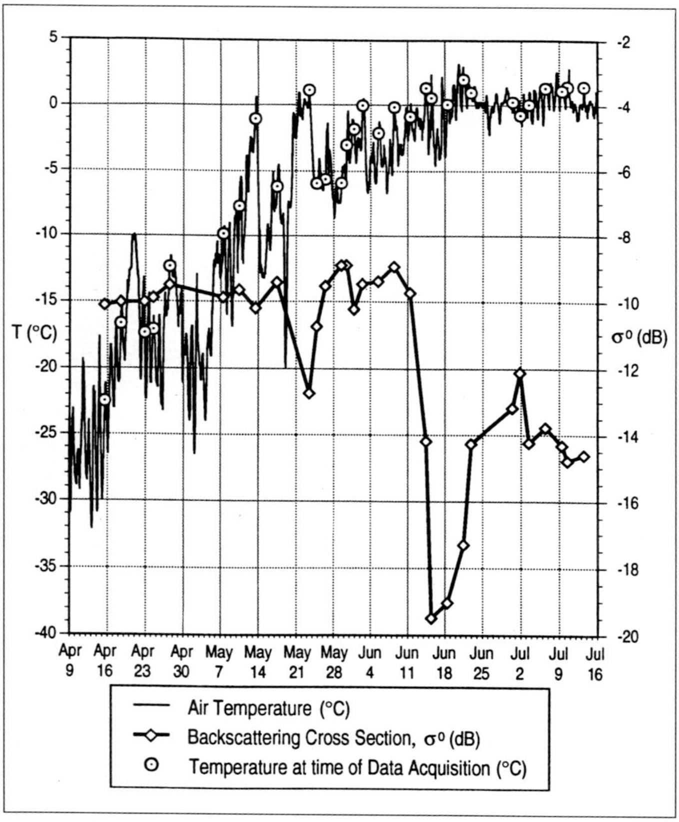


**Fig. 1.** Near-surface air temperature from a drifting buoy in the Beaufort Sea, and the 5.3-GHz, VV-polarized backscattering cross-section of multiyear ice near the buoy as observed using the ERS-1 SAR, during the period mid-April to July, 1992. The *circled dots* on the temperature record denote the temperatures at the time of SAR data acquisition

**Fig. 2.** **a** ERS-1 SAR image of a representative ice region on June 2, 1992, near the buoy that provided the data in Fig. 1. Temperatures were still well below freezing. **b** A similar image of the same ice scene on 15 June 1992, just after the onset of melt. Equal gray levels in separate images connote equal backscattering cross-sections. The *white rectangular regions* define sampling boxes used in previous signature-change study. (Copyright ESA 1992)

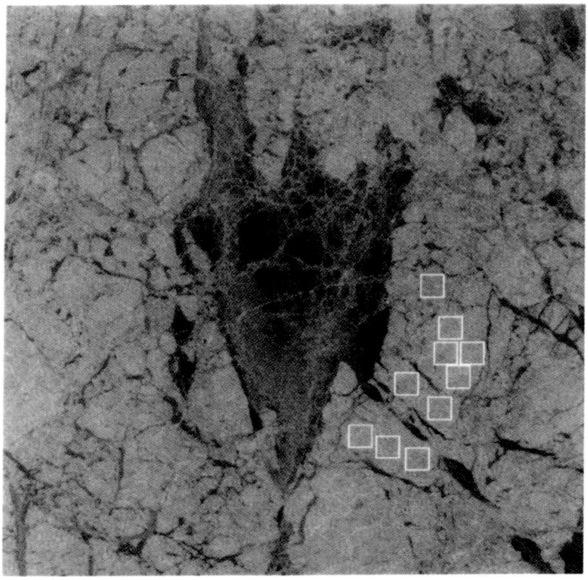



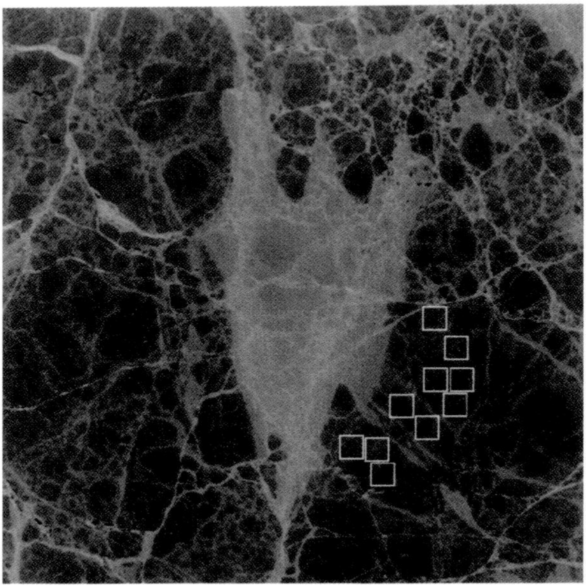

as 2% renders previously (dry) microwave-transparent snow quite lossy; thus illumination of the strongly scattering bubbly upper layer of multiyear ice is reduced, and scattered returns are attenuated by their passage through the snow. In the earliest stage of melt onset, this is all the microwave physics that matters. The fact that the strong scatterer beneath the snow happens to be a bubbly layer is not essential – any strong backscatterer would produce similar phenomenology because the shielding by newly wet snow is the source of the change. The difficulty in sensing melt onset on first-year ice at 5.3-GHz is the weakness of winter first-year ice backscattering. The drop in backscattering due to shielding of an already weak scatterer is difficult to observe, especially from space. (Note, however, that these comments apply only to the earliest stage of melt onset; scattering from both first- and multiyear ice increases to high levels within roughly 2 weeks of melt onset, probably because the snow surface becomes water-logged and rough. Utilization of the information contained in this latter signature variation is a topic of current research in several groups.) Spaceborne observation of melt onset on first-year ice might well be facilitated by observation at a wavelength where winter first-year ice cross sections are higher. We return to this point in the next section.

Two considerations directly influence the design of machine automatable algorithms for melt onset detection. First, the drop in backscattering cross sections varies slightly between multiyear ice floes, but is large – typically 7–10 decibels – for virtually every individual floe (Winebrenner et al. 1994). Thus it is not necessary to monitor the backscattering cross-section history of individual multiyear floes to observe melt onset; it is only necessary to derive from imagery information on the temporal backscattering evolution of typical multiyear ice pixels. Second, near-surface air temperatures over Arctic sea ice are highly correlated over distances on the order of at least 100 km (Rigor and Muñoz, "Statistics of Surface Air Temperature Observed in the Arctic", manuscript in preparation, 1996). Thus it is necessary only to sample Lagrangian regions of size on the order of 100 km with SAR imagery to produce a map of melt onset dates with useful spatial resolution.

For regions covered predominantly by multiyear ice at the time of melt onset, these considerations suggest a simple algorithm. Divide the region of interest into fixed (Eulerian) grid cells of the minimum dimension consistent with high temporal resolution of sampling with SAR imagery around the time of melt onset (preferable sizes will be on the order of 100 km or less). Assign each SAR image to a grid cell, and compute for each image a histogram of pixels with backscattering cross-sections in narrow ranges. Because the most probable surface cover is multiyear ice (in the selected regions), simply track the backscattering cross section value of the peak (mode) of the histogram through time; the time series of such peak values is a reasonable proxy for multiyear ice backscattering cross sections under the stated conditions. Set a threshold such that when the proxy for multiyear ice cross-sections crosses the threshold from above, melt onset is inferred to have taken place within the entire cell. This algorithm was automated and run using all ERS-1 SAR imagery over the Beaufort Sea collected by the Alaska SAR Facility (ASF) during spring 1992, with the result shown in Fig. 3. This particular map used cells of 200 km on a side on the grid defined for Special Sensor Microwave/Imager (SSM/I) brightness temperatures. The algorithm required operator intervention in one cell, denoted with a numeral 2 in the map, where an early "false start" to melt onset occurred – backscattering dropped briefly in late May but then

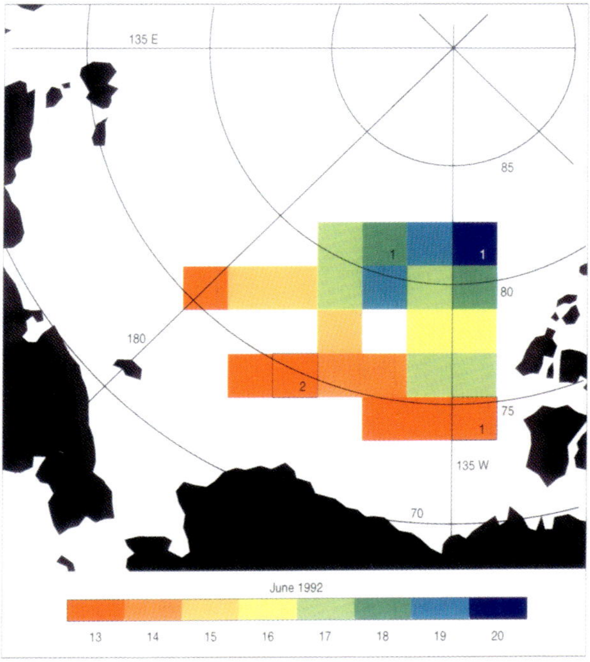


**Fig. 3.** Map of the dates of melt onset estimated from ERS-1 SAR data and the algorithm of Winebrenner et al. (1994), for the spring of 1992 in the Beaufort Sea. The cells are square and measure 200 km on a side; they coincide with groups of cells on the SSM/I grid. *Unshaded cells* indicate a lack of sufficient data or that the area contains less than 50% multiyear ice

quickly returned to winter levels until mid-June, when backscattering began a typical summer progression. The inferred melt onset dates shown in Fig. 3 agree with independent estimates of those dates from buoy data, to within 4 days (Winebrenner et al. 1994). In cells where temperatures repeatedly reach 0 °C and refreeze during spring, the algorithm will pick the first date of melting as the melt onset date. The largest single source of uncertainty in satellite-derived melt onset dates in the study of Winebrenner et al. (1994) was the temporal lag between SAR images sampling each grid cell; thus there is the prospect of reducing such uncertainty in future studies.

The fundamental phenomenology underlying mapping of autumn freeze-up, i.e., the transition from temporally erratic, sometimes high but often low backscattering to high, stable backscattering, is illustrated in Fig. 4, and its visual manifestation in SAR imagery is shown in Fig. 5. The physics of the transition is more difficult to model in a precise, quantitative way, but there is little doubt that the freezing of liquid water and consequent rise in backscattering from the bubbly upper layer of refreezing ice is the primary physical cause of changes in backscattering from ice floes (Carlstrom and Ulander 1993; Beaven and Gogineni 1994). In the case of freeze-up, however, areally averaged backscattering often displays a peculiar overshoot phenomenon, in which cross sections initially increase to 1–2 dB above winter values around the time of the decrease in temperatures from near to below freezing (for the final time in the year). The cross sections then gradually settle back to typical winter multiyear ice values within 7–10 days.

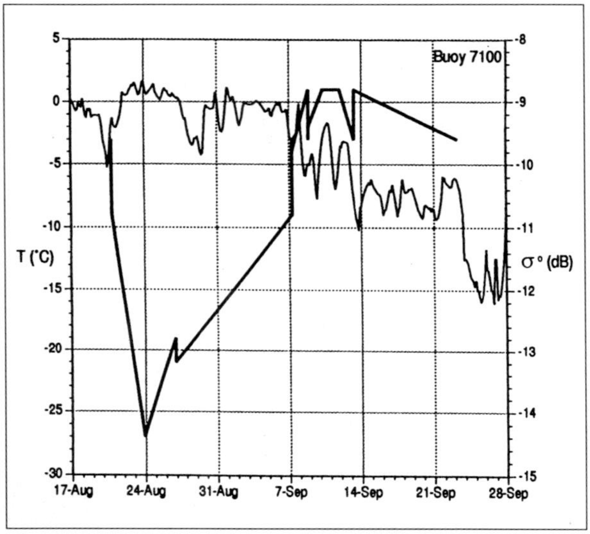


**Fig. 4.** Near-surface air temperature at the buoy that provided data for Fig. 1, and the mode of the histogram of nearby backscattering cross-sections (at 5.3 GHz, VV-polarization, measured using the ERS-1 SAR), during autumn 1992

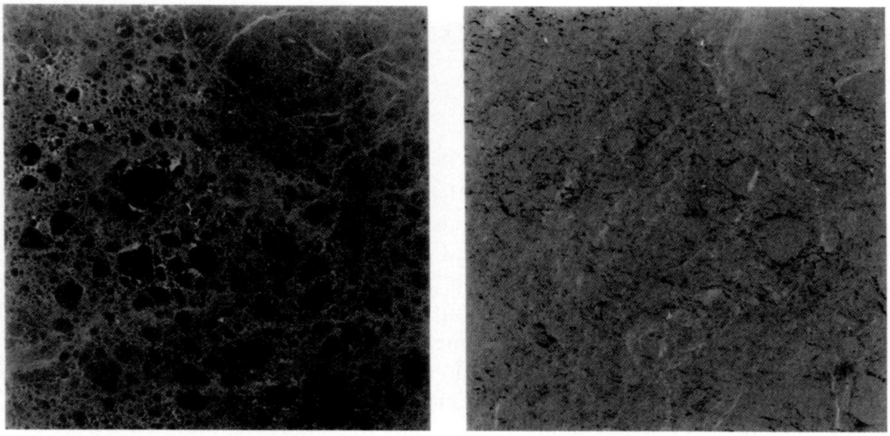

**Fig. 5.** **a** ERS-1 SAR image of a representative ice region on September 6, 1992, near the temperature-reporting buoy of Figs. 1 and 4. Temperatures were near freezing at the time of image acquisition. **b** Similar image data from another region near the same buoy on September 8, 1992, after temperatures had dropped below freezing. Equal gray levels in separate images connote equal backscattering cross sections. (Copyright ESA 1992)

This observation led to the following automatable algorithm for retrospective estimation of freeze-up dates. Using the method of SAR image sampling on the fixed grid from the melt onset algorithm, and limiting consideration to regions in which the areal coverage by summer ice is greater than 50% at the end of the summer, consider time series of histograms of backscattering cross sections from the images sampling each grid cell. Again, track the peaks (mode) of the histograms in time (which, by construction, indicate typical ice backscattering values). When that peak value crosses a predetermined threshold, $X$, and remains continually above that threshold for a predetermined number of days, $N$, estimate the time of freeze-up to have been some number of days, $M$, prior to the threshold crossing. An example based on ERS-1 SAR observations of the Beaufort Sea (as acquired by the ASF) is shown in Fig. 6, with $X$ = -10.5 dB, $N$ = 10 days, and $M$ = 0 days. We estimate the uncertainty in freeze-up dates on this map to be ±4 days, due largely to the available temporal sampling of grid cells (though an additional uncertainty may be caused by the overshoot phenomenon) (Winebrenner et al. 1996). This, as well as the rather sparse collection of cells within which we can estimate freeze-up dates, is an indication of the difficulty of sampling using only the autumn 1992 ERS-1 SAR images collected by the ASF. We expect that data from later years during the life of the satellite can provide denser sampling, but the severity of the problem is noteworthy.

Nonetheless, the construction and testing of melt onset and freeze-up mapping algorithms shows that melt season lengths can be estimated to a useful accuracy (perhaps 5%) using SAR data and machine automatable algorithms. Limitations of the current data and algorithms include the limitation to regions largely covered by multiyear ice and the temporal and spatial coverage limitations of early SAR data. Though the latter problem may be solved simply by a greater collection of SAR or ScanSAR data, this has yet to be demonstrated (and in the case of ScanSAR introduces the complication of varying incidence angles). In the next section, we present observations that may point the way toward an easing of these limitations.

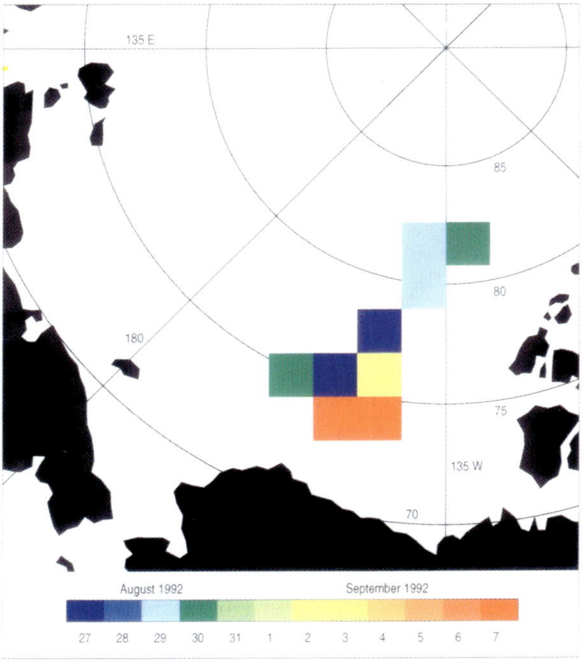


**Fig. 6.** Map of the dates of freeze-up estimated from ERS-1 SAR data and the algorithm of Winebrenner et al. (1996)

## 7.3 Enhanced-Resolution Scatterometer Observations of Freeze-Up

Spaceborne scatterometers are stable, well-calibrated instruments for measuring backscattering cross sections, albeit at spatial resolutions coarser than the scales of cross section (and geophysical) variability inherent in some geophysical targets. (For a discussion of relevant operational details of spaceborne scatterometry, including the matter of normalization to a standard incidence angle, see Long et al. 1993).

We are motivated to investigate the utility of scatterometers for observing seasonal transitions on sea ice for at least two reasons. The first is a prospective advantage stemming from the shorter wavelength (higher frequency) at which many spaceborne scatterometers operate, relative to that of Radarsat or the ERS-1 and -2 SARs and scatterometers. At the frequency (5.3 GHz) of those latter instruments, backscattering from winter first-year sea is weak. Hence it is difficult to observe the melt onset transition on first-year ice and to automate data interpretation in areas covered mostly by first-year ice (see the previous section). However, the operating frequency of the Seasat scatterometer was 14.6 GHz, and that of the new NASA scatterometer (NSCAT) is 14.0 GHz. Ground-based observations near those frequencies (Onstott 1992) indicate that winter backscattering from first-year ice is considerably higher than at 5.3 GHz (though the physical reason for this is not clear as of this writing). Dry snow would be expected, on theoretical and observational grounds to be slightly less transparent at the higher frequencies than at 5.3 GHz, but a given thickness of wet snow would be expected to be a more effective shield (Hallikainen and Winebrenner 1992). Therefore we would expect a much stronger drop in 13- to 15-GHz backscattering cross sections at melt onset than is observed at 5.3 GHz. This would make backscattering observations at the higher frequencies far more easily usable for melt onset observation in areas such as the Siberian shelves, where first-year ice predominates. This would significantly expand the geophysical utility of melt onset mapping based on backscattering.

The typical spatial resolutions of space borne scatterometers, however, and of the Seasat, NSCAT and ERS-1 scatterometers in particular, are nominally 50 km. As Carsey (1985) and Wismann et al. (1996) have shown, even this resolution is of some use on sea ice, but there is mixing and a loss of signature information because sea ice characteristics and signatures vary on horizontal scales as short as meters. This is not to say that information on all spatial scales down to meters is always useful for all purposes – experience with low-resolution imagery of sea ice indicates that sea ice imagery with 5-km resolution shows much more information than 50-km imagery, especially if the 5-km imagery comes at an acceptable price in terms of spatial coverage, temporal resolution, and radiometric fidelity. As we discuss below, such resolution is evidently available. Moreover, while the resolution of SAR provides a way to usefully focus on a specific ice type (see the previous section), there remains a strong practical motivation to distill the information in a high-resolution image quickly into a smaller volume. We therefore have a second motivation, in addition to the promise of shorter wavelength observations, to explore further the potential of enhanced-resolution scatterometry – scatterometer observations may complement or even, in some cases, substitute for SAR data. It seems likely that whatever is learned in this effort will contribute insight useful in building better automated SAR, as well as scatterometer, algorithms.

Long et al. (1994) have in fact begun this process using ERS-1 scatterometer data from the Weddell Sea and an adaptation of the resolution enhancement method of Long et al. (1993). Heuristically, Long's Scatterometer Image Reconstruction with Filtering (SIRF) operates by combining scatterometer observations from footprints whose backscatter values are correlated because of they share area in common. The aim is to reconstruct the underlying, fine-scale map of backscattering cross-section, given sufficient knowledge of the overlap geometries. In practice, the scatterometer observations are acquired over some time interval, during which the target could change; any such temporal variation is treated as "noise" in the algorithm. The SIRF algorithm is based on an iterative, multivariate modification of multiplicative algebraic reconstruction; the standard version of the latter is a special case of maximum-entropy reconstruction. In cases where footprint overlap is insufficient to completely determine all matrix elements in the estimation problem, a maximum-entropy-based assumption is used to specify those elements. This is, in effect, an assumption concerning the autocorrelation structure of the underlying backscattering map at subresolution scales (Jaynes 1982; see also pages 420–426 of Percival and Walden 1993), namely that the correlation is that of the most random, least predictable structure on those scales that is consistent with the observations at larger scales. The assumption is likely to be more accurate when noise is limited (Jaynes 1982). The result is a map of backscattering cross-section estimates on a finer grid than the original data, consistent with the assumptions on that underlying map and "noise" inherent in the method.

In the case of sea ice, ice motion and possible temporal changes limit the period of time over which observations of a given ice region may be accumulated, and therefore the fineness of the resulting grid of backscattering cross-sections relative to the original data resolution. In the case of the study of Long et al. (1994) in the Weddell Sea, this interval was 7 days and allowed enhancement from 50 km to 25 km. In the case of the Seasat scatterometer, differences in predelivery processing of the data allowed greater enhancement, to 9 km, while accumulating data over 8-day intervals.

From the preceding discussion, it is clear that high-frequency scatterometer observations of melt onset as well as freeze-up would be valuable. However the lifetime of the Seasat, together with the unavailability of NSCAT data as of this writing, limit our high-frequency observations for now to freeze-up during 1978. The promise of higher-frequency observations is strikingly evident, nevertheless, in the measurements we have. Figure 7 shows temperature, precipitation, and backscattering cross section time series from 1978, the former measured at Mould Bay in the Canadian Archipelago, the latter from a time series of resolution-enhanced, 14.6-GHz backscattering images based on data from the Seasat wind scatterometer. Because the location of the backscatter measurements is just west of the temperature measurement site, small temporal offsets between events in the temperature and backscatter records might be expected. Nonetheless, the highly variable backscattering characteristic of summer conditions quickly gives way to high, stable backscattering after September 15 (Julian day 258), giving a clear indication of freeze-up. Even rather short, minor melting events just prior to freeze-up seem to be reflected in the backscattering record. This appears to be consistent with the pattern seen in SAR studies at 5.3 GHz (see the previous section and Winebrenner et al. 1996).

Figure 8 shows a series of 22 wide-area images including the Alaskan coast and the Beaufort and Chukchi Seas, with a pixel spacing of 9 km. The series begins on July 8

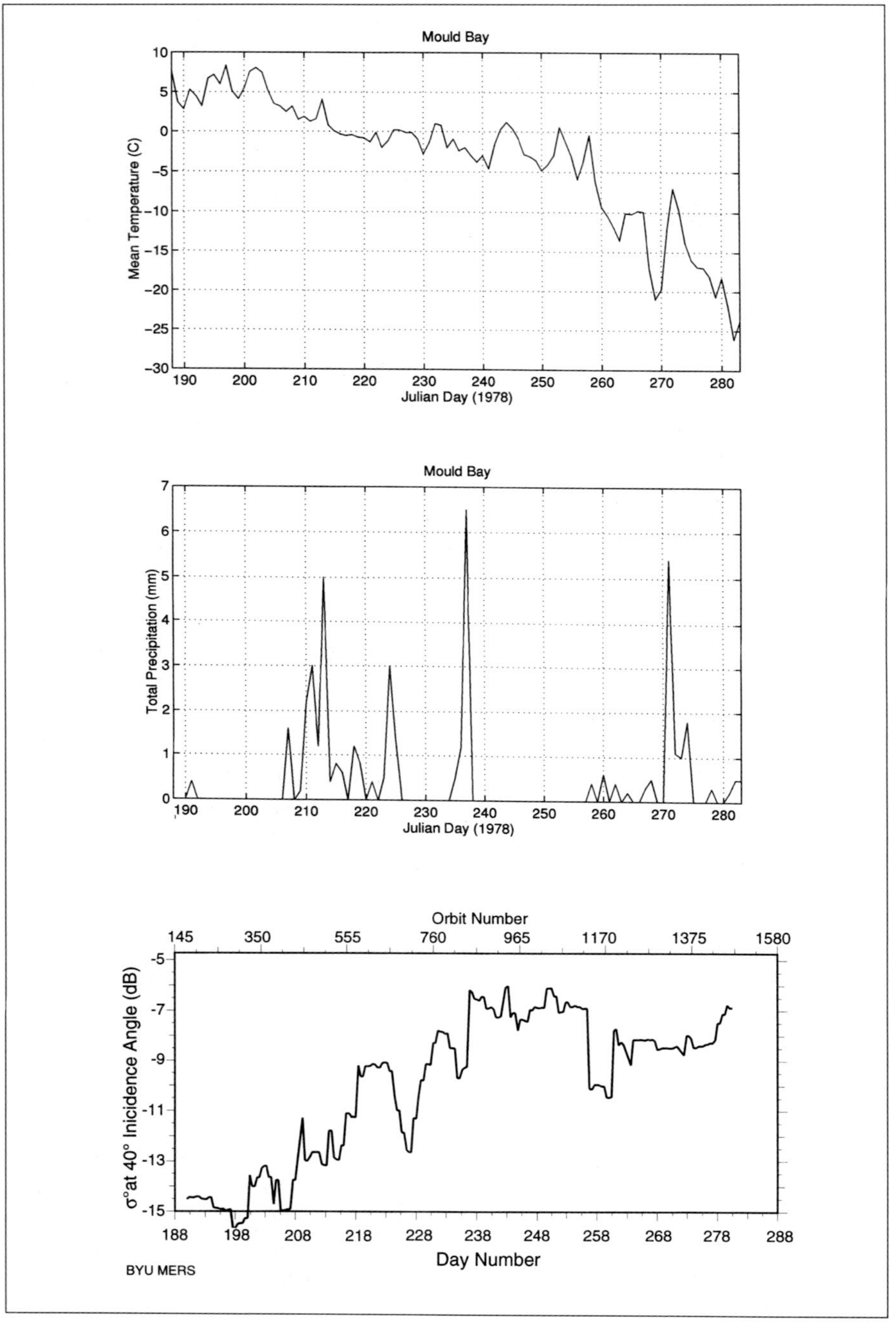


**Fig. 7.** **a-c.** Time series of **a** temperature and **b** precipitation at Mould Bay, and **c** 14.6-GHz backscattering from sea ice just to the west, as measured with the Seasat scatterometer, during autumn 1978

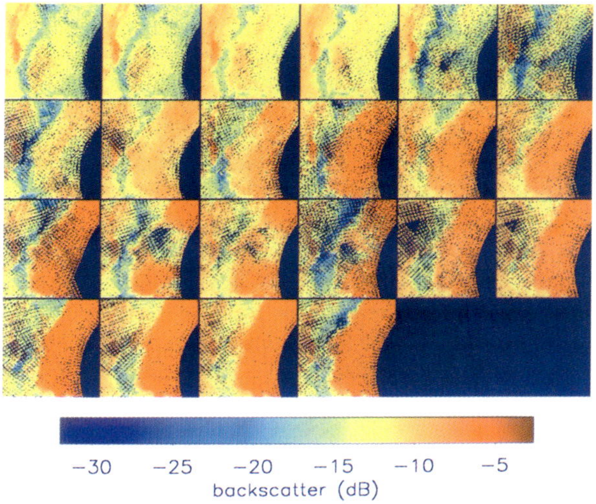


**Fig. 8.** A time series of 22 enhanced-resolution Seasat scatterometer images, each resulting from data accumulation of 8 days, at time intervals 4 days apart (data accumulations for adjacent images overlap by 4 days). The time series begins on July 8, 1978 and ends on October 8, 1978. Each image is presented in a polar stereographic projection with center coordinate 70°N, 225°E, and measures approximately 1468.5 km on its horizontal axis and 1530.8 km on its vertical axis. Pixel spacing is 8.9 km. The coast of Alaska is visible in most images as the relatively high backscatter region along the *left borde* r. Th on the right of each image is the region in which no data are available due to the limiting latitude of the Seasat orbit

(Julian day 189), and continues to October 7 (Julian day 280). The images are spaced 4 days apart; the 8-day data accumulation intervals for each pair of images overlap by 4 days. The north-to-south progression of stable, high backscattering marking freeze-up is very clear, at least for regions known independently to be covered in winter largely by multiyear ice. Note that freeze-up either has not occurred or is not so clearly visible in the area near the Alaskan coast at the end of the image series.

This response at 14.6 GHz in the Arctic stands in contrast to that observed with 5.3 GHz scatterometry in the Antarctic (Drinkwater et al. 1994). In the latter case, freeze-up is marked much more subtly by a slight decrease in backscattering at the reference incidence angle (40°), but the change in the gradient of backscattering with incidence angle is pronounced.

It is clear in any case that 14.6 GHz (or, almost equivalently, 14.0 GHz) backscattering and enhanced-resolution scatterometry of Arctic sea ice can provide freeze-up, and very likely melt onset, information. It is likely that such information can be obtained on first-year as well as multiyear Arctic sea ice using the higher frequencies. The wide geographic coverage of scatterometer data make them complementary, and perhaps in some cases a substitute for, SAR data. Moreover, the use of scatterometry involves observations at a number of incidence angles and referencing those observations to a common incidence angle (Long et al. 1993); experience in learning to use scatterometry data may therefore be highly valuable in learning to use wide-swath, multi-incidence angle (i.e., ScanSAR) data as well. Hence there is considerable motivation to expand attempts to automate seasonal transition date mapping on sea ice to include scatterometer and higher-frequency SAR observations.

## 7.4 Future Work

Clearly much work on, and involving, automated mapping of seasonal transitions on sea ice remains to be done as of this writing. Methods and observations to date are no more than prototypes; neither mature technology nor geophysically useful data sets yet exist.

We believe the single most pressing need, from both a geophysical and an algorithm-development standpoint, is for observational studies outside the popular Beaufort Sea region. There is evidence that SAR signatures of sea ice vary significantly across the Arctic basin, and the adjustment of algorithm parameters or entire algorithms may prove necessary in the face of wider observations. From a geophysical standpoint, it is Arctic-basin-scale variations in melt onset and freeze-up timing and melt season length that are of most interest. We hope that data records now available from ERS-1- and -2-receiving stations in addition to the Alaska SAR facility can be used to address this need.

As we have shown in the previous section, scatterometer data and backscattering observations at 13–15 GHz hold considerable promise for extending seasonal transition mapping. However, the physics of the relevant phenomenology, especially strong winter backscattering from first-year ice, is not well understood. We cannot say how broadly, in a geographical sense, higher frequency data may be useful without a better understanding of why the phenomenon arises, and therefore when, if ever, it is likely not to arise. A physical understanding of backscattering also aids the multiple-incidence angle processing necessary in constructing enhanced resolution scatterometer imagery and ScanSAR imagery (e.g., RADARSAT).

Finally, from a purely geophysical standpoint, seasonal transition dates have been mapped so far, by whatever means, for very few years in the Arctic. Progress on questions about the role of the Arctic in global climate and possible climate change cannot be addressed without decadal and multidecadal time series. We therefore hope that the next few years will bring the publication and availability of such time series. We believe that with such availability, the role of automated interpretation of satellite observations investigating polar geophysical questions will be greatly strengthened.

## References

Aagaard K, Carmack EC (1994) The Arctic Ocean and climate: a perspective. AGU Geophys Monogr 85

Barber DG, LeDrew EF, Flett DG, Shokr M, Falkingham J (1992) Seasonal and diurnal variations in SAR signatures of landfast sea ice. IEEE Trans Geosci Remote Sens 30, no 3: 638–642

Beaven SG, Gogineni SP (1994) Shipborne radar backscatter measurements from Arctic sea ice during the fall freeze-up. Remote Sens Rev 9: 3–25

Carlstrom A, Ulander LMH (1993) C-band backscatter signatures of old sea ice in the central Arctic during freeze-up. IEEE Trans Geosci Remote Sens 31, no 4: 819–829

Carsey F (1985) Summer Arctic sea ice character from satellite microwave data. J Geophys Res 90, no C3: 5015–5034

Drinkwater MR, Early DS, Long DG (1994) ERS-1 investigations of Southern Ocean sea ice geophysics using combined scatterometer and SAR images. Proc Int Geosci Remote Sens Symp IGARSS '94, August 8–12, Pasadena, California, pp 165–167
Early DS, Long DG, Drinkwater MR (1994) Comparison of enhanced resolution images of Greenland from ERS-1 and Seasat scatterometers. Proc Int Geosci Remote Sens Symp IGARSS '94, August 8–12, Pasadena, California, pp 2382–2384
Ebert EE, Curry JA (1993) An intermediate one-dimensional thermodynamic sea ice model for investigating ice-atmosphere interactions. J Geophys Res 98, no C6: 10085–10109
Hallikainen M, Winebrenner DP (1992) The physical basis for sea ice remote sensing. In: Carsey FD (ed) Microwave remote sensing of sea ice, 478, American Geophysical Union, Washington, DC
Jaynes ET (1982) On the rationale of maximum-entropy methods. Proc IEEE 70, no 9: 939–952
Livingstone CE, Onstott RG, Arsenault LD, Gray AL, Singh KP (1987) Microwave sea ice signatures near the onset of melt. IEEE Trans Geosci Remote Sens 25, no 2: 159–173
Long DG, Drinkwater MR (1994) Greenland ice-sheet surface properties observed by the Seasat-A scatterometer at enhanced resolution. J Glaciol 40, no 135: 213–230
Long DG, Hardin PJ, Whiting PT (1993) Resolution enhancement of spaceborne scatterometer data. IEEE Trans Geosci Remote Sens 31, no 3: 700–715
Long DG, Early DS, Drinkwater MR (1994) Enhanced resolution ERS-1 scatterometer imaging of southern hemisphere polar ice. Proc Int Geosci Remote Sens Symp IGARSS '94, August 8–12, Pasadena, California, pp 156–158
Maykut G, Untersteiner N (1971) Some results from a time dependent thermodynamic model of sea ice. J Geophys Res 76: 1550–1575
Onstott RG (1992) SAR and scatterometer signatures of sea ice. In: Carsey FD (ed) Microwave remote sensing of sea ice, 478, American Geophysical Union, Washington, DC
Onstott RG, Grenfell TC, Mätzler C, Luther CA, Svendsen EA (1987) Evolution of microwave sea ice signatures during early summer and midsummer in the marginal ice zone. J Geophys Res 92, no C7: 6825–6835
Parkinson C (1992) Spatial patterns of increases and decreases in the length of the sea ice season in the north polar region 1979–1986. J Geophys Res 97, no C9: 14377–14388
Percival DB, Walden AT (1993) Spectral analysis for physical applications. Cambridge University Press, Cambridge, UK
Scharfen G, Barry RG, Robinson DA, Kukla G, Serreze MC (1987) Large-scale patterns of snow melt on Arctic sea ice mapped from meteorological satellite imagery. Ann Glaciol 9: 200–205
Schwartz K, Jeffries MO, Li S (1994) Using ERS 1 SAR data to monitor the state of Arctic Ocean sea ice between spring and autumn 1992. Proc Int Geosci Remote Sens Symp IGARSS '94, August 8–12, Pasadena, California, pp 1759–1762
Smith LC, Forster RR, Isacks BL, Hall DK (1997) Seasonal climatic forcing of alpine glaciers revealed with orbital synthetic aperture radar. J. Glaciol, in press
Winebrenner DP, Nelson ED, Colony R, West RD (1994) Observation of melt onset on multiyear Arctic sea ice using the ERS 1 synthetic aperture radar. J Geophys Res 99, no C11: 22425–22441

Winebrenner DP, Holt B, Nelson ED (1996) Observation of autumn freeze-up in the Beaufort and Chukchi Seas using the ERS 1 synthetic aperture radar. J Geophys Res 101, no C7: 16401–16419

Wismann V, Cavanie A, Hoekman D, Woodhouse I, Boehnke K, Schmullius C (1996) Land surface observations using the ERS-1 windscatterometer. Final Report for European Space Agency Contract 11103/94/NL/CN, Institute for Applied Remote Sensing, Wedel, Germany

# CHAPTER 8

# Satellite Microwave Radar Observations of Antarctic Sea Ice

M.R. Drinkwater

## Contents



Analysis of SAR Data of the Polar Oceans
Edited by C. Tsatsoulis and R. Kwok

## 8.1 Introduction

Historical data on Antarctic sea ice extent and concentration have traditionally been derived from visible and near-infrared images acquired by the polar-orbiting National Oceanic and Atmospheric Agency's (NOAA) meteorological satellites, using the Advanced Very High Resolution Radiometer (AVHRR), and more recently by the Defense Meteorological Satellite Program's Operational Linescan System (OLS) (Massom 1991). The limitation of these systems is that the majority of energy imparted to the Antarctic sea-ice system is transferred during frequent episodic storm bursts, caused by fast-moving low pressure systems (McPhee et al. 1996). Since the Southern Ocean sea-ice cover is completely bounded at its lower latitude limit by open ocean, these "polar lows" transport large amounts of moisture (contained in warm air masses) over the outer ice cover. The result is that most, if not all, noteworthy periods of wind- and temperature-driven dynamic changes in the ice cover are accompanied by periods where the region is blanketed by cloud, and when the atmosphere is inherently more electromagnetically opaque. During storms, the probability with which the area is cloud covered is extremely high, thereby ruling out use of visible or near-infrared images as a practical method of monitoring the associated changes in ice conditions. Instead, Nimbus-7 Scanning Multichannel Microwave Radiometer (SMMR) and DMSP Special Sensor Microwave/Imager (SSM/I) have been the primary workhorses to build up a microwave record of Antarctic sea-ice characteristics. Similar problems, however, occur in passive microwave retrievals of sea-ice concentration, and the algorithms are called into question during these periods of change. Oelke (1997) discovers that the influence of water vapor in the atmosphere alone can modify the ice concentration retrievals by fractions exceeding 10%, and that retrievals of ice concentration must compensate for the atmospheric water vapor and liquid water contents.

Prior to 1991, the only active microwave radar data to be acquired in the Southern Ocean were collected during the Shuttle Imaging Radar-B (SIR-B) mission of the space shuttle in 1984 (Carsey et al. 1986; Martin et al. 1987). Seasat in 1978 acquired no synthetic aperture radar (SAR) data due to the lack of a direct broadcast receiving station in Antarctica, and due to the brevity of its lifetime (lasting only 3 months long before a fatal power failure). Despite the brief Arctic success of Seasat, a period of almost 15 years passed before the second radar remote sensing satellite mission was flown. For the first time, the launch of the European ERS-1 in July 1991 enabled compilation of the first detailed picture of the C-band microwave backscattering characteristics of Southern Ocean sea ice. The launch of an identical ERS-2 in April 1995 provides continuity in this series of Antarctic radar observations.

In this chapter, a variety of microwave data sets are presented and the results of these combined data applied in investigations of the spatial and temporal characteristics of sea ice. Since most field data collection and experiments coinciding with the ERS missions occurred within the South Atlantic quadrant of the Southern Ocean, this study focuses particularly upon that Weddell Sea region.

## 8.2 Obtaining Spaceborne Radar Data in Antarctica

### 8.2.1 Antarctic SAR Data

The following sections briefly describe the data from currently operating SAR satellites, which can be received and recorded in Antarctica. Figure 1 shows the present locations of three operational Antarctic receiving stations.

#### 8.2.1.1 *ERS-1 and ERS-2 SAR*

Since 1991, the European Space Agency (ESA) ERS-1 and ERS-2 satellites have acquired microwave radar image data of the Weddell Sea. The C-band imaging SAR mode of the active microwave instrument (AMI) on board ERS allows vv-polarized backscatter ($\sigma^0_{vv}$) data to be collected (i.e. wavelength $\lambda = 5.6$ cm). Presently, these are the only operational instruments capable of acquiring high-resolution (approximately 30 m), all-weather (day or night), frequent repeat, 100 km-swath Antarctic sea-ice images (Drinkwater 1995b). Neither spacecraft has SAR tape-recording capability and data access is limited to the areas of direct broadcast to existing operational and ESA-approved ERS-1 Antarctic ground receiving stations (ESA Earthnet 1992).

#### 8.2.1.2 *J-ERS-1 SAR*

Scientific data from the L-band (1.2-GHz) Japanese Earth Resources Satellite (J-ERS-1 or "Fuyo") SAR have been made available to approved investigators by the National Space Development Agency (NASDA) of Japan. Currently, hh-polarized backscatter ($\sigma^0_{hh}$) Antarctic data can only be acquired within the Syowa and McMurdo station mask areas shown in Fig. 1. These SAR data may be ordered at "research" cost from RESTEC (Remote Sensing Technology Centre; i.e., the distribution agent for NASDA) through a research data distribution and purchase agreement.

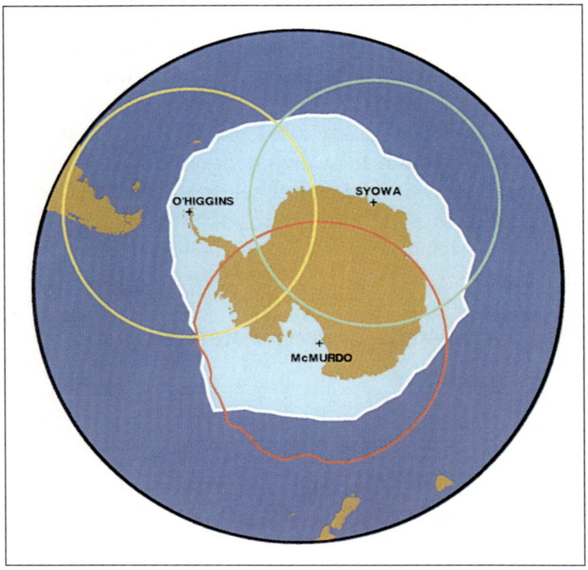


**Fig. 1.** Operational Antarctic ERS, J-ERS and RADARSAT receiving stations, together with their data acquisition masks

### 8.2.1.3
### *RADARSAT SAR*

The Canadian RADARSAT satellite, launched by NASA on November 4, 1995, carries a C-band hh-polarized SAR. The expected life span of the satellite is 5 years. Presently, the Canadian Space Agency's (CSA) RADARSAT [1] SAR is the only satellite radar offering more than simply 100 × 100 km images. It offers a wide variety of flexible beam modes of which the most unique, the SCANSAR mode, is the only source of wide-swath (500 km), high-resolution (approximately 100 m) ice cover information for the Southern Ocean. However, the only CSA-approved Antarctic receiving station operational and equipped to downlink and record RADARSAT data is McMurdo. RADARSAT also has on-board tape recorders which will allow a limited amount of data (wide-swath "narrow" mode) to be recorded and downlinked elsewhere, other than simply at the Antarctic receiving stations. This, in principle, allows data access in any planned location in the Southern Ocean. Until now, no data have been acquired for approved Antarctic investigators. Further information on the RADARSAT mission and its operating modes can be found at the RADARSAT URL (http://radarsat.space.gc.ca).

### 8.2.1.4
### *ERS SAR Backscatter Calibration Accuracy*

SAR data products received and processed by the German Processing and Archiving Facility (D-PAF) must be treated differently than most US processed ERS data, since they

1 At time of writing RADARSAT is in orbit undergoing commissioning. Data are expected to come online in mid 1997.

are distributed in a full resolution (i.e., 12.5 m pixel spacing), uncalibrated image form. According to a study by Bally and Fellah (1995) for these three-look SAR images (i.e. ESA SAR precision image products), the single pixel (12.5 m) 90% radiometric confidence interval is bounded by a ±4.5 dB interval. Thus, the probability that the measured intensity lies within a ±4.5 dB error bound is 90%, whereas the radiometric accuracy and stability errors are within specifications of a fraction of a decibel (Laur et al. 1993).

To calculate the SAR backscattering coefficient ($\sigma^0_{SAR}$) of a distributed target which corresponds to an area of sea ice ($N$ pixels in extent), averaging is performed. Block averaging is generally first used to create a speckle-filtered image, with the new equivalent number of looks modified by the area of block averaging together with the target area (i.e., number of pixels averaged from target). The resulting confidence interval for an 8 × 8 block-averaged image with 100 m pixel spacing exceeds 90% for ±1 dB bounds. With further target averaging of a sample box of over 250 pixels (i.e., 16 × 16 pixel box), a confidence level exceeding 90% may be achieved with error bounds of ±0.5 dB. Over 99% probability of errors less than ±0.5 dB may be reached with samples of greater than 500 pixels (i.e., 23 × 23 pixel box). Thus, for most purposes, having already used a box-filter (8 × 8 window) to reduce the image to 100 m pixel spacing, it is generally only necessary to derive samples from target box areas exceeding 8 × 8 pixels to exceed the ±0.5 dB error bound at 99% confidence. All samples presented in this paper are from target areas exceeding this minimum threshold area.

### 8.2.2 Antarctic SAR Receiving Stations

It is normally only feasible to acquire high-resolution SAR images within direct broadcast range or line-of-sight to a ground receiving station, due to the high data rates. Prior to the launch of ERS-1 in July 1991 there were no satellite receiving stations in Antarctica equipped to downlink or record SAR data. Since 1992, a chain of contiguous SAR data reception areas has been developed to support the SAR data requirements of a number of nations. Currently, McMurdo (US), O'Higgins (German), and Syowa (Japanese) are all equipped and capable of downlinking and recording SAR image data. Figure 1 illustrates that almost complete spatial coverage of the mean maximum seasonal ice extent is possible when all three stations are operating together.

The first SAR station to open in Antarctica was the German Antarctic Receiving Station (GARS) at the Chilean General Bernardo O'Higgins station, located on the northern end of the Antarctic peninsula (at 63.32°S, 57.90°W). GARS has been collecting ERS-1 data since October 1991, but its data reception is limited to 1–2 month "campaign-style" operating periods when it is manned. These periods typically coincide with German Antarctic experiments, or periods when the German icebreaker R.V. *Polarstern* is operating in the Weddell or Bellingshausen Seas. To date the O'Higgins receiving station has enabled the most comprehensive SAR data records of Antarctic ice to be collected.

Presently, SAR data are transported out of Antarctica by ship or aircraft (via the Chilean Teniente Marsh Base), thereby resulting in delays before SAR image processing takes place. However, a high-speed satellite link is being developed which will (in late 1997) enable McMurdo SAR data to be relayed back to the US for processing at the Alaska SAR Facility (for further information see ASF URL: http://www.asf.alaska.edu). The recent addition of RADARSAT with its tape-recording capacity has increased the flexibility for recovering information in inaccessible locations such as the Southern Oceans.

### 8.2.3
### Scatterometer Imaging

The nonimaging wind scatterometer mode of the AMI on board ERS (EScat) enables continuous radar coverage of Antarctica along a 500-km swath during the period when the SAR is switched off. Low bit-rate EScat data (with an intrinsic resolution of approximately 50 km) do not require direct station transmission and are tape recorded and downlinked to more accessible ESA ground stations (ESA Earthnet 1992). Scatterometer $\sigma^0_{vv}$ data can therefore be collected at times and in regions where direct SAR downlink is impossible. Extensive daily coverage of the ERS-1 scatterometer is particularly valuable due to the lack of data reception in the AMI SAR mode when Antarctic receiving stations are closed. This is especially true in the Weddell Sea, because of the "campaign-style" operation of GARS and limited access to Syowa.

A method is proposed using data acquired in the EScat mode for mapping sea-ice characteristics over the entire Southern Oceans ice cover. Recent advances in image generation from scatterometer data (Drinkwater et al. 1993a) enable complementary, medium-scale resolution images to be generated from EScat orbital swath data. This method allows construction of enhanced-resolution (approximately 20 km) scatterometer images (with pixel spacing of 12–14 km) from the backscatter data record, and by virtue of the wider swath and larger coverage allows mapping of the entire sea-ice cover in a few days (Drinkwater et al. 1993a; Long et al 1993). The advantage of EScat image data is that they form an uninterrupted sequence of C-band backscatter of Antarctic sea ice from December 1991 onwards. Not only are these data critically necessary to fill in areas of sparse temporal and spatial radar data coverage, but they are also complementary to the restricted incidence angle data (20–26° incidence) of the SAR. The contrasting 20–60° incidence angle diversity of these C-band scatterometer data enables much greater capability for discriminating ice types around Antarctica than by either SAR or passive microwave alone. Coupled with higher frequency passive microwave and finer-resolution SAR, these data enable a better understanding of the time sequence and seasonal history of Antarctic sea-ice formation, drift, deformation and decay.

## 8.3
## Sampling Issues

### 8.3.1
### Temporal and Spatial Coverage

One of the priorities for scientific studies of polar ice and snow is high spatial and temporal coverage, due to the short timescales of variability in snow and ice conditions. Exploitation of radar data, particularly in studies related to the mapping and monitoring of the sea-ice cover, was one of the main driving forces behind inclusion of the SAR instrument on board ERS-1 (Drinkwater 1995a). A large number of sea-ice studies are currently being performed using Arctic Ocean SAR data received at northern hemisphere ground stations the Alaska SAR Facility, Fairbanks (and Gatineau, Canada; Kiruna, Sweden; and Tromsø, Norway). Together these SAR receiving stations enable the majority of the northern-hemisphere sea-ice cover to be mapped using SAR, especially at the latitudinal limit of the orbit where the SAR swaths converge. Contrastingly less coverage is possible of the more extensive Antarctica sea-ice cover due to the latitudinal extent of the sea ice and the limited operating times of GARS and Syowa. The result is a high-resolution SAR image database which is at best discontinuous in space and time. Alternative methods have therefore been sought using other forms of EScat data, in order to fill the temporal active microwave record in space and time.

### 8.3.2 Filling the Gaps

The key advantage of the ERS scatterometer is that it operates whenever the SAR is switched off, continuously retrieving information from the Weddell Sea without the necessity of a local receiving station. Its wider swath provides more frequent and broader incidence angle ($20^\circ \leq \theta \leq 60^\circ$) coverage in a given location. This low-bit-rate (LBR) data source is essential to fill in areas of sparse temporal and spatial SAR coverage of the Antarctic ice cover. Until now, the intrinsic low resolution of the raw measurement data is the main reason why scatterometer data have not been used in such applications. Nevertheless, the approach described by Drinkwater et al. (1993a) produces weekly average backscatter maps of the entire Southern Ocean sea-ice cover at a resolution higher (approximately 20 km) than present conventional alternatives such as the SSM/I passive microwave radiometers. EScat images form an uninterrupted sequence of vv-polarized C-band backscatter ($\sigma^0_{EScat}$) of Antarctic sea ice from the beginning of the ERS-1 mission. Since these backscatter images are based on multiple azimuth and incidence angle observations, these data provide greater capability for discriminating ice types and separating of ice from ocean. Coupled with higher-frequency passive microwave and finer resolution SAR, these data enable a better understanding of the time sequence and seasonal history of Antarctic sea-ice formation, drift, deformation and decay. Furthermore, medium-scale-resolution EScat images essentially provide the large-scale context within which the high resolution SAR images may be interpreted.

## 8.4 Ice Classification Issues

Recognition of Antarctic sea-ice types in C-band microwave images, as a proxy for ice thickness, is limited by the lack of a historical sea-ice signature database equivalent to that existing for the Arctic (Drinkwater 1995a). Only since ERS-1 has flown have surface experiments focused on the acquisition of a microwave database for comparison with satellite measurements. These provide the foundation for preliminary attempts to recognize and categorize ice types observed in SAR and scatterometer images. Due to the logistical constraints of working in accessible regions, most of these measurements have been focused in the Weddell Sea. Results from these demonstrate that Antarctic ice appears considerably different from its Arctic counterpart, due in part to its relatively higher salinity and younger mean age and thickness (Drinkwater et al. 1993a).

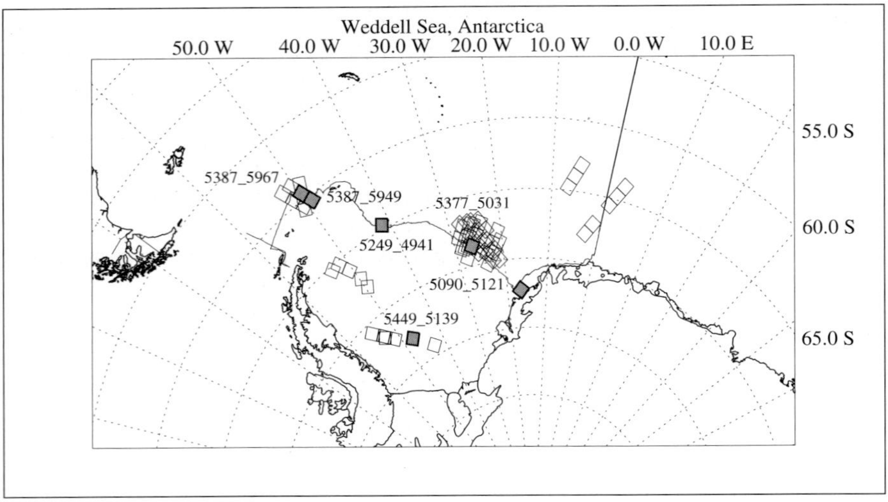


**Fig. 2.** Location of ERS-1 SAR image frames used in the Weddell Sea backscatter analysis. Backscatter distribution data from the *bold* SAR frames are given in Table 1

## 8.4.1 Backscatter Characteristics of Antarctic Ice

Since the launch of ERS-1, C-band radar backscatter characteristics of Antarctic sea ice have been investigated during a number of Weddell Sea surface experiments (Drinkwater et al. 1993b, 1995a; Hosseinmostafa et al. 1995; Lytle et al. 1996). Most, however, have focused on the smaller-scale influences of sea-ice physical properties upon surface-scatterometer measurements, without placing them in the context of satellite radar measurements. In this section, a synthesis is provided of small- and large-scale measurements, using the shipborne track and ERS SAR image locations shown in Fig. 2. Examples of field observations and SAR measurements are provided together with corresponding surface validation from the 1992 Winter Weddell Gyre Study (WWGS'92) (Drinkwater and Haas 1994; Lemke 1994).

### 8.4.1.1 *Shipborne Scatterometer Signatures*

Field backscatter data were acquired in 1992 using a shipborne microwave radar, during WWGS'92. A C-band (4.3-GHz) frequency-modulated continuous-wave radar scatterometer was operated from the port rail of R.V. *Polarstern* to obtain shipborne measurements of the microwave scattering properties of Antarctic sea ice in mid-winter (Drinkwater et al. 1995a). The radar had dual polarization, enabling like- (vv) and cross-polarization (hv) data to be acquired at a variety of incidence angles ($15^\circ \leq \theta \leq 70^\circ$). When the ship was stationary by a large, uniform ice floe, the instrument was scanned in the vertical plane to obtain independent samples of $\sigma^0_{Ship}$ as a function of incidence angle and polarization. In support, detailed surface measurements were made within the footprint of the shipborne radar, after each scan was performed, "surface information" comprising measurements of physical and chemical properties of snow and ice together with structural information (Drinkwater and Haas 1994). The objective of the 1992 winter scatterometer measurements was to provide validation data for ERS-1 SAR observations and to obtain a detailed microwave backscattering catalogue of various ice conditions. Overlapping data acquisitions were planned and made possible by the GARS as part of the International Space Year Project (Lemke 1994).

### The Ice Margin and Pancake Ice Floes

The Antarctic outer ice margin or marginal ice zone (MIZ) is typically characterized as a high-energy wave environment, where wave propagation into the ice edge plays a dominant role in determining the style of ice formation during winter ice advance and the characteristics of floes during ice-edge recession. The MIZ is typically broad in its extent during ice-edge advance. In 1992, during WWGS'92, a high amplitude swell was experienced along the Greenwich meridian and frazil and pancake young-ice growth was observed for hundreds of kilometers from the MIZ into the seasonal ice zone. Figure 3a shows a photograph taken in June 1992 of these swell-rocked pancakes in the

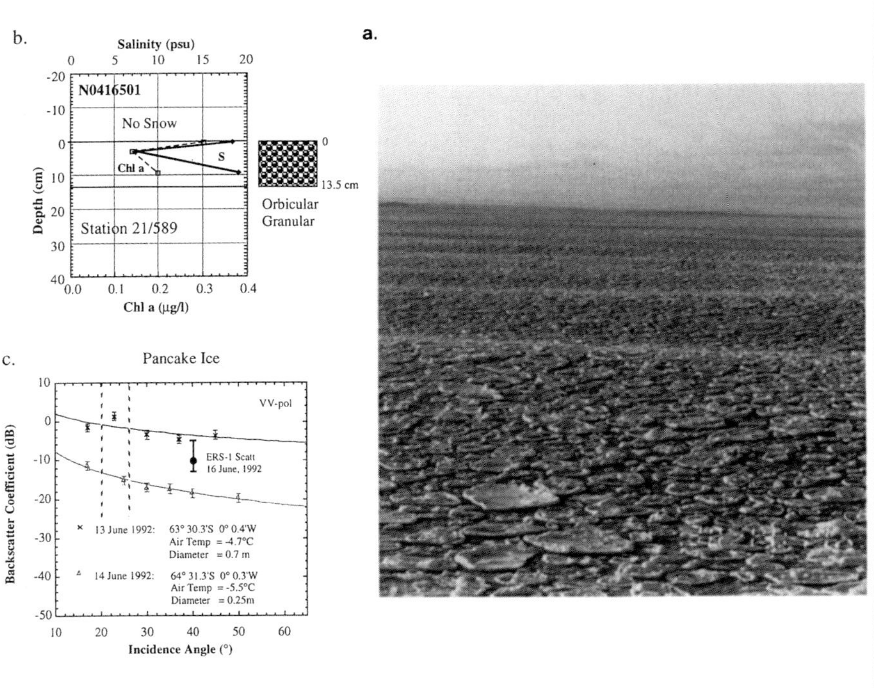


**Fig. 3.** **a** Photograph of pancake ice field on June 13, 1992, **b** sea-ice sample at this location at site N0416501 (after Drinkwater and Haas 1994), and **c** shipborne scatterometer scan at this site location, together with contrasting scan from June 14 and $\sigma^0{}_{Escat}$ measurement in the same location

location of Maud Rise. In contrast, after the peak ice extent the ice edge is typically in a recessional mode, and the MIZ usually comprises brash ice and a mixture of small wave-broken floes and wave-washed piles of rubble.

In contrast to the situation for nilas and young ice forms found in calm environments, pancakes were found to be almost ubiquitous at the ice margin (along the Greenwich meridian) during ice edge advance in 1992. The occurrence of vast expanses of pancakes and/or streamers with densely packed subresolution floes results in an extremely characteristic MIZ signature (Drinkwater et al. 1993a; Gohin 1995; Early and Long 1997). Typically riding on the surface of waves, small wave-washed, porous pans often have a high salinity (Fig. 3b) and sufficiently high permittivity that there is relatively strong backscatter near nadir. The rate at which $\sigma^0_{Ship}$ falls with incidence angle, however, appears dependent on the size and packing density of the pancakes, the amount of open water between them, and how wet or deformed their surfaces are. Figure 3b shows the salinity characteristics of a pancake sampled on June 13, 1992, and Fig. 3c contrasts the June 13 radar signature with another site from the following day. The June 13 signature is greater by 10 dB throughout the incidence angle range, with values exceeding -5 dB at 40° incidence or more, due in part to the denser packing of pans and their rough, deformed, and often rafted surfaces. In cases of smaller pancakes spaced by open water, as on June 14, $\sigma^0_{Ship}$ falls more rapidly. Gohin (1995) reports an intermediate MIZ signature (in his Fig. 15) which appears a composite of backscatter curves originating over combinations of open water and dispersed floes, or even low-concentration patches of pancakes.

Considerable azimuthal anisotropy is also noted in pancake ice margins in the ERS scatterometer data (Early and Long 1997), and is likely due to the fact that the pancakes damp out smaller gravity and capillary waves, leaving only long-wavelength swell waves. This results in a dominant wave-propagation direction sensed by the scatterometer in some circumstances. Future attempts to fit the functional form of the wind relationship may enable wave propagation direction to be derived.

**Nilas**

Large expanses of recently formed nilas (< 10 cm thick) have rarely been observed during surface shipborne experiments, largely because Weddell Sea experiments have focused on the ice margin and swell-influenced regions of the seasonal ice pack. Consequently, the total fraction of observed ice formed under calm, nondynamic conditions is small, and the opportunity to sample large areas other than in recently opened leads or polynyas with scatterometers is rare. Some young ice was observed during WWGS 92, which is reported in Drinkwater and Haas (1994). When observed, nilas typically had little or no snow cover other than frost flowers. The typical signature includes large backscatter coefficients at near nadir-angles, with a high gradient of $\sigma^0_{Ship}$, falling to values less than -20 dB at high incidence angles. A large sample of calibrated SAR pixel values shown as a thick black global probability distribution function in Fig. 4 (from all image locations in Fig. 2) testifies that the likelihood of finding large fractions of new ice and nilas with values below -15 dB is rare in the central Weddell Sea, and extensive nilas areas are generally only found in locations of persistent divergence such as coastal or ice-shelf polynya systems.

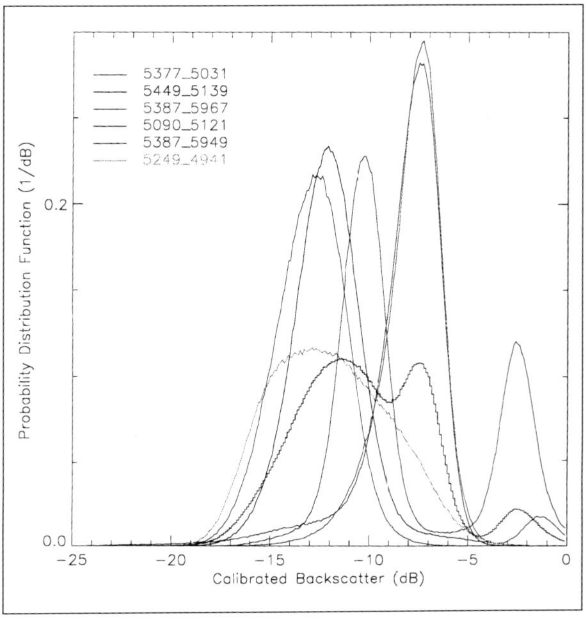


**Fig. 4.** SAR backscatter probability distribution functions (pdf's) for each individual scene and for the entire grouped winter SAR data set (see Fig. 2 and Table 1 for SAR scenes) as the stepped thick black histogram

### White Ice Floes

The mean thin first-year ice or white ice (in the range 30–50 cm thick) signature shown in Fig. 5a is typical of ice observed in the divergent central Weddell Sea region and is an average of seven individual site measurements made from July 1 to 14, 1992 in the location of scene 5377–5031 (Fig. 2). The signature of this relatively smooth, light snow-covered (1–6 cm snowdepth) ice has $\sigma^0{}_{Ship}$ values between -24 and -26 dB in the ERS SAR incidence angle range $20^\circ \leq \theta \leq 26^\circ$, an intermediate gradient of around 0.49 dB/degree from $20^\circ \leq \theta \leq 60^\circ$ incidence, and a neutral cross-polarized backscatter curve. Large expanses of white ice were scoured by strong winds and had snow depths of less than 5 cm. Lytle et al. (1996) have shown previously that snow-free young ice (10–30 cm thick) may result in extremely low backscatter values in an equivalent incidence angle range, thereby increasing as the ice ages (in the absence of deformation) and roughens. These results nevertheless imply that smooth, level white ice with a bare ice surface or shallow snow cover alone does not play a dominant role in driving the backscatter to values shown in the probability distribution functions (pdf's) in Fig. 4, and suggest alternatively that areas of congealed pancakes or ridged portions of sea ice drive up measured SAR values in this extensive region.

### Smooth First-Year Ice Floes

The mean signature of smooth, level (undeformed) uniform first-year ice floes shown in Fig. 5b is constructed from nine individual scatterometer samples, each site of which exceeded 60 cm ice thickness. A distinguishing feature of this signature is that $\sigma^0{}_{Ship}$ exceeds that of the white ice signature at all angles. The surface roughness of such level ice is generally greater than those white ice forms, while also displaying an older and deeper layered snowcover. ERS SAR backscatter values of equivalent ice floes are

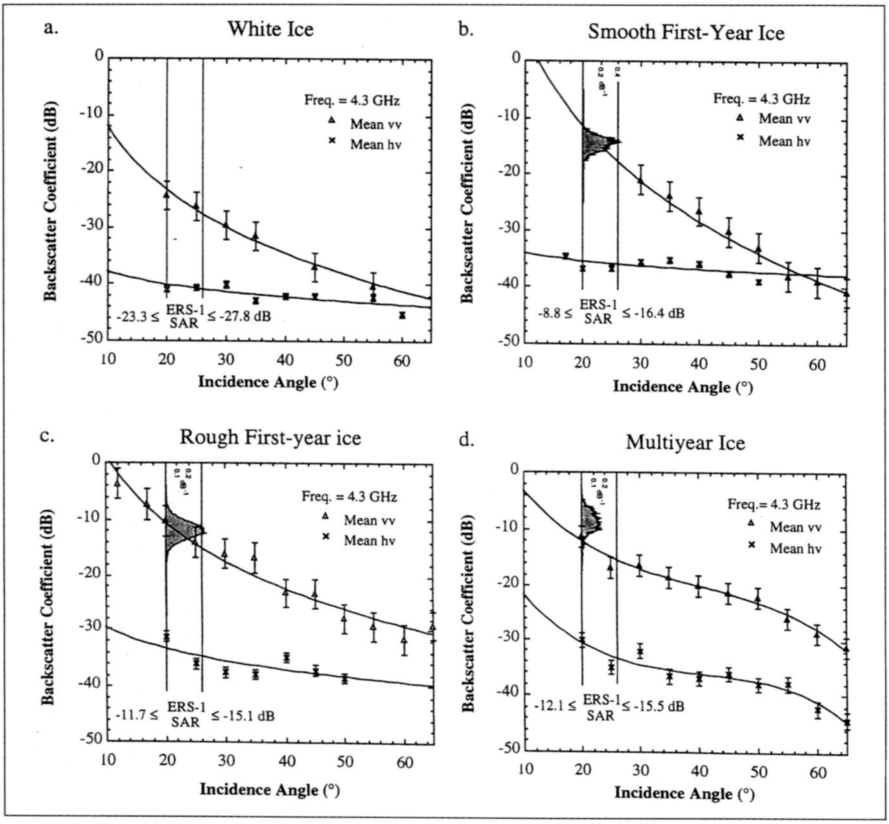


**Fig. 5.** Mean WWGS'92 shipborne scatterometer ice signatures, from **a** white ice, **b** smooth first-year ice, **c** rough first-year ice, and **d** multiyear ice. Curves are fitted exponentially and error bars indicate the standard error of the data. *Vertical lines* delineate the incidence angle range of ERS SAR (20° ≤ θ ≤ 26°) and overlapping *shaded* pdf's show the probability distribution from SAR pixel samples in the locations of shipborne scatterometer measurements. A range of observed SAR values is shown for all sampled images for each ice type

expected to fall in the range $-8.8 \leq \sigma^0_{Ship} \leq -16$ dB, based on the surface scatterometer measurements, but this is heavily dependent on the physical surface characteristics.

A typical medium thick (67 cm) first-year ice sample acquired on July 13, 1992 at 65.99°S, 33.53°W had interleaved bands of frazil and columnar sea-ice crystal growth and 12 cm of layered snow. The snow-ice interface temperature was -13° C, despite air temperatures lower than -30° C, and an accompanying salinity at the ice surface of 14 psu. Angular depth-hoar crystals developing under the strong thermal gradient in the basal snow were of significantly higher salinity than the ice surface itself, having typical values of around 27 psu. This highly saline, rough scattering interface at the base of the snow probably accounts for extremely high backscatter values in the range 20–26° in Fig. 5b. Likewise, the remainder of the snow had significant salt content, and airborne spray from nearby leads is thought to be a contributor to the mean background level of surface snow salinity of 2 psu.

**Rough First-Year Ice Floes**

Well-deformed, ridged, or rubbled first-year ice floes are difficult to measure using a shipborne scatterometer, because a complete incidence angle scan is not possible from a fixed viewing position when the local topography is punctuated by piles of ice blocks. Shipborne scatterometer measurements were limited to five sites where surface roughness was relatively evenly distributed, and where the signature was not significantly biased by one or more features within the scan. The result is the mean signature in Fig. 5c. Values of $\sigma^0_{Ship}$ are lower at near normal incidence, and fall between -11.7 and -15 dB for the SAR incidence angle band. The gradient in $\sigma^0_{Ship}$ is approximately -0.52 dB/degree in the 20–60° range, but the small sample size and averaging of individual ridges or local roughness elements cause the variability in measurement points around the logarithmic fit. Cross-polarized backscatter is greater and more variable for these rough ice floes, especially in the 20–40° range, and is explained by second-order scattering effects such as multiple scattering from blocks.

**Multiyear Ice Floes**

According to the field scatterometer data in Fig. 5d, multiyear ice floes (largely second-year ice) are practically indistinguishable from rough first-year ice forms in the ERS incidence angle range. Undeformed, old floes were distinguished on the basis of snow depth, salinity, and thickness as second-year ice. Mean net annual thermodynamic ice growth is estimated to be around 1.5 m in the absence of dynamic thickening, based on the lifecycle, mean snow depth, and annual net freezing rate in the Weddell Sea (Fischer 1995). However, perennial ice floes observed during WWGS'92 comprised massive undeformed ice with considerable snow accumulation. Snow depths exceeding 0.75 m were commonly measured (Massom et al. 1997) while ice thickness ranged from around 2 m upwards. Shipborne scatterometer $\sigma^0_{Ship}$ values from these floes fall between -12 and -15.5 dB, and the low gradient of -0.36 dB/degree observed in the 30–50° incidence angle range is caused by snow volume scattering from layers and distinctive ice lenses and ice pipes (Massom et al. 1997). To account for the volume scattering characteristics exhibited in Fig. 5c, a modified third order polynomial fit is applied instead of a single parameter exponential, to accommodate the flattening at 25° and rolloff beyond 55°. The distinctive plateau and higher backscatter value in the range 40–50° may be used in 40°-incidence EScat images as a method for discriminating high concentrations of thick, snow-covered perennial ice.

#### 8.4.1.2
#### *Regional Winter C-band Backscatter Variability*

The map of the Weddell Sea region in Fig. 2 indicates (in bold) the regional locations of several winter SAR image frames listed in Table 1. Corresponding image pdf's of calibrated backscatter coefficients are shown in Fig. 4. Each pdf comprises individual pixel values from 8 × 8 averaged and calibrated full-resolution SAR.PRI images (originally 8000 × 8000 pixels), resulting in $10^6$ pixel samples per 100 × 100 km frame.

A global pdf histogram, indicated as the thick black stepped line in Fig. 4 (shown as a bar-histogram with 0.1 dB bin width), shows the combined pdf from all images indicated in Fig. 2, including those listed in Table 1. This summary pdf indicates a trimodal distribution of backscatter $\sigma^0_{SAR}$, with distinct components forming peaks at –11.5, –7.5, and –2.5 dB. Individual image pdf's indicate that the highest backscatter mode comprises pixels containing glacial ice (i.e., ice shelf). Image 5090–5121 in Fig. 2 was acquired in the vicinity of R.V. *Polarstern* as she made a transect along the periphery of the Riiser-Larsen ice shelf. This scene contains a significant number of ice-shelf pixels, falling in the range –5 to +5 dB: air temperatures were around –15° C at the time of SAR imaging. Similar characteristics are observed in the upper peak in the histogram of image 5387–5967. This image captures a large iceberg drifting in the outer MIZ northeast of the tip of the Antarctic peninsula. Its backscatter values exceed those of the ice shelf in the previous example, due in large part to its lower latitude and warmer air temperatures. Glacial ice, therefore, comprises a distinct portion of the upper limb of the backscatter distribution in the Weddell Sea, and may conveniently be used in winter to find tabular icebergs of significant size, either in the marginal ice zone or within the interior ice pack. Nontabular Antarctic icebergs do not always present a uniform target area to the SAR, and rotation and drift with respect to the imaging swath often result in a reduction in contrast between the iceberg and its surrounding background.

The next brightest component distribution is described by the pdf of two contiguous image frames (5387–5949 and 5387–5967) along an orbit crossing the western Weddell Sea MIZ (Fig. 2). The peaks in their image histograms overlap almost identically, but for a small transition from the swell-disturbed marginal ice zone to the outer ice edge. Mixtures of brash ice and small first-year ice floes together with occasional multiyear ice floes characterized the swell-rocked ice margin at the time of SAR imaging (Haas et al. 1992). Air temperatures and windspeeds at the time of imaging were recorded on board R.V. *Polarstern* as between –5 and 0 °C, and 5 and 10 m $s^{-1}$, respectively. These mixtures have an extremely distinctive signature, with $\sigma^0_{SAR}$ values ranging between -10 and –5 dB, with a mode at –7.3 dB. Rough surface scattering, occurring from sub-resolution floes, has a distinctively uniform texture and bears a close resemblance to wave-disturbed northern hemisphere MIZ ice signatures observed by airborne SAR in the Labrador Sea (Drinkwater and Squire 1989; Livingstone and Drinkwater 1991). The further into the MIZ one traverses, the higher the probability of observing large undeformed first-year ice floes and new ice in leads. These pixels appear as a toe of darker material in the histogram of image 5387–5949, extending from –12 dB down to around –19 dB.

Signatures similar to the previous MIZ cases (in terms of backscatter amplitude) may be encountered in the central ice pack, but with the obvious distinction of visible ice-floe outlines. Image 5249–4941, shown in Fig. 6 (overlapping the path of R.V. *Polarstern*), contains a typical mixture of perennial and seasonal ice, and was acquired in the outflow path of old ice from the Weddell Sea. The mixture histogram combining these two distinctive end-members appears as a broad pdf in Fig. 4. Shipborne observations, noted by Haas et al. (1992) as *Polarstern* traversed the region encompassed by this image (less than 24 h after SAR imaging – see Table 1), indicated a closed ice cover (100% concentration) comprising level first-year ice punctuated by distinct ridges and rougher multiyear ice in fractions up to 30%. First-year ice was extremely rubbled and ridged, particularly around the perimeter of thick multiyear ice floes. The broad pdf comprises two overlapping distributions without distinctive peaks, due to the continuum of states of the ice ranging from patches of smooth, relatively undeformed first-

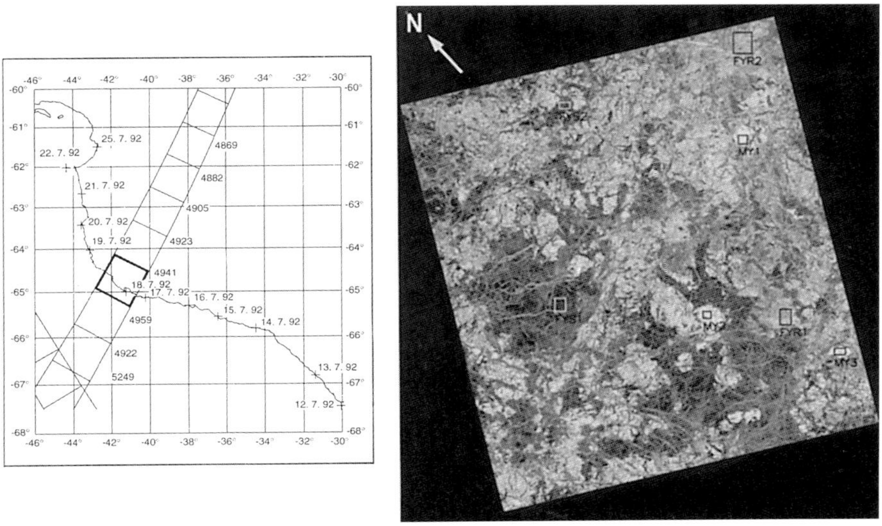


**Fig. 6.** ERS-1 SAR image from orbit 5249, frame 4941, taken on July 17, 1992 (Copyright ESA 1992) and located in Fig. 2. *Boxes* highlight regions from which component backscatter distributions are plotted in Sect. 8.4.1.2. The *map* marks the location of the image with respect to the track location of the research icebreaker *Polarstern*

**Table 1.** List of ERS-1 SAR image frames used in regional backscatter data analysis [a]

| Image (orbit/frame) | Date and time (UTC) | Center latitude | Center longitude |
|---|---|---|---|
| 5090 5121 | 6 July 92: 09:01:35.786 | -72.263 | 343.424 |
| 5249 4941 | 17 July 92: 11:34:19.254 | -64.706 | 318.550 |
| 5377 5031 | 26 July 92: 10:12:16.392 | -68.614 | 333.541 |
| 5387 5949 | 27 July 92: 03:11:05.712 | -60.406 | 310.928 |
| 5387 5967 | 27 July 92: 03:11:20.807 | -59.581 | 310.165 |
| 5449 5139 | 31 July 92: 10:56:54.090 | -72.934 | 312.629 |

a See Fig. 2 for geographic locations in the Weddell Sea

year ice to old, deformed snow-covered floes. Individual SAR image samples identified in Fig. 6 and tabulated in Table 2 are shown later and used to illustrate the distribution of backscatter in small-scale areas.

The least deformed, or level first-year ice is observed in images 5377–5031 from the central Weddell Gyre region, and 5449–5139 from the southern Weddell Sea (Belgrano Bank) region (see Fig. 2). Both yield distinctive pdf modes between −12 and −13 dB. The former is a largely divergent region, in which vast expanses of undeformed white ice

**Table 2.** Summary of backscatter samples from ERS-1 SAR image 5249/4941 (Fig. 6)

| Ice type | Sample no | Mean/median (dB) | Std. deviation (dB) |
|---|---|---|---|
| Smooth first-year | FYS 1 | -16.09/-16.10 | ±0.99 |
| | FYS 2 | -15.46/-15.50 | ±1.05 |
| Rough first-year | FYR 1 | -10.14/-10.30 | ±1.83 |
| | FYR 2 | -9.54/-9.5 | ±1.69 |
| Multiyear ice | MY 1 | -5.72/-5.50 | ±1.88 |
| | MY 2 | -6.81/-6.50 | ±2.27 |
| | MY 3 | -5.92/-5.50 | ±2.03 |

alternated with snow-covered stony fields along the transect of *Polarstern* (Haas et al. 1992). July air temperatures remained stable around −27 °C and ice formation was almost immediate under divergent conditions. The SAR image indicates no open water and an extremely low ridge density. The example from the southern Weddell (image 5449–5139) in contrast contains a larger proportion of deformed seasonal ice (with higher ridging density), and a small fraction of grounded icebergs and perennial ice exiting the basin northwards (Viehoff and Li 1995). These higher backscatter components of the scene account for the higher pdf mode for image 5449–5139 in Fig. 4, and the tail extending to values exceeding −5 dB. Based on the predominant drift direction, the origin of the old ice and icebergs found in this region appears to be the Filchner ice shelf front, or the location of grounded icebergs A22 and A23. Large 5- to 10-km-diameter, high backscatter (−7 to −2 dB) multiyear ice floes are clearly conglomerates of distinctive old ice floes cemented together by a matrix of what was likely deformed, snow-covered seasonal ice. Floes have broken away and drifted north-westward in the distinctive streamer of bright material (Drinkwater 1997). Similarly, the iceberg chain described by Viehoff and Li (1995), which originated as one large iceberg from the Filchner ice shelf, subsequently grounded on the broad General Belgrano Bank. Ice-motion data obtained from SAR, scatterometer, and buoys provide complementary evidence for this assertion, but further supporting evidence is provided in the form of tracks of deformed sea-ice created as the sea ice drifts past the location of each of these icebergs (see Viehoff and Li 1995, Fig. 9). The ice cover had drifted in an extremely constant fashion northwards past the icebergs, leaving linear tracers of deformed material extending 100 km in the direction of drift. At a typical drift speed of 16 cm $s^{-1}$ north from the iceberg barrier, these features represent a period of around 1 week. Furthermore, the persistence of such bands (in lee of the icebergs) in time as well as in space (≥ 100 km) indicate a consistent ice drift direction and relatively homogeneous ice motion field in this location.

Examples from images 5377–5031 and 5449–5139 indicate that the least deformed first-year ice generally comprises the leftmost peak in the global pdf (indicated as the black stepped histogram in Fig. 4). Old ice and marginal ice, on the other hand, fill an overlapping range in backscatter to form the central peak in the global pdf. As expected, heavily ridged or deformed first-year ice falls squarely in the middle of the range, and comprises the main peak in image 5090–5121. The roughness of sea ice in the location of this SAR image is further described by laser altimeter flights made from *Polarstern*, with mean ridge heights of the order of 1.2 m and a mean ridge spacing of 54 m (Dierking

1995). Therefore, for the central ice pack, linear mixing of the two end members (1) smooth undeformed ice and (2) thick perennial ice accounts for a continuum of $\sigma^0_{SAR}$ values between –19 and –4 dB. Except for the confusion existing between multiyear ice and MIZ signatures, the value of SAR backscatter appears closely related to the amount of deformation and ridging density of the ice, or broadly speaking, ice thickness. Although backscatter values may be encountered in the range –20 to –25 dB, for newly growing areas of nilas and gray ice (in leads or ice shelf polynya systems), large expanses of such low-backscatter material are rarely encountered in the Weddell Sea.

### 8.4.1.3
### *Snapshot Signature Summary*

The SAR image example shown in Fig. 6, provides the largest mixture of ice ages and snow depths, due to its prime location in the perennial ice outflow region (Drinkwater 1997). The corresponding pdf for frame 5249–4941 in Fig. 4 confirms a broad peak comprising a variety of pixels spanning the range of possible white, first-year, and multiyear ice $\sigma^0_{SAR}$ values. A number of samples are extracted from this image to generate Table 2. Values describe the signatures of the primary ice components, each of which is extracted on the basis of the scattering characteristics and of their homogeneity. Smooth and rough first-year and multiyear ice mean $\sigma^0_{SAR}$ values range from –16.1 dB to –5.5 dB. In total, a further 60 samples (each of 100 or more pixels) were collected in the same manner from the bold SAR images shown in Fig. 2, to generate a summary of snapshot winter backscatter characteristics of these primary ice types. All except the southernmost of the images were acquired in regions in which WWGS'92 surface samples and ice cores were collected (Drinkwater and Haas 1994). Figure 7 compares and

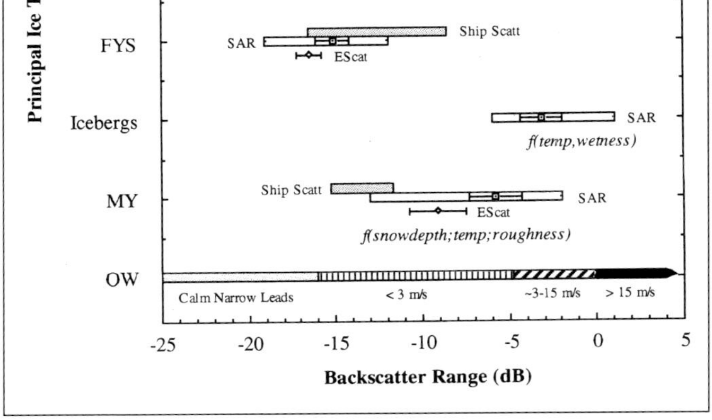


**Fig. 7.** Summary plot of C-band Weddell Sea ice and ocean signatures, indicating the ranges and/or standard deviation of SAR, and shipborne ($\sigma^0_{Ship}$) and EScat ($\sigma^0_{EScat}$) backscatter coefficient measurements of each broad ice category. *MIZ* Marginal ice zone, *FYR* rough first-year ice, *FYS* smooth first-year ice, *MY* multiyear ice, *OW* open water

contrasts the ranges and statistics of these data with matching summary statistics inferred from the mean field scatterometer $\sigma^0_{Ship}$ curves (i.e., shaded pdf's) in Fig. 5. Additional comparison data are included for open water backscatter coefficients, and MIZ scattering from brash ice or pancakes.

In Fig. 7, the MIZ provides one of the largest ranges of backscatter coefficient, due to the influence of floe size, wave environment, and variable air temperature. Depending on the season and location, frazil ice growth and pancake formation may be favored. However, brash ice also provides extremely high–values due to its similar characteristics in terms of subresolution floes and floe packing. A good analogy may be drawn between the Weddell and the Labrador Sea MIZ, which was observed by SAR during the LIMEX experiments (Drinkwater 1989; Livingstone and Drinkwater 1991). In Fig. 7 SAR and EScat values overlap due to the broad range in scattering conditions described in Sect. 8.4.1.1.

Field sampling of smooth first-year ice floes is inherently biased, since an icebreaking ship normally seeks the path of least resistance. Although this is generally true, the ship's route during WWGS'92 was based largely upon large-scale information provided by AVHRR satellite images and did not always enable sampling of the smoothest first-year ice. The benefit of SAR images is that they enable the very smoothest ice forms to be easily found and selected. Samples shown in Fig. 7 show that the measurement ranges plotted from the shipborne scatterometer and SAR do not overlap particularly well. In other words, shipborne radar measurements of smooth first-year ice indicate a higher range of backscatter coefficients than samples selected from the SAR image data. Similarly, the ship navigated around multiyear ice floes or heavily ridged ice regions visible in the AVHRR images. As a consequence, the multiyear ice which was sampled is at the least deformed extreme of such old ice floes. Haas et al. (1992) document the level nature of most of the multiyear ice floes sampled. In particular, $\sigma^0_{Ship}$ values in Fig. 7 indicate that the field scatterometer measurements are biased toward the lowermost end of the multiyear and rough first-year backscatter ranges.

Other than for smooth first-year ice areas, shipborne scatterometer measurements are generally lower at 40° incidence than equivalent EScat image data extracted in the same locations. Spatial resolution and the averaging over a number of resolution cells influences mean EScat backscatter measurements. Intervening ice motion (typically 1–2 pixels over the 6-day Escat averaging window) and resulting mixtures of backscatter elements combine to give a relatively higher $\sigma^0$ than that of any individual ice component. In addition, the larger roughness elements dominate the scattering in the higher incidence angle range, yet shipborne scatterometer measurements cannot easily measure the signature of such nonuniform or deformed ice floe surfaces.

Open water signatures are generated in Fig. 7 for comparison, using the *CMOD4* algorithm (Stoffelan and Anderson 1993). As expected, the range of backscatter values under differing wind conditions brackets the entire range of sea-ice signatures. Nevertheless, by skillfully using image context (i.e., knowledge regarding the shape of leads within the ice pack) together with ice dynamics information from ice tracking, it is possible to distinguish open water areas within the pack. In any case, observing an open lead within the Weddell Sea in winter is rare, and for the most part new leads freeze within a matter of hours after opening.

## 8.4.2 The Future of Ice Classification

The advent of a global dataset from ERS and NSCAT scatterometers enables uninterrupted C- and Ku-band radar coverage of the Arctic and Antarctic sea-ice cover to be generated. These synoptic-scale images can be exploited to address global ice classification in conjunction with ice dynamics issues. Recent work by Beaven and Gogineni (this volume) fusing active and passive datasets indicates that radar in combination with SSM/I passive microwave image data increases the amount of independent information and adds dimensionality to the problem of unambiguous identification of classes of ice.

### 8.4.2.1 *Scatterometer Ice Classification*

Early attempts at one-dimensional classification of Antarctic ice have used simple supervised clustering analyses on the basis of EScat backscatter statistics. Backscatter coefficients at 40° incidence appear to characterize the large-scale deformation and surface roughness of the sea ice (Drinkwater et al. 1993a). Simple slicing of an EScat image using the values given in Table 3 yields results shown in Fig. 8. Figure 8a is generated for a 6 day period beginning on June 16, 1992 (day 168), and Fig. 8b for a similar period starting at July 31, 1996 (day 213). Each part of Fig. 8 indicates the track of R.V. *Polarstern* during WWGS'92 together with the sea-ice sample sites marked by symbols. Physical descriptions of individual samples of ice are provided by Drinkwater and Haas (1994), together with shipborne scatterometer scattering coefficients (i.e., $\sigma^0_{Ship}$).

Figure 8a coincides temporally with the first four sea-ice samples acquired along the Greenwich meridian (marked by "+" symbols). The classification routine identifies a large patch of pancake ice formation corresponding directly to the location shown in Figure 3. Table 3 indicates that pixel values fall in the range $-11 \leq \sigma^0_{EScat} < -6$ dB. Later in the WWGS'92 experiment, as *Polarstern* traversed the Weddell Sea, large expanses of undeformed level white ice were observed ($-20.0 \leq \sigma^0_{EScat} < -14.0$ dB), and the ship track was intentionally modified to avoid a large region of deformed ice, which was identified in real time using on-board received AVHRR data. Level first-year ice regions (FYS) are confirmed at sites marked by squares. The ship track bends around a large patch of deformed, rough first-year ice which originated along the eastern coast of the Weddell Sea and drifted progressively northwestwards from day 168 onwards. The ice outflow region in the northwest Weddell Sea is characterized by rough first-year and multiyear ice forms. In the classified image, the margin between level and thicker, deformed first-year ice closely fits field observations of a boundary between characteristics in first-year ice, where sample sites of thicker, deformed first-year ice are marked by diamonds. At the time of imaging, the ship was preparing to exit the sea ice, in the northwestern Weddell Sea MIZ, and circles symbols indicate mixed multiyear and brash ice samples in the location of orbit 5387 shown in Fig. 2 (Drinkwater and Haas 1994). This region of high backscatter corresponds with brash and wave-broken pieces of deformed first-year sea ice and a small concentration of deep snow-covered old ice, and the algorithm classifies the region as pancakes and/or multiyear ice.

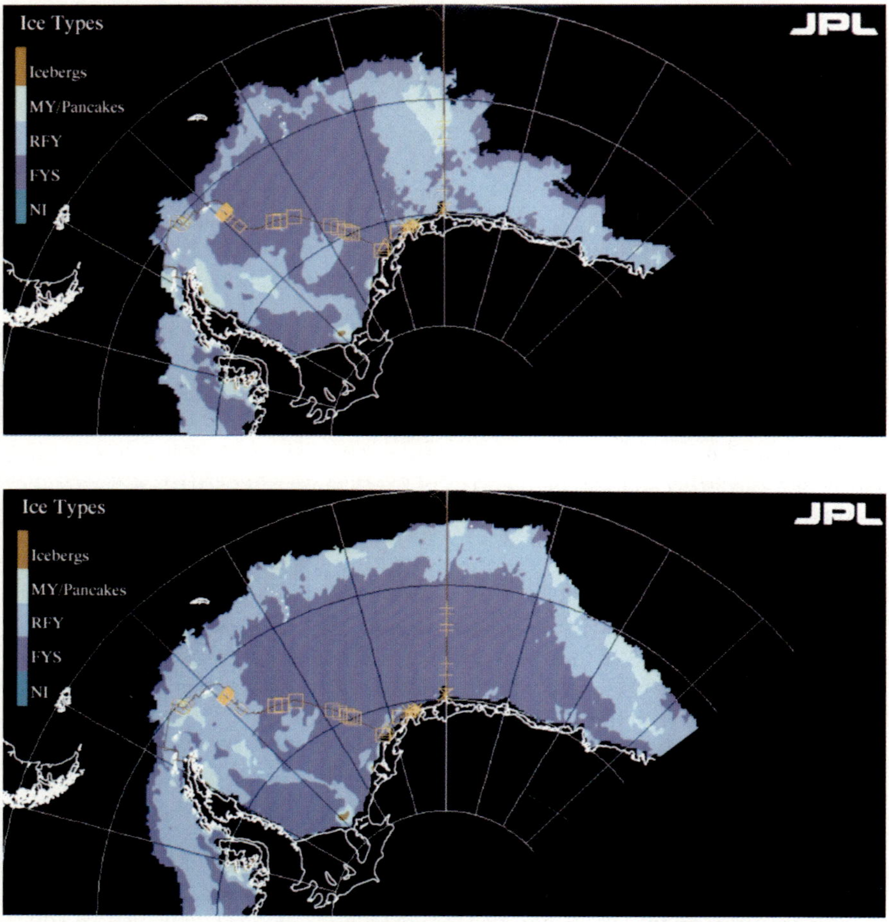


**Fig. 8.** Classified weekly mean SIRF *A* images in the ERS-1 scatterometer mode, for **a** June 16 (i.e., days 168–173), and **b** July 31 (i.e., days 213–218), 1992. (ERS-1 Scatterometer data copyright ESA)

**Table 3.** Summary of winter C-band EScat backscatter ranges ($\theta = 40°$ incidence) of main categories of sea ice in the Weddell Sea

| Ice type | Backscatter range |
|---|---|
| Icebergs | $-6.0 \leq \sigma^0 \leq 0.0$ |
| Multiyear/pancakes | $-11.0 \leq \sigma^0 < -6.0$ |
| Rough first-year | $-14.0 \leq \sigma^0 < -11.0$ |
| Smooth first-year | $-20.0 \leq \sigma^0 < -14.0$ |
| Nilas | $-32.0 \leq \sigma^0 < -20.0$ |

#### 8.4.2.2
#### *Multichannel and Multisensor Data Fusion*

Further preliminary work to classify sea ice in ERS scatterometer data is being undertaken by Early (unpublished) using isodata clustering algorithms on three-dimensional ERS-1 scatterometer image data sets. Three dimensions are constructed from six-day *A* and *B* images and an isotropy image, where the primary *A* image is a standard EScat enhanced product comprising a weekly mean 40° incidence angle normalized backscatter coefficient image, the *B* image is the gradient in scattering coefficient in the 20–60° incidence angle range, and the isotropy image characterizes the azimuthal variance in scattering coefficient and combines information acquired using different beams of the scatterometer instrument. As illustrated by the section above, the main problem in one-dimensional ice classification using EScat data is the separation of bright returns from pancake ice, multiyear ice and rough first-year and brash ice. While the former and latter ice types are typically found seasonally in a specific area, it is difficult to unambiguously distinguish perennial ice near the ice margin. The primary advantage of this three-dimensional method is that pancakes and brash in the MIZ can be distinguished from multiyear ice in high concentrations. Similarly, rough ice with isotropic scattering characteristics can be distinguished from bright patches of pancake growth, where wave effects create a large degree of anisotropy.

Multidimensional algorithms can further address the problems of overlap between the backscatter coefficient distributions by using multispectral approaches which merge microwave radar and radiometer data. Beaven and Gogineni (this volume), for instance, merge SAR and passive microwave data in an attempt to improve sea ice classification. Future studies will attempt to fuse ERS, NSCAT, and SSM/I data to improve ice classifications, especially in situations where atmospheric water vapor limits the accuracy of passive microwave retrievals.

## 8.5
## Antarctic Sea Ice Dynamics

One of the primary scientific goals of employing high-resolution satellite radar is to generate SAR ice kinematics data products with which to develop a spatial and temporal picture of Antarctic ice drift and opening and closing in response to the momentum transfer (Drinkwater 1995b). The benefit of radar is that it can image the ice night and day, such that a continuous database of images may be acquired for ice motion tracking processing. Until now, ice-motion studies utilized satellite-tracked buoy data as a means of characterizing the drift and dynamics primarily in the Weddell Sea. As such, Argos/GPS buoys are also the only source of detailed independent information on ice kinematics. Their drawback, on the other hand, is that they do not provide information on the spatial characteristics of drift unless they are carefully deployed in a network.

### 8.5.1
### Length Scales of Antarctic Sea-Ice Drift

Vihma et al. (1996) and Kottmeier and Sellmann (1996) have investigated the spatial correlation function of drift in the Weddell Sea using various spatial arrangements of buoy data in 1992. Corresponding length scales are computed from European Centre for Medium Range Weather Forecasting (ECMWF) analysis fields of atmospheric pressure and geostrophic wind. The length scales of sea ice drift and atmosphere appear to vary seasonally and the ice-drift field is especially dependent upon location and primary drift direction with respect to the coast.

#### 8.5.1.1
#### *Coastal Regime*

In the northwards drifting perennial ice regime of the western Weddell Sea, ice motion is forced by both wind and currents. Periodic deviations from this relatively constant drift pattern are caused by passing cyclones or "polar lows." Because of the large fractions of thick ice, ice conditions most closely resemble those of the Arctic basin, and thus the ice drift correlation length scales may be expected to vary seasonally and interannually with ice concentration. In 1992, Argos buoys were spaced at intervals in the east-west direction with respect to the Antarctic peninsula, thereby allowing characterization of the east-west correlation functional form. Vihma et al. (1996) find longitudinal ($u_{\mathrm{ice}}$) and lateral ($v_{\mathrm{ice}}$) integral correlation length scales of 700 and 550 km, respectively. Comparative results of Thorndike (1985) for the Arctic Beaufort Sea are 800 and 600 km for $u_{\mathrm{ice}}$ and $v_{\mathrm{ice}}$ respectively. In contrast the east-west integral correlation length scale of the scalar atmospheric pressure was 890 km during this period, while the geostrophic wind integral correlation length scales were $u_{\mathrm{G}} = 1050$ and $v_{\mathrm{G}} = 330$ km, respectively. These results indicate that the region of predominantly northward ice motion off the continental shelf exhibits a large decorrelation in velocity in the east-west direction, and that there is confirmed existence of a large gradient in northward velocity components with distance from the Antarctic coast. Indeed, the synoptic pressure pattern also reflects a large gradient in velocity in this same direction.

#### 8.5.1.2
#### *Central Basin*

Kottmeier et al. (1992) investigated seasonal and spatial variations in the correlation length and integral length scales of ice motion to find lateral integral correlation length scales ranging between 270 and 540 km and longitudinal integral correlation length scales of between 490 and 680 km. Generally, the correlation length scales become smaller with greater mobility of the ice, or vicinity to the ice margin, and the smallest values correspond to summer ice drift near the ice edge in ice concentrations less than 70%. Thus, less compact summer ice tends to have smaller length scales than, for instance, winter ice or ice drifting close to the coastline, as above. In the winter, length scales found by Kottmeier et al. (1992) in the central Weddell are somewhat smaller than those observed in the western Weddell. Though the basin-scale circulation in the Weddell and Beaufort Seas is of a similar scale (approximately 1500 km), the difference between these central Weddell length scales and those obtained in the Arctic Basin by Thorndike (1985) suggest that the ice drift differences are caused largely by smaller scales of forcing. Ice motion data derived from satellite radar images must be used to test this hypothesis.

### 8.5.2
### Motion Spectra and Temporal Covariance with Wind

Ice drift, with respect to previously measured buoy motion in the Antarctic ice cover, is known to be highly dependent upon the drifter location with respect to the coast, the primary characteristics of the ocean circulation, and often the water depth. A large distinction can be made between (a) buoys in deep water under relatively free-drift conditions, and (b) buoys drifting in shallow water coastal regimes. The former implies largely divergent situations which characterize large portion of the Antarctic ice cover, and where buoys generally suffer less from internal ice resistance to drift. The latter, in contrast, occurs on the continental shelf in situations in which periodic high-frequency tidal currents play a significant role in driving shear motions within the ice cover, and where internal ice stresses can be transferred from the coast to the location of drift during ice convergence events.

#### 8.5.2.1
#### *Buoys Under Relatively Free Drift*

Velocity spectra of several freely drifting deep-water buoys in the Weddell Sea are analyzed in conjunction with geostrophic wind data to investigate the correlation and spectral relationships between forcing and response. Velocity components of several mid-winter WWGS '92 buoys (with 6 h resolution data) are computed from 1-month July records in 1992, and concurrent geostrophic wind velocity components estimated from an optimally interpolated pressure field (ECMWF analysis data fitted to weighted buoy pressure measurements).

Figure 9 shows the results of spectral analysis of the magnitude of the complex ice and wind velocity vectors. Fig. 9a and Fig. 9b indicate the temporal autocorrelation functions of this scalar ice and wind speed, respectively, and show a temporal e-folding length scale of less than one day. The cross-covariance function in Fig. 9c demonstrates clearly that the wind and ice speeds are highly correlated on timescales of ±1 day and that there is a slight lag of around 12 h or so between the wind forcing and ice drift response. This confirms that the drift adjusts in a matter of hours to wind forcing, explaining the short response time exhibited by the large-scale ice motion in the example shown later in Fig. 11. Similarly, Bartlett-window-filtered power spectra (normalized by the velocity variance) in Fig. 9d and Fig. 9e show clearly that ice drift speed contains the most energy at periods above 100 hours, and the cross amplitude spectrum in Fig. 9f shows that the spectra are almost identical in characteristics for this 1 month period. Accompanying coherency and phase spectra of the complex velocities illustrate how the correlation between the wind and ice drift increases rapidly to a peak exceeding 0.8 at periods of 45–70 hours (i.e., 2–3 days). This corresponds to the frequency of passing low pressure systems during this winter period, while in contrast the coherency rises to a plateau exceeding 0.8 at periods exceeding 5 days and is comparable to the results of Kottmeier and Sellmann (1996).

The phase spectrum reveals that ice consistently drifts to the right of the geostrophic wind at all periods of forcing with respect to the upper level winds, yet with some scatter. July radiosondes launched from R.V. *Polarstern* in the vicinity of the buoy array indicate a relatively constant turning angle of 44° between surface (10 m) and upper

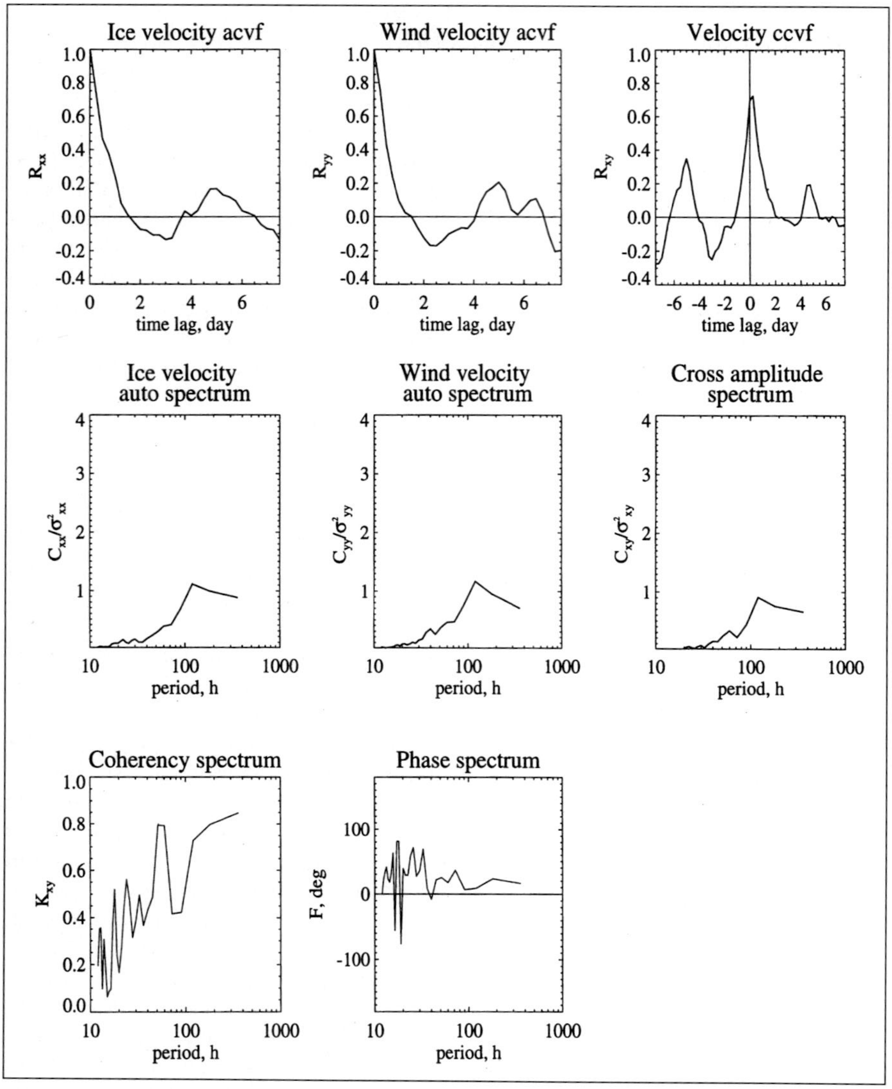


**Fig. 9. a-h.** Auto-covariance functions and normalized variance spectra of Argos-derived ice drift and wind speeds for free drifting buoy in the central Weddell Sea during WWGS'92. For further details, see text

level (900 mB) winds (i.e., ageostrophic angle of the atmospheric boundary layer). Thus ice drifts with an angle of approximately 25° to the left of the surface wind in this July period. This turning angle is fairly typical of previous results observed by Martinson and Wamser (1990) for drifters in the eastern Weddell Sea in 1986.

### 8.5.2.2
### *Coastal Shallow Water Buoy Drift*

In Figure 10, a comparative correlational analysis is made for the 5 month long drift of Ice Station Weddell (ISW) in the western Weddell Sea in 1992 (ISW Group 1993), together with hourly winds. ISW was located on an old ice floe just off the continental shelf break at 54° W in the region of perennial ice (Drinkwater 1997). Its motion, in contrast,

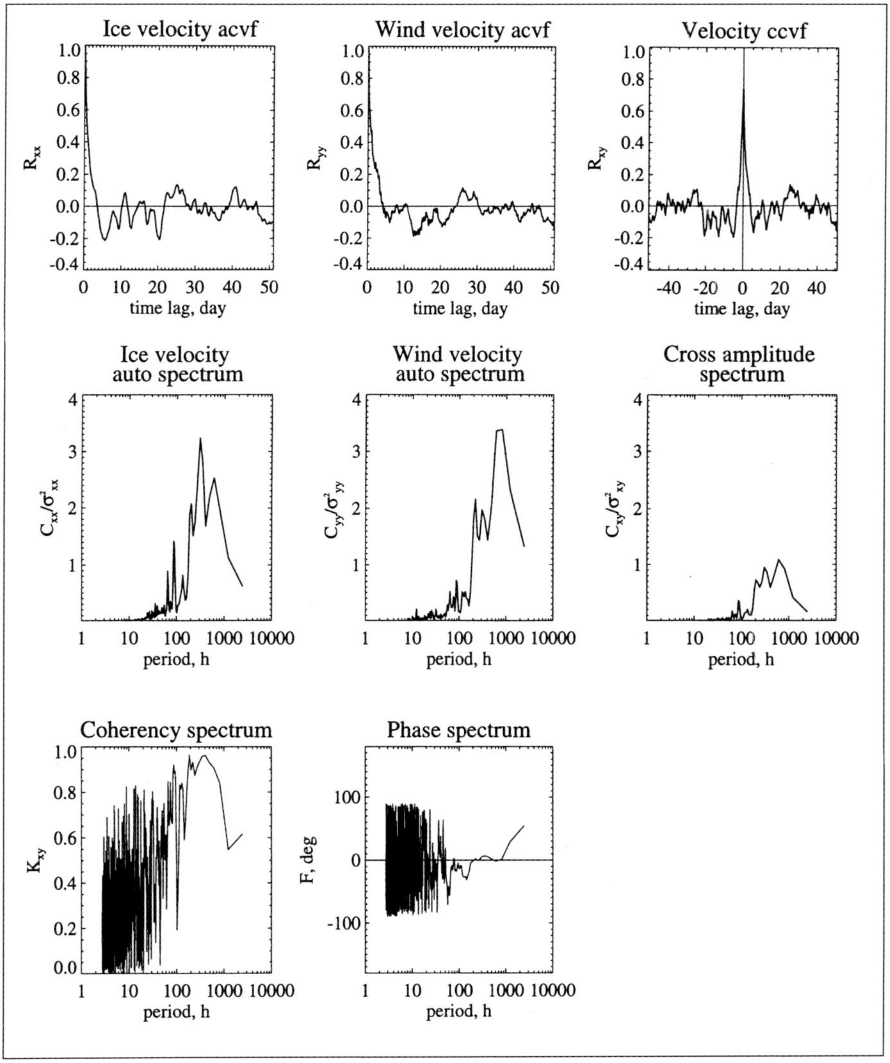


**Fig. 10. a-h.** Auto-covariance functions and normalized variance spectra of GPS-derived ice drift and wind speeds for Ice Station Weddell. For further details, see text

is highly influenced by tidal currents and high-frequency motions. Hourly GPS positions afford the possibility to investigate correlations on shorter timescales than Argos buoys, since the positional errors which accompany hourly interpolated Argos positions are unacceptably large. This limits spectral analysis in Argos buoy data to typically 3-h or greater sampling intervals, since in practice typically only around 18–20 Argos fixes are obtained daily for each buoy in the Southern Oceans.

This combined hourly GPS and wind record indicates that ice on the continental shelf also responds largely to wind fluctuations on timescales of 120 h (i.e., 5 days) or more. Both the ice and wind auto-spectra indicate a predominant peak at around 500–1000 h (i.e. 20–40 days), which represents a dominant response of drift speed to wind fluctuations on timescales associated with synoptically driven pressure systems. Therefore, the monthly mean pressure situation over the Weddell Sea is responsible for the primary forcing of northwards drift in the western limb of the Weddell Gyre. The spectrum for ISW drift speed in Fig. 10g shows a relatively high degree of coherency for fluctuations of a period exceeding 100 h, but the phase spectrum indicates a mean of zero with a high degree of variance in the angle between the wind and drift directions for periods up to around 12 h. This indicates that the ice does not drift in a consistent fashion on short timescales. Short term fluctuations in coherency and phase (below 12 h) appear to be caused largely by aliasing of motions induced by tidal currents and breakdown of the relationship with wind. Though the geostrophic wind is the predominant forcing mechanism on long timescales, short-term fluctuations in velocity caused by internal ice forces and tides must therefore be considered with respect to satellite observations of ice drift in these regions.

In order to resolve the specific tidal influence, one must focus on nonnormalized ice velocity components. Power spectral density results indicate strong diurnal (M2) and semi-diurnal (K1 and O1) tidal current forced ice velocity fluctuations, with strong peaks at periods close to 12 and 24 h. Current meters suspended beneath ISW also demonstrate similar fluctuations (Muench and Gordon 1995). This result compares favorably with the findings of Viehoff and Li and recent studies of Vihma et al. (1996) and Kottmeier and Sellmann (1996). It may be expected, therefore, that tidal fluctuations on timescales of 1 day or less have an important impact on the variance in velocity over the period of satellite repeat orbits. Exact repeat orbits will suffer no serious impact from tidally induced ice-drift velocity variance, whereas image pairs acquired at fractions or non-multiples of 24 h, suffer from aliasing of tidally induced velocity components. The impact of tides is discussed in more detail in the context of SAR results later.

### 8.5.3
### Large-Scale Sea-Ice Tracking Using Visible Wavelength Data

Several remote sensing methods have been used to track sea-ice drift in Antarctica, the most common of which utilizes visible or near-infrared wavelength data. During the WWGS '92 experiment a shipborne high resolution picture transmission (HRPT) receiving station was used to collect data from the Advanced very High Resolution Radiometer (AVHRR). During this period data from the NOAA-9, NOAA-10, NOAA-11 and NOAA-12 satellites were received (Lemke 1994). Because of missing daylight during the austral winter, overpass information is limited to the infrared channels (3, 4, and 5). Similarly, cloud-free data are relatively sparse. Despite these restrictions, some

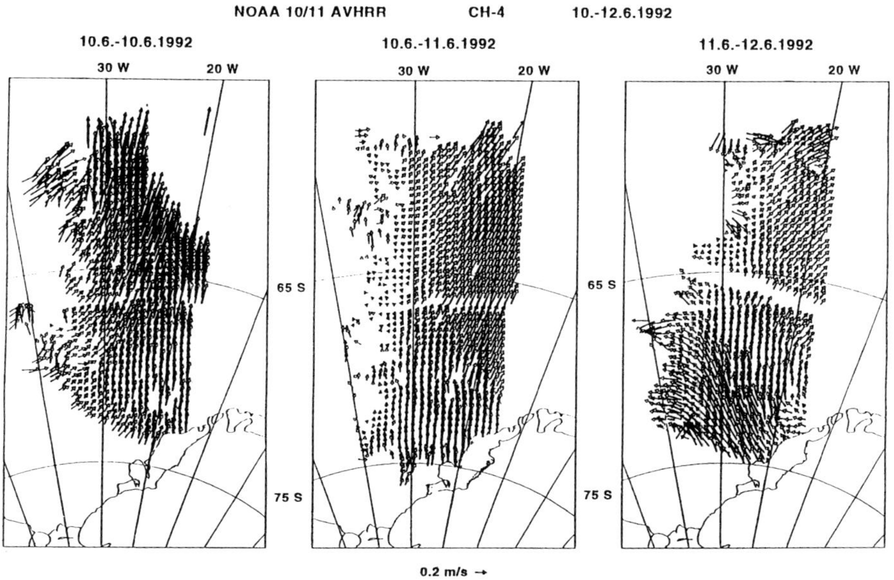


**Fig. 11.** a-c. Examples of AVHRR channel-4 derived sea-ice velocity fields for the 1992 periods **a** 07:27 – 21:31 h GMT on June 10, **b** 21:31 h on June 10 to 21:07 h GMT on June 11, **c** 21:07 h on June 11 to 07:51 h GMT on June 21

170 data sets were recorded throughout this experiment and used for ice motion tracking. Figure 11 shows an example of 3 consecutive ice-motion pairs, from images acquired in June 1992 (10th June: 07:27 GMT and 21:31 GMT; 11 June: 21:07 GMT; and 12 June: 07:51 GMT). These large-scale data (courtesy of the late Thomas Viehoff) indicate that the temporal adjustment to wind forcing is rapid, and especially that the sea-ice drift adjusts to the mesoscale patterns in the wind forcing within a short period of time (and especially in the absence of onshore flow). In Figure 11 this is illustrated by a burst of off ice-shelf winds, which began late on 11 June, which cause a rapid adjustment from predominantly northerly drift conditions to north-westerly drift.

Unfortunately the extent of large-scale examples such as Figure 11 is limited, as is the western extent of the Figure in panels a and b. Together with the restriction of sparse temporal coverage of a given region, this makes AVHRR ice tracking somewhat limited as a source of ice-drift information. This drawback, therefore, is the primary motivation for using microwave data for ice-motion tracking in Antarctica.

## 8.5.4
## Small-Scale Sea-Ice Tracking from SAR

Recent preliminary studies by Thomas et al. (1995) and Viehoff and Li (1995) have used satellite radar data to manually chart regional patterns of ice drift in the Weddell Sea in response to winds and currents. Recent work by Drinkwater and Kottmeier (1994) introduced ice-motion tracking from mesoscale SAR data as an alternative method for deriving kinematic measurements of the Weddell ice cover. In contrast to the Arctic ice motion studies using ERS SAR (Kwok et al. 1990), the capability to track Antarctic ice using automated SAR motion tracking had not been demonstrated. The reason was because in the Arctic, large fractions of multiyear ice provide high contrast targets which can be successfully tracked in time using radar images. Weddell Sea ice in comparison exhibits large expanses of low contrast, level first-year ice, and a lack of distinct features with which computer algorithms may successfully track the ice. Here it is demonstrated that SAR images are effective at resolving features which may be used throughout the year in tracking both perennial and seasonal Antarctic ice floes.

#### 8.5.4.1
#### *Perennial Ice Drift*

Perennial ice in the Antarctic shares some similarities with Arctic multiyear ice, in that it is relatively thicker than its seasonal counterpart and generally supports a high snow load (Massom et al. 1997). Old ice may be expected to produce a relatively bright backscatter signature at C-band during all seasons of the year, as demonstrated by examples in Sect. 8.4.1. During 1992, a series of SAR image acquisitions were planned in conjunction with the drift of ISW. ISW was deployed on a perennial ice floe in the south-western Weddell Sea (Drinkwater and Lytle 1996) and positioned over the continental shelf-break at 71.5°S, 50°W (ISW Group 1993).

Figure 12 shows a result from tracking snow-covered perennial ice floes in the vicinity of ISW during the February – March drift phase of this 1992 experiment. Overlapping image pairs acquired along 1- and 3-day-spaced crossing orbits (during the so-

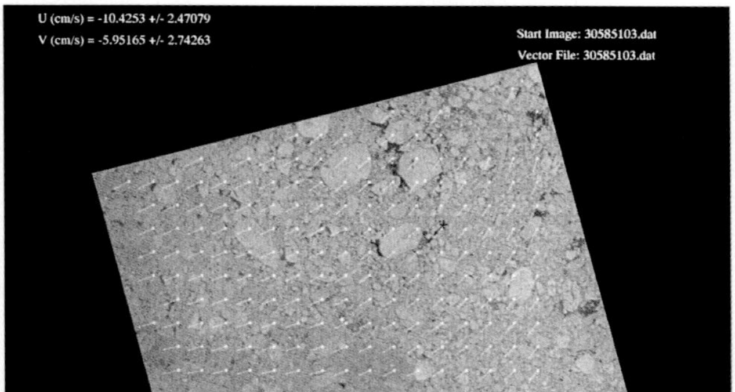


**Fig. 12.** Three-day ice motion vectors superimposed onto an ERS-1 SAR image for March 15, 1992 (image copyright ESA 1992). The scale indicates the distance (km) of the vector displacements and the mean u and v components of ice velocity are given at *upper left*. For further details, see text

called "ice phase" orbit of ERS-1) were correlated to produce ice motion vectors using the algorithm described by Kwok et al. (1990). Figure 12 shows the 5-km-gridded ice-drift displacement field superimposed on the ERS-1 SAR reference image (orbit 3058 frame 5103) acquired at 11:34 GMT on February 15, 1992 at 71.59°S, 53.04°W. Spots indicate the gridded starting positions of tracked features, and the scaled vector arrows indicate the distance traveled in the 17 h period separating the pair of tracked SAR images. The location of ISW at the time of the image is indicated by the black "+", and the scaled velocity vector, derived from the corresponding instantaneous GPS locations of ISW, indicated by an arrow. Ice motion vectors around ISW imply strong cyclonic motion, and mean and standard deviations of the $u$ and $v$ velocity components are -10.43±2.5 and 5.95±2.7 cm $s^{-1}$, respectively. Partial derivatives of velocity over the entire tracked portion of the scene indicate a mean divergence of 0.15± 0.63 % $d^{-1}$, a mean vorticity of -0.45± 0.70 % $d^{-1}$, and mean shear of 2.01±1.02 % $d^{-1}$.

In total, 15 pairs of SAR images were processed to ice drift over ISW between February 7 and March 15, 1992. Over this period, the mean sea-ice drift speed around ISW was 7.45±1.37 cm $s^{-1}$ on a bearing 323°, while the corresponding mean wind speed was 5.0 m $s^{-1}$. In summary, there was a net mean divergence of only 0.6% over the period of satellite observations, indicating that net opening of the ice cover was minimal. Indeed both opening and closing events were observed, each effectively redistributing the ice area. Corresponding rotation and shear values were $-0.023\pm0.3\times10^{-6}$ $s^{-1}$ and $0.65\pm0.5\times10^{-6}$ $s^{-1}$ respectively (i.e., 0.2±2.5% $d^{-1}$ and 5.6±4.3% $d^{-1}$), indicating shear to be the primary deformation mechanism.

#### 8.5.4.2
#### *Seasonal Ice Drift*

During 1992, several buoys were deployed in the eastern Weddell Sea as part of the WWGS'92 experiment (Lemke 1994), in a region comprised of level, undeformed white ice. Simultaneous overlapping SAR acquisitions were planned in July 1992 such that the buoys could be used as validation for the SAR motion retrievals. Under divergent conditions in the central Weddell Sea, and consequently relatively free drift, the majority of ice-drift vectors observed by SAR in the vicinity of the buoy array follow a pattern expected from Ekman dynamics. Figure 13 shows 3-day SAR-observed drift vectors together with the corresponding WWGS'92 buoy drift tracks marked as dotted Lagrangian trajectories. The superimposed pressure field is optimally interpolated using a polynomial scheme which combines weighted buoy pressure measurements together with 1000-mB ECMWF pressure analysis fields. This technique was originally employed by Kottmeier and Sellmann (1996) but is hybridized here by extending and constraining the fitting procedure to broaden the geographic limits to include synoptic-scale features and the SAR-tracked domain. Corresponding gridded geostrophic winds are indicated in Figure 13 together with optimally interpolated climatological mean geostrophic currents derived from original data supplied by Ross (in preparation) and Kottmeier and Sellmann (1996).

Figure 13a shows a 3-day period with a low pressure system to the southeast and relatively constant surface atmospheric pressure conditions over the buoy array (marked by diamonds). ERS-1 SAR ice tracking from two consecutive parallel descending orbits captures the east northeast sea ice drift, which aligns itself approximately along the iso-

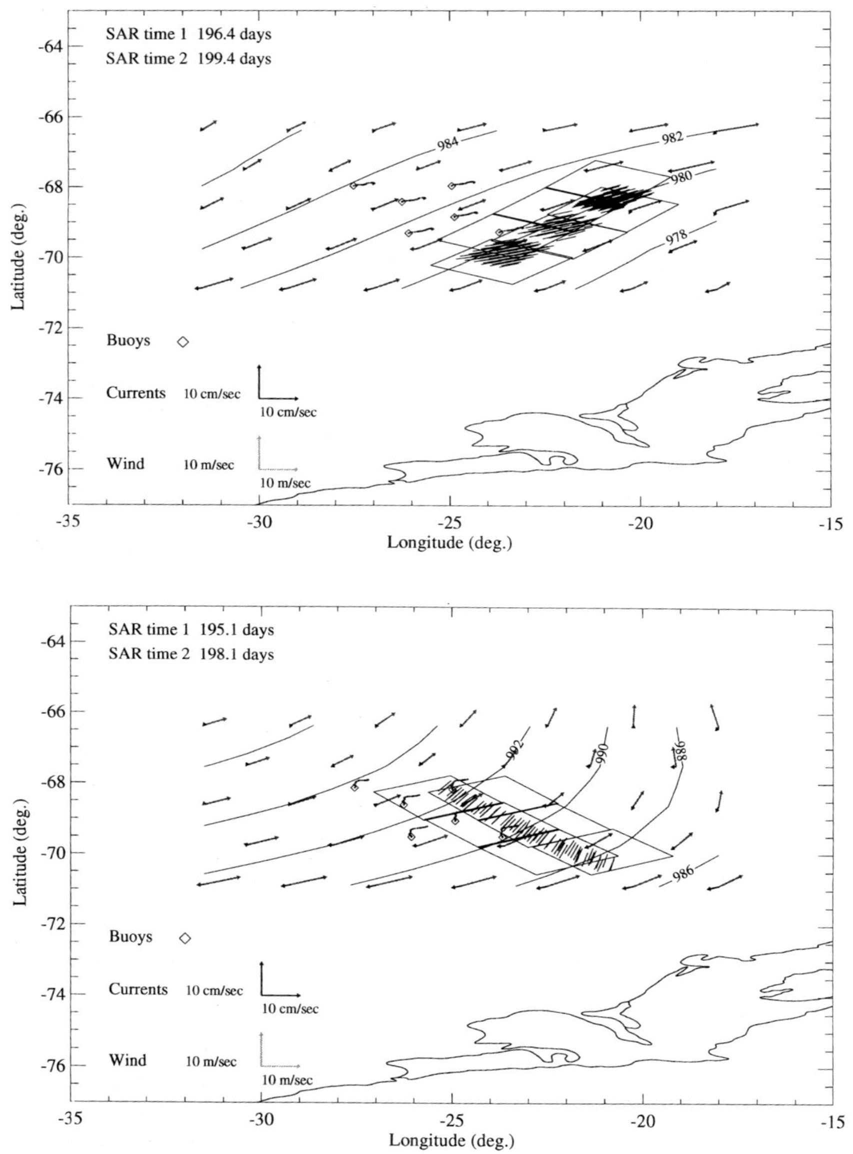


**Fig. 13. a,b.** SAR-derived ice motion superimposed on the WWGS'92 buoy drift array, for periods **a** July 14–17, 1992, and **b** July 13–16, 1992. Gridded geostrophic winds and mean climatological geostrophic currents are shown together with the optimally interpolated ECMWF pressure field. Buoy start locations are shown by *diamonds*. For further details, see text

bars. The indications are that sea-ice motion in this region can for the most part be considered in "free drift" (Martinson and Wamser 1990), especially during divergent conditions when ice interaction forces are negligible. As the low-pressure system in Fig. 13 moves to the east, winds become more southerly, causing increasing divergence as high pressure is reestablished. The weak magnitude of the mean coastal geostrophic current, flowing in the opposite direction in Fig. 13a in the location of the buoys, appears to have little impact other than moderating the drift velocity. More deformed coastal ice in SAR drift examples further to the southeast appear to respond more to basal current stresses due to stronger coastal currents which reach 5 cm $s^{-1}$ in magnitude.

Episodic bouts of divergence and convergence are observed primarily in response to storm impulses at the leading and trailing edges of low pressure systems, respectively. Figure 13b shows the mean synoptic and ice-motion situation prior to the example in Fig. 13a. In Figure 13b, the passing trough of low pressure causes a rapid adjustment in the buoy drift, each buoy exhibiting a 90° direction change in direction. Geostrophic winds are observed to change from southerlies to westerlies as a consequence of the eastward passage of the low-pressure trough, and the along track SAR motion vectors show a convergence in response to winds towards the Antarctic coast.

During the July drift of the WWGS'92 buoys, the sea-ice drifts at radiosonde balloons launched from R.V. *Polarstern* close to the location of buoy deployment region show a mean ageostrophic turning angle of around 44° between the upper-level and surface winds. The mean angle between the geostrophic wind and the SAR-measured ice-drift vectors is observed to vary considerably in the range ±20°, depending on the pressure characteristics in the region, as found by Kottmeier and Sellmann (1996). In general, the sea ice in this region drifts to the left of the surface wind vector with a mean turning angle of around 25°, and thus the ice drift vectors for the most part are closely aligned with the isobars.

### 8.5.4.3
### *Sea-Ice Deformation*

The advantage which SAR has over large-scale ice motion derived from low resolution satellite images is that the deformational response of the ice may be characterized. Figure 14 illustrates a product derived from the SAR ice-velocity measurements shown in Fig. 12. Ice deformation is graphically described in terms of the ratio of strain-rate invariants $E_I$ and $E_{II}$, calculated from the velocity derivatives on the 5 km grid-cell scale. The angle of change, θ, expresses the style of deformation [$\theta = \tan^{-1}(E_{II}/E_I)$], where $E_I$ quantifies positive divergence (or negative convergence) and $E_{II}$ the rate of shearing. The result is a graphical illustration of whether the motion is accompanied by either predominantly divergence (i.e., $\theta \approx 0$), shear ($\theta \approx 90°$), or convergence ($\theta \approx 180°$). It may be noted for the field of cyclonic motion recorded in Fig. 12 that a large proportion of the deformation represents shear-style motion (indicated by dark blue and pink). Only one single purple 5 km grid cell located to the northwest of the ice camp (marked by "+") illustrates a situation tending to pure convergence. It clearly begins as an area with a large fraction of open water. Such sea-ice deformation data are used in validating ice rheologies presently employed in coupled ice-ocean-atmosphere models (Stern et al. 1995).

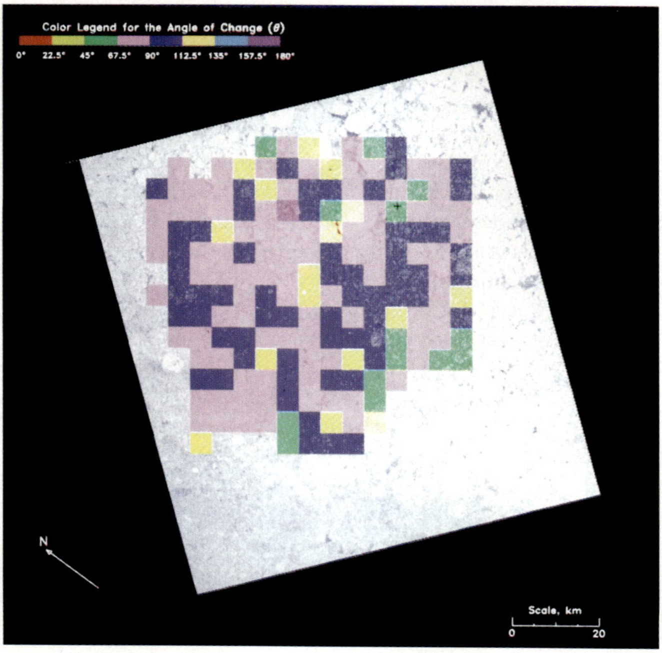


**Fig. 14.** Color-coded display of deformation angle θ [i.e., $\tan^{-1}(E_{II}/E_I)$] superimposed upon the SAR image shown in Fig. 12. For further details, see text

## 8.6
## Algorithm Issues

Various difficulties in implementing algorithms using Antarctic radar data may be anticipated on the basis of experience with problems encountered in Arctic applications. The performance of existing algorithms must be tested in the Antarctic in order to evaluate their skill and accuracy. To date, the main area in which existing algorithms have been tested is ice tracking. Examples are shown here which evaluate the precision and accuracy of ice tracking in Antarctica. Various other scenarios which warrant further study are also briefly illustrated in subsequent sections.

### 8.6.1
### Antarctic Ice-Motion Tracking Performance

#### 8.6.1.1
#### *SAR Motion Tracking Validation*

Further work to quantify the accuracy of automated sea-ice tracking algorithms used on SAR image pairs has recently been undertaken by the present author. Positions of Argos drifter and GPS receivers in the vicinity of ISW are compared with SAR ice-tracked motion vector information to establish the success of algorithms developed for the winter Arctic (Kwok et al. 1990). The period of consecutive ERS-1 SAR imaging of the ISW camp floe lasted between day 38 (February 7) through day 75 (March 15) in 1992. Results, such as those shown in Figs. 12 and 13, demonstrate that algorithms originally developed for the winter Arctic can be used without serious problems in both ISW summer perennial ice or relatively featureless smooth Weddell Sea winter first-year ice. In order to establish the accuracy of the tracking algorithms, however, positions from a fixed GPS receiver at the ISW camp are compared with SAR ice-tracked drift vectors. GPS positional fixes were not logged until after day 57, preventing accurate interpolation of latitude and longitude positions of the camp earlier than this date. The period of highest-frequency GPS measurements overlaps with six consecutive SAR ice velocity image "pairs" (i.e., 12 consecutive images), allowing comparisons to be made between day 57 and 75. Figure 15 shows the correlation between mean velocity components derived from SAR (such as that in Fig. 12) and those derived from instantaneous GPS ISW locations (interpolated to the SAR imaging times). SAR-derived $u$ and $v$ velocity components are rotated into the local coordinate system to match calculated GPS velocities.

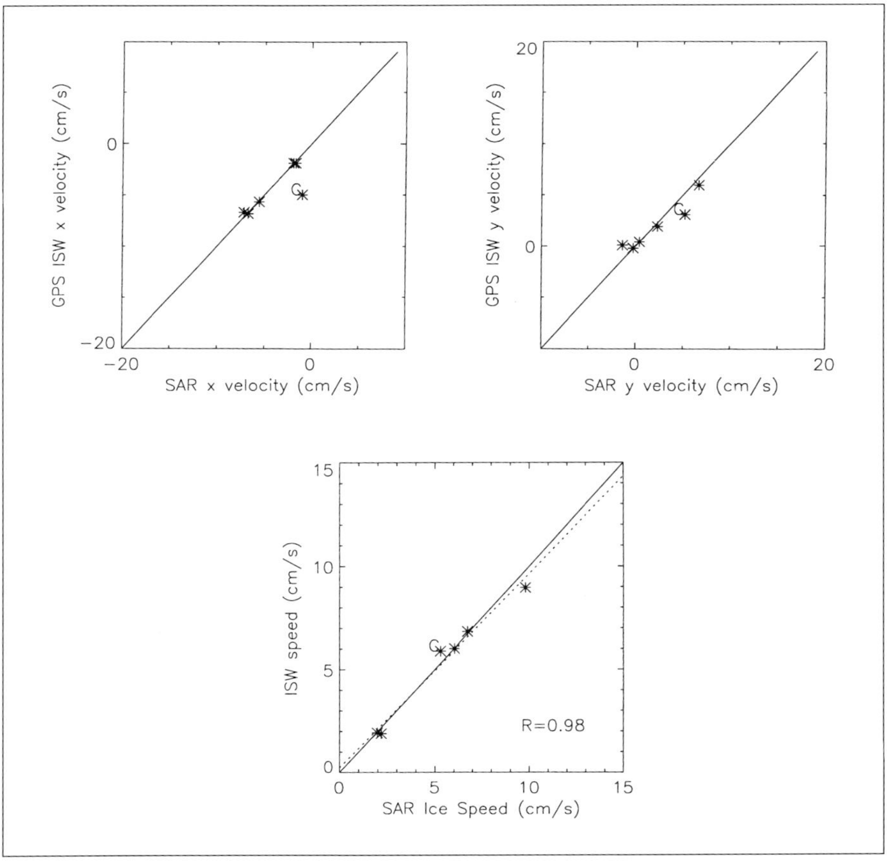


**Fig. 15.** Comparison of SAR-tracked 3-day velocities with equivalent 3-day velocities computed from instantaneous GPS locations of Ice Station Weddell. *Curly arrows* signify points extracted from velocity fields exhibiting a high degree of vorticity. For further details, see text

In terms of different mean ice drift speeds Fig. 15 indicates a sample ranging from 2 to 10 cm $s^{-1}$. Figure 15a shows the west-east (i.e., x or u component) and Fig. 15b south-north (i.e. y or v) components of velocity, while Fig. 15c indicates the magnitude of the velocity. Ice drift is largely westward and northward during this period. Though a small sample, the plot shows a high degree of accuracy in deriving both the direction and the magnitude of ice drift. Although spatial mean SAR velocities are used in the regression analysis, as opposed to values of the SAR-derived velocity interpolated to the ISW floe location, the result in Fig. 15c indicates a correlation of 0.98 (illustrated by the dotted regression line). The rms error in SAR motion tracking is computed to be less than 0.5 cm $s^{-1}$ for 3-day motion tracking. This translates into an error of approximately 100 m, or equivalently 1 pixel, after accounting for error variance generated by temporal interpolation of GPS locations and spatially averaged mean SAR velocity vectors.

The largest outlying vector in Fig. 15 is identified throughout by a curly arrow, indicating that the motion field displays a large amount of rotation or vorticity. As Fig. 12 shows, the mean velocity components of such a field are not sufficiently representative of any single point in space for a good comparison to be made. On the other hand, it is felt that spatial interpolation of the velocity field is not representative of true inter-floe motions, and so further attempts are not made to try to improve this comparison.

#### 8.6.1.2
#### *Time-Series Observations of Ice Floes*

Results of processing all 3-day or 1-day repeat overlapping "pairs" into ice kinematics information are shown as a velocity time series in Fig. 16 together with surface wind speed and the GPS monitored drift of the ISW camp floe. In Figure 16a ISW measurements of the wind speed are shown as a dotted line. Prior to completion of the meteorological mast on day 57, once- or twice-daily wind speeds were recorded. The 3-day smoothed solid line indicates Finnish Salargos buoy (no. 5908) wind speed measurements prior to day 57 (courtesy Timo Vihma), whereafter the solid line is a 3-day running mean of the hourly means recorded by the ISW met station anemometer.

Figure 16b and Fig. 16c show the mean SAR instantaneous velocity components joined by a dotted trend line. Horizontal bars indicate the period separating the tracked image pair, and the vertical bar indicates one standard deviation about the mean velocity vector (spatially averaged over a 100 × 100 km scene). An overlapping solid line indicates the 3-day smoothed velocity of ISW and begins upon installation of a fully functional GPS receiver at ISW (on day 57). Hollow diamond symbols represent the velocity computed from distances traveled between instantaneous GPS locational "fixes" at SAR imaging times (shown in Fig. 15). Where diamonds overlap mean SAR velocity measurements, the spatially averaged SAR velocity is representative of a single GPS velocity vector. In a spatial context, where the mesoscale velocity field does not represent translational motion, the ISW GPS velocity diamond does not overlap the mean SAR velocity. Thus, in the case of day 70 the discrepancy shown in Fig. 16b,c is explained by strong anticyclonic or cyclonic motion, and is similar to the example in Fig. 12. A mean velocity of a spatial field of cyclonic or anticyclonic flow is not representative of any single vector within that field, and such spatial means cannot be directly compared to the "ensemble-averaged" ISW GPS-derived velocity. Nevertheless, smoothed 3-day velocity components indicate responses to sustained bursts of wind shown in the solid line in Fig.

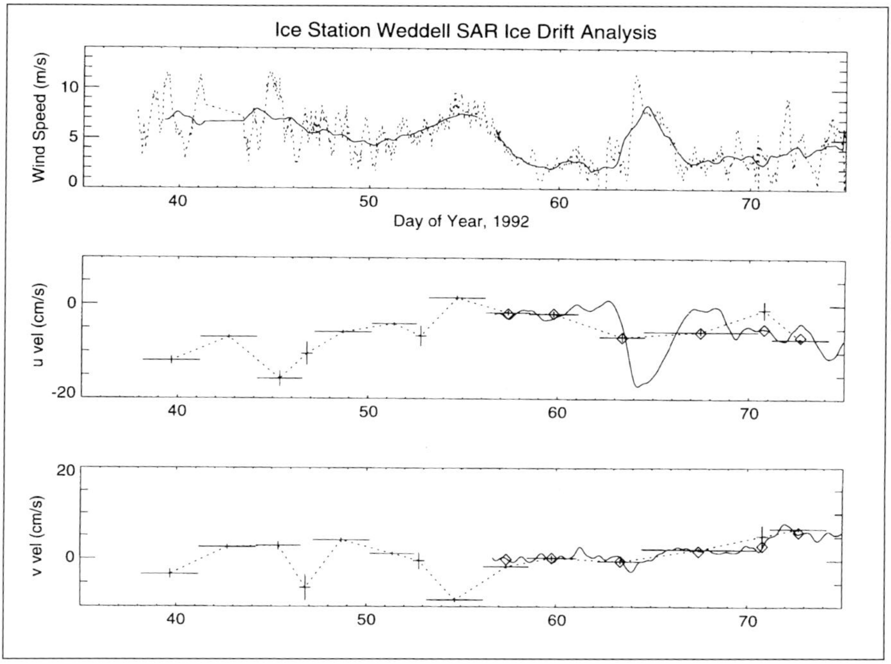


**Fig. 16.** **a-c.** Time series of ERS-1 SAR-tracked ISW ice motion results. **a** Measured surface wind speed (*dotted*) and 3-day smoothed windspeed, **b** SAR- (*dotted*) and GPS-derived (*solid*) u component of velocity (east is positive), **c** SAR (*dotted*) and GPS (*solid*, 3-day running mean) v component of velocity (north is positive). For further details, see text

16a. Additionally, though not in "free-drift," SAR-reported drift velocity components (dotted line in Fig. 16b,c) clearly respond to the 3-day filtered wind forcing. The largest discrepancy between SAR velocity and 3-day smoothed ISW velocity occurs as a result of the wind burst from day 63 to day 67, when ISW is pushed to the west at -15 cm $s^{-1}$. In the corresponding SAR image record, the closest pair of image acquisitions was at 6 days, and thus the minimum in ISW u velocity (solid line) is missed. A higher temporal resolution in SAR imaging is required to capture such deviations in ice motion.

Generally, examples such as Fig. 16 indicate that the sea-ice drift field adjusts to the wind. The temporal frequency of fluctuations in the field of ice velocity vectors closely matches those in wind velocity. Temporal sampling at 3-day intervals or less is concluded to be necessary if the small-scale response to synoptic storms is to be studied in more detail in an open basin such as the Weddell Sea. The shorter timescale examples such as the 1-day repeat orbits plotted at days 46, 52, and 71 capture more variance in ice velocity (indicated by the vertical bars for u and v velocity components). Spectral results in buoy data in Sect. 8.5.2.2 confirm this phenomenon to be attributable to the aliasing of velocity components induced by tidal motions. Importantly, therefore, exact 3-day repeat orbits do not alias diurnal or semidiurnal (which are close to 12 h period) tidal motions, whereas nonmultiples of 12 h contain significant tidal variance in this location. Ultimately, a trade-off is necessary between resolution in temporal sampling and the sampling interval over which ice floes are tracked. The best solution appears to be 3-day SAR ice-motion tracking at regular, exact 1-day intervals, but issues such as the SAR data volumes (for required coverage of the entire Weddell Sea) then become paramount. RADARSAT ScanSAR image coverage may be the only future high-resolution SAR solution which can approach this sampling requirement.

### 8.6.2
### Temporal Changes in Ice Characteristics

The majority of algorithm problems occur as a consequence of large temporal variations in ice characteristics. Time series of ERS microwave radar data demonstrate that within ice class backscatter signatures vary both seasonally and interannually in the Weddell Sea. The primary distinction, therefore, between seasonal and perennial ice covers can only be made by separating data acquired during the austral winter ice maximum and summer minimum, or alternatively by continuously tracking perennial ice floes identified during the summer months through time and space. During February and March of 1992, planned SAR acquisitions over ISW resulted in over 35 calibrated images over this perennial ice zone region. Other, previously described winter SAR data acquisitions were planned during WWGS'92 in locations where R.V. *Polarstern* was traveling across the Weddell Sea, and in a number of additional instrumented locations. The result in 1992 is a total collection of more than 50 images (exceeding $3.5 \times 10^9$ pixels) in regions where sea-ice cover was either measured or for which independent sources of information are available. Two examples of seasonal backscatter variability are illustrated below. In the first, all ERS-1 SAR images in winter and summer states are separated and grouped, and their backscatter statistics plotted in Fig. 17. In the second, in Fig. 18, follow a number of ice parcels in space, to illustrate the progression in ice signatures (after Drinkwater and Lytle 1997).

**Fig. 17.** Seasonal SAR backscatter pdf's. Bimodal peaks in winter distribution comprise first-year or seasonal (*FY*) and multiyear (*MY*) or perennial ice.

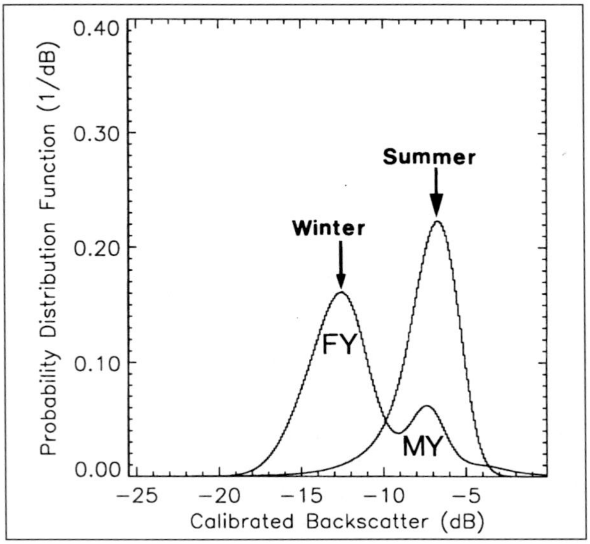

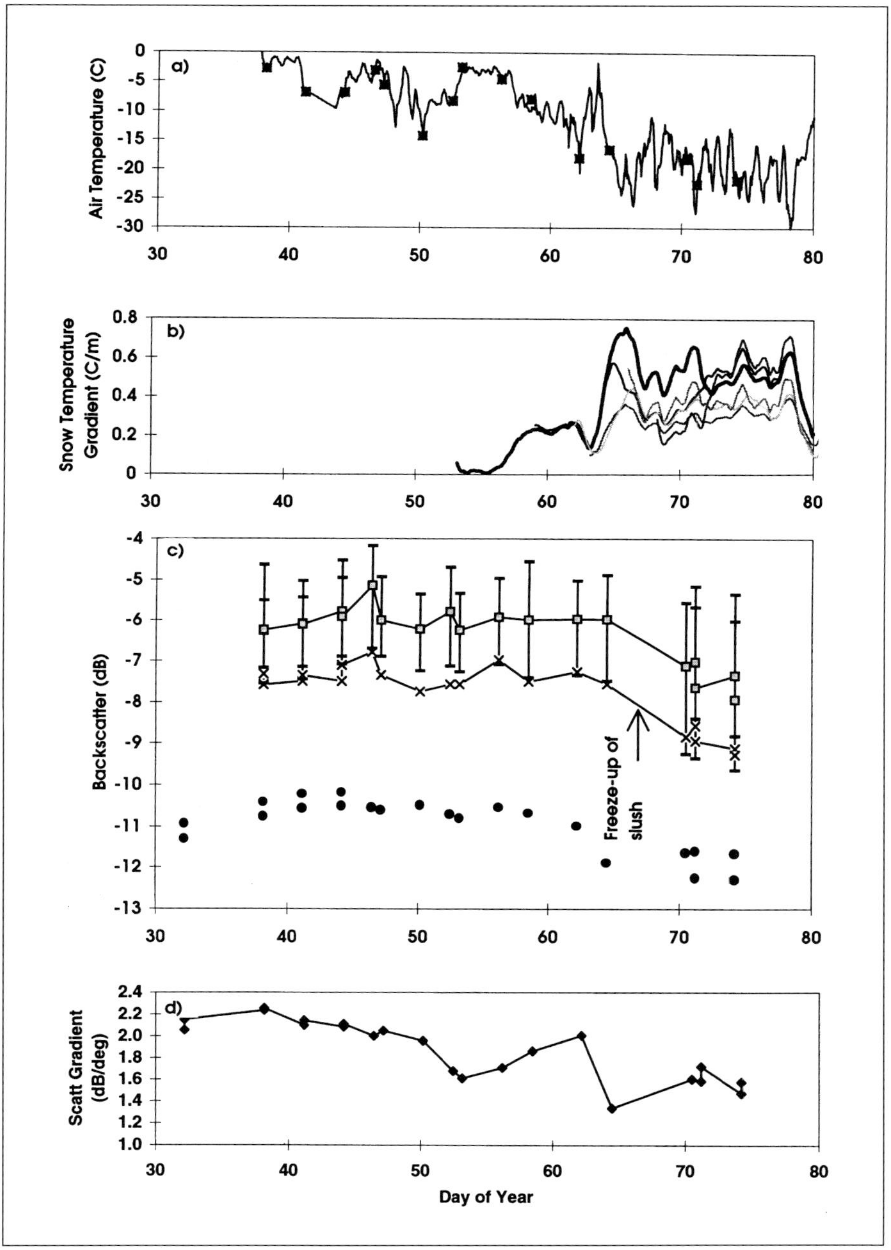


**Fig. 18.** **a-d.** Time series of **a** surface air temperatures from the Finnish Salargos buoy; **b** snow-temperature gradients at thermistor sites on snow-covered perennial ice floes; **c** □ or $\bar{\sigma}^{\circ}_{MY}$ of multiyear floes (*vertical* bars indicating range of values), × or $\bar{\sigma}^{\circ}_{SAR}$ (i.e,. global SAR image means), and ● $\bar{\sigma}^{\circ}_{Scatt}$ or 40° incidence normalized backscatter coefficient; **d** ◆ EScat backscatter gradient $M$ (linear regressed) at 40° incidence. For further details, see text (after Drinkwater and Lytle 1997).

#### 8.6.2.1
#### *Summer and Winter SAR Backscatter Characteristics*

If contrasts in midwinter and midsummer conditions are first considered, then a collection of all statistics of calibrated backscatter images may be made on the basis of the ice growth or melt seasons. Since the Weddell Sea ice cover is largely composed of seasonal ice, the residual ice cover at the end of summer typifies the various states of perennial ice, all ice surviving the summer melt being defined as multiyear ice. All the grouped pixel values are combined into either summer and winter pdf's in Fig. 17 to describe the differences between the summer and winter perennial, or the annual ice signatures. The summer pdf shows a mode around -6 dB, with a small tail to higher values comprising highly deformed ice and icebergs. Ice concentrations typically exceed 95% in the perennial ice pack and the contribution from wind-roughened open water is negligible (Drinkwater and Lytle 1997). At the lower end of the summer pdf is a longer tail extending to –15 dB and beyond. This includes level, undeformed multiyear ice and snow-covered first-year ice thick enough to survive the summer melt.

In winter, the pdf becomes bimodal with the appearance of seasonal ice. As time progresses, advection of an increasing fraction of multiyear ice northwards out of the basin reduces the probability of multiyear ice, thereby producing the imbalanced bimodal midwinter distribution shown in Fig. 17. While a secondary peak remains, with a mode at around –7 dB, the main peak now occurs at around –12.5 dB and comprises more prevalent forms of level, first-year ice. The low end of the pdf, at values of –15 dB and below, indicates the least deformed first-year ice. A winter tail extending down to the noise floor of the SAR (approximately –25 dB) includes very small proportions of new and young ice forms, and recent results from (Morris and Jeffries 1997) in the Bellingshausen Sea indicate that typical undeformed new ice appears with a mean backscatter value of –23± 0.5 dB in small areas of coastal polynyas and flaw leads.

#### 8.6.2.2
#### *Time-Series Observations of Backscatter*

Early observations of the temporally changing C-band backscatter response within fixed regions have been made in Drinkwater et al. (1993b). However, temporal monitoring of the backscatter signature of moving floes has not previously been attempted. Studies focusing specifically on multiyear ice in summer and winter SAR images show that its backscatter coefficient changes seasonally, and the mode of the multiyear ice pdf in Fig. 17 is observed to shift slightly from around -6 dB to -7 dB between warm and cold seasons (i.e., around 25% change in image scattering intensity). Surface measurements made at ISW (Lytle and Ackley 1996) illustrate that perennial ice retains a deep snowcover on its surface. A large proportion of the ISW camp floe and surrounding ice floes experienced summer flooding at the base of the snow.

Figure 18 illustrates the temporal change of the SAR backscatter signature of tracked perennial ice floes in the vicinity of ISW as they cooled during the summer-autumn transition period. As the cold wave (i.e., 0 °C isotherm) propagates downwards through the snow, areas of flooded or saturated basal snow (slush) cool and refreeze. In response to this process, C-band microwave backscatter values shown in Fig. 18 decrease (after Drinkwater and Lytle 1997). The period of cooling air temperatures at ISW (Fig. 18a) is shown in parallel to thermistor profiles at a variety of sites with differing snow depth (Fig. 18b). Thermistors indicate that protracted cold air temperatures after day 63 enable the snow layer to develop a strong thermal gradient. This cools the basal snow sufficiently quickly that the slush freezes around day 67.

Large numbers of individual multiyear ice floes were tracked in ERS-1 SAR images to derive the microwave backscatter trend shown in the upper trace of Fig. 18c. For comparison, global mean SAR-image and EScat-image $\sigma^0_{vv}$ values measured over the identical 100 × 100km SAR-imaged areas, each report essentially the same decreasing trend. The largest decrease in $\sigma^0_{vv}$ is observed before and after the freeze-up of the slush. Thus, the change from rough surface scattering from snow-covered slush to rough surface scattering from dry snow-covered perennial ice appears to be responsible for the observed shift in the multiyear ice peak in Fig. 17.

#### 8.6.2.3 *Meltponding*

Additional effects of seasonal melting and transformation at the surface of perennial ice floes are noted here as a consequence of summer processes and periodic winter excursions in air temperature due to storm systems. Large warm excursions in air temperature have been observed to have considerable impact upon the microwave backscattering characteristics of sea ice, in response to the changing surface snow and ice characteristics (Drinkwater et al. 1995a; Massom et al. 1997). Though the appearance of classical meltponds in the Weddell Sea appears to be checked largely by cold dry southerly winds and the latitude or extent of the southward wandering of the zero degree seasonal isotherm, a recent summer experiment in 1995 recorded surface meltponding in the Weddell Sea (Low 1995). The appearance and expression of meltponding in SAR images is discussed in more detail by Low (1995) by use of coincident SAR and aerial photography. Ongoing studies of the spatio-temporal expression of seasonal melt in the EScat image time series (1992 to the present day) will reveal the context of these SAR-observed features in the long-term record. Furthermore, these studies are expected to confirm whether extreme melt events noted in 1992 and 1995 are associated with climatic anomalies such as the 4-year period in sea-surface temperature and reduced poleward wind stress observed by White and Petersen (1996).

## 8.7 Conclusions

ERS satellite data have for the first time been extensively studied and applied in studies of sea-ice dynamics and characteristics in Antarctica. The main result of analysis of these data is an understanding of how microwave signatures and their time-varying properties can be used as a basis for understanding geophysical changes in the Antarctic sea-ice cover. Examples show that satellite radar is a valuable addition to the arsenal of tools now available for building up the long-term observational database necessary for addressing climate-related issues. Furthermore, algorithms which have been applied to these data will be refined and ultimately lead to data products including ice kinematics fields and area flux information, with which to address mass- or freshwater balance questions in the Southern Ocean.

The most promising application of radar data in Antarctica is undoubtedly the ability to observe ice dynamics in time-sequential data sets. Further work to realize automated algorithms for processing these images to large-scale ice-kinematics products will be undertaken in the near future, such that these data sets become widely available to more general users of dynamic information. Ice-drift statistics alone are an important advance in understanding how the wind-aided advection of ice influences regional heat and freshwater budgets of the Southern Ocean.

Observation and recognition of particular ice characteristics and types is critical to the interpretation and evaluation of new microwave image products in Antarctica. This chapter provides the first broad picture of the backscatter characteristics or "signatures" observed by C-band satellite radars, and serves as a foundation for a catalogue of Southern Ocean sea-ice signatures for regions in which in-situ measurements are frequently made. Examples illustrated in this study indicate that Antarctic ice can often look considerably different from its Arctic counterpart, warranting future studies to expand knowledge of its regional and seasonal variability. A caveat must be added, however, in relationship to the utility of more general automated ice-classification schemes in Antarctica, especially in regions and seasons where products cannot be characterized in terms of their uncertainty. Indeed, the concept and requirement for proxy classes of ice thickness should be revisited, and future algorithms to classify Antarctic ice should be carefully crafted. Future schemes should utilize the present framework of large-scale measurements by passive microwave and scatterometers, together with drawing upon quantitative information derived from direct observations of the kinematics. Together these information sources provide the basis for a more valuable geophysical product than simply ice-type classification.

Coupled regional ocean-ice-atmosphere models will initially benefit from Southern Ocean radar data through more accurate parameterizations of the ice rheology and resulting motion and deformation fields. In turn, they will be used to evaluate the direct relationships between the changing basin-wide distribution of sea ice and the flux of momentum. Accurate monitoring of Southern Ocean ice formation, drift, deformation, and divergence is of additional importance to indirect estimation of surface fluxes of heat, freshwater and salt.

Mesoscale coverage of ERS-1 and NSCAT can now be used together with the meteorological and oceanographic data to generate regional surface flux estimates in varying locations in the Southern Ocean. These are presently being compared with the regional distribution of warm and cold water regimes observed in oceanographic data, to map the spatial arrangement of ice surface flux environments in relation to the ocean and atmospheric circulation patterns. Resultant estimates will be compared with simulated fluxes in one- and two-dimensional coupled sea-ice models to provide keys to the principal sea-ice factors regulating the surface fluxes in ice-covered regions. Furthermore, the advantage of the coverage which the scatterometer image data provide is to facilitate investigations of links between the Antarctic circumpolar wave and temporally varying attributes of the sea-ice cover. For the first time, high-resolution global coverage of sea ice enables teleconnections between mid- and high-latitude processes to be studied in Antarctica in a consistent fashion.

**Acknowledgments.** ERS-1 data were supplied as part of the European Space Agency-supported AO2.USA.119 project. Enhanced resolution ERS-1 scatterometer data were processed by David Long and David Early of Brigham Young University as part of a collaborative project. SAR data are processed and supplied by the German Processing and Archiving Facility (D-PAF), and special thanks go to Jörg Gredel and Birgit Schättler for their product support and Klaus Reiniger for GARS receiving-station support. WWGS'92 buoy data were kindly supplied by Christoph Kottmeier as part of an ongoing collaborative analysis project and Timo Vihma is thanked for use of the Salargos buoy data. Mark Drinkwater performed this work at the Jet Propulsion Laboratory, California Institute of Technology, under contract to NASA, with the support of Robert H. Thomas of Code YSG, NASA.

## References

Bally P, Fellah K (1995) Evaluation of the accuracy of the backscattering coefficient measurement in SAR data products. ESA/ESTEC Earth Sciences Division Tech Note, July

Carsey FD, Holt B, Martin S, McNutt L, Rothrock DA, Squire VA, Weeks WF (1986) Weddell-Scotia Sea marginal ice zone observations from space, October 1984. J Geophys Res 91, no C3: 3920–3924

Dierking W (1995) Laser profiling of the ice surface topography during the Winter Weddell Gyre Study 1992. J Geophys Res 100, no C3: 4807–4820

Drinkwater MR (1989) LIMEX '87 Ice surface characteristics; implications for C-band SAR backscatter signatures. IEEE Trans Geosci Remote Sens 27, no 5: 501–513

Drinkwater MR (1998) Active microwave radar observations of Weddell sea ice. In: Jeffries MO (ed) Antarctic sea ice processes, interactions and variability, Antarctic Research Series American Geophysical Union, in press

Drinkwater MR (1995a) Airborne and satellite SAR investigations of sea-ice surface characteristics. In: Ikeda M, Dobson F (eds) Oceanographic applications of remote sensing, CRC Press, Boca Raton, Florida, pp 345–364

Drinkwater MR (1995b) Applications Of SAR measurements in ocean-ice-atmosphere interaction studies. In: Ikeda M, Dobson F (eds) Oceanographic applications of remote sensing, CRC Press, Boca Raton, Florida, pp 391–406

Drinkwater MR, Haas C (1994) Snow sea-ice and radar observations during ANT X/4: summary data report. Berichte aus dem Fachbereich Physik 53. Alfred-Wegener-Institut für Polar- und Meeresforschung, Bremerhaven, Germany

Drinkwater MR, Kottmeier C (1994) Satellite microwave radar- and buoy-tracked ice motion in the Weddell Sea during WWGS'92. Proc Int Geosci Remote Sens Symp IGARSS '94, August 8–12, Pasadena, California, pp 153–155

Drinkwater MR, Lytle VI (1997) ERS-1 SAR and field-observed characteristics of austral fall freeze-up in the Weddell Sea, Antarctica. J. Geophys. Res., vol. 102, C1, pp 1101–1117.

Drinkwater MR, Squire VA (1989) C-band SAR observations of marginal ice zone rheology in the Labrador Sea. IEEE Trans Geosci Remote Sens 27, no 5: 522–534

Drinkwater MR, Long DG, Early DS (1993a) Enhanced resolution scatterometer imaging of Southern Ocean sea ice. ESA J 17: 307–322

Drinkwater MR, Hosseinmostafa R, Dierking W (1993b) C-band microwave backscatter of sea ice in the Weddell Sea during the winter of 1992. Proc Int Geosci Remote Sens Symp IGARSS '93, August 18–21, Tokyo, Japan, pp 446–448

Drinkwater MR, Early DS, Long DG (1994) ERS-1 investigations of Southern Ocean sea-ice geophysics using combined scatterometer and SAR images. Proc Int Geosci Remote Sens Symp IGARSS '94, August 8–12, Pasadena, California, pp 165–168

Drinkwater MR, Hosseinmostafa R, Gogineni SP (1995a) C-band backscatter measurements of winter sea ice in the Weddell Sea, Antarctica. Int J Remote Sensing 16, no 17: 3365–3389

Drinkwater MR, Fischer H, Kreyscher M, Harder M (1995b) Comparison of seasonal sea-ice model results with satellite microwave data in the Weddell Sea. Proc Int Geosci Remote Sens Symp IGARSS '95, July 10–14, Florence, Italy, pp 357–359

Early DS, Long DG (1997) Azimuth modulation of C-band scatterometer $\sigma^{o}$ over Southern Ocean sea ice. IEEE Trans Geosci Remote Sens, vol. 35, no. 5, pp 1201–1209.

ESA Earthnet (1992) ERS-1 System. ESA Special Publication SP-1146, ESA Publications Division, ESTEC, Noordwijk, The Netherlands

Fischer H (1995) Comparison of an optimized dynamic-thermodynamic sea-ice model with observations in the Weddell Sea. Rep Polar Res 166. Alfred-Wegener-Institut für Polar- und Meeresforschung, Bremerhaven, Germany

Gohin F (1995) Some active and passive microwave signatures of Antarctic sea ice from mid-winter to spring 1991. Int J Remote Sens 16, no 11: 2031–2054

Haas C, Viehoff T, Eicken H (1992) Sea ice conditions during the Winter Weddell Gyre Study 1992 ANT X/4 with R.V. *Polarstern*: shipboard observations and AVHRR imagery. AWI Berichte aus dem Fachbereich Physik 34. Alfred-Wegener-Institut für Polar- und Meeresforschung, Bremerhaven, Germany

Hosseinmostafa AR, Lytle V, Jezek KC, Gogineni SP, Ackley SF, Moore RK (1995) Comparison of radar backscatter from Antarctic and Arctic sea ice. Electromagn Wave Applic 9: 421–438

Ice Station Weddell Group (1993) Weddell sea exploration from ice station. EOS 74, no 11: 121, 124–126

Kottmeier C, Sellmann L (1996) Atmospheric and oceanic forcing of Weddell sea ice motion. J Geophys Res 101, no C9: 20809–20824

Kottmeier C, Olf J, Frieden W, Roth R (1992) Wind forcing and ice motion in the Weddell Sea region. J Geophys Res 97, no D18: 20373–20383

Kwok R, Curlander JC, McConnell R, Pang SS (1990) An ice-motion tracking system at the Alaska SAR Facility. IEEE J Oceanic Eng 15, no 1: 44–54

Laur H, Meadows PJ, Sanchez JI, Dwyer E (1993) ERS-1 SAR radiometric calibration. Proc of the CEOS SAR Calibration Workshop, ESTEC Noordwijk, Netherlands, ESA Publication WPP-048, pp 257–281

Lemke P (ed) (1994) The expedition ANTARKTIS X/4 of R.V. *Polarstern* in 1992. Rep Polar Res 140. Alfred-Wegener-Institut für Polar- und Meeresforschung, Bremerhaven, Germany

Livingstone CE, Drinkwater MR (1991) Springtime C-band SAR backscatter signatures of Labrador Sea marginal ice: measurements vs modelling predictions. IEEE Trans Geosci Remote Sens 29, no 1: 29–41

Long DG, Hardin PJ, Whiting PT (1993) Resolution enhancement of spaceborne scatterometer data. IEEE Trans Geosci Remote Sens 31, no 3: 700–715

Low D (1995) The validation of ERS-1 summer SAR data for Antarctic summer sea ice. M.Sc. Thesis, University of Dundee, UK

Lytle V, Ackley S (1996) Heat flux through sea ice in the western Weddell Sea: convective and conductive transfer processes. J Geophys Res 101, no C4: 8853–8868
Lytle VI, Jezek KC, Gogineni SP, Hosseinmostafa AR (1996) Field observations of microwave backscatter from Weddell sea ice. Int J Remote Sens 17, no 1: 167–180
Martin S, Holt B, Cavalieri DJ, Squire V (1987) Shuttle imaging radar B (SIR-B) Weddell Sea ice observations: a comparison of SIR-B and scanning multichannel microwave radiometer sea ice concentrations. J Geophys Res 92, no C7: 7173–7179
Martinson DG, Wamser C (1990) Ice drift and momentum exchange in the winter Antarctic pack ice. J. Geophys. Res. 95: 1741–1755
Massom RA (1991) Satellite remote sensing of polar regions. Lewis Publishers, Boca Raton, Florida
Massom M, Drinkwater MR, Haas C (1997) Spatial and temporal distribution of winter snowcover properties on sea ice in the Weddell Sea. J Geophys Res 102, no C2: 1101–1117
McPhee MG, Ackley SF, Guest P, Huber BA, Martinson DG, Morison JH, Muench RD, Padman L, Stanton TP (1996) The Antarctic zone flux experiment. Bull Am Meteorol Soc 77, no 6: 1221–1232
Morris K, Jeffries MO (1998) Sea ice characteristics and seasonal variability of ERS-1 SAR backscatter in the Bellingshausen Sea. In: Jeffries MO (ed) Antarctic sea ice processes, interactions and variability, Antarctic Research Series American Geophysical Union, vol. 74.
Muench RD, Gordon AL (1995) Circulation and transport of water along the western Weddell Sea. J Geophys Res 100, no C9: 18503–18515
Oelke C (1997) Atmospheric signatures in sea-ice concentration estimates from passive microwaves: modelled and observed. Int J Remote Sens 18, no 5: 1113-1136
Ross H (in preparation) Ein massenerhaltendes hydrographisches Inversmodell mit verallgemeinerter Vertikalkoordinate angewandt auf den Atlantischen Sektor des Südlichen Ozeans. PhD Thesis, Alfred-Wegener-Institut für Polar- und Meeresforschung, Bremerhaven, Germany
Stern HL, Rothrock DA, Kwok R (1995) Open water production in Arctic sea ice: satellite measurements and model parameterizations. J Geophys Res 100, no C10: 20601–20612
Stoffelan A, Anderson DLT (1993) ERS-1 scatterometer data characteristics and wind retrieval skill space at the service of our environment. Proc 1st ERS-1 Symp, November 4–6, 1992, Cannes, France, pp 41–47
Thomas JP, Turner J , Lachlan-Cope TA, Corcoran G (1995) High resolution observations of Weddell Sea surface currents using SAR sea-ice motion vectors. Int J Remote Sensing 16, no 17: 3409–3425
Thorndike AS (1985) Kinematics of sea ice. In: Untersteiner N (ed) The geophysics of sea ice, Plenum, New York, pp 489–549
Viehoff T, Li A (1995) Iceberg observations and estimation of submarine ridges in the western Weddell Sea. Int J Remote Sens 16, no 17: 3391–3408
Vihma T, Launiainen J, Uotila J (1996) Weddell Sea ice drift: kinematics and wind forcing. J Geophys Res 101, no C8: 18279–18296
White WB, Petersen RG (1996) An Antarctic circumpolar wave in surface pressure wind temperature and sea-ice extent. Nature 380: 699–702

# CHAPTER 9

# Alaska SAR Facility: The US Science Center for Sea Ice SAR Data

F. Carsey, R. Harding, C. Wales

## Contents



## 9.1 Introduction

In 1985 NASA approved the Alaska SAR Facility (ASF) at the Geophysical Institute of the University of Alaska Fairbanks (UAF). The implementation was to be a joint effort between UAF and the Jet Propulsion Laboratory. The objective of ASF was and is to supply scientific and operational data users with calibrated, timely satellite SAR data. In 1991 ASF began to receive data from the First European Remote Sensing Satellite (ERS-1), in 1992 from the Japanese First Earth Resources Satellite (JERS-1), in 1995 from the ERS-1 follow-on, ERS-2, and in 1996 from the Canadian RADARSAT. In 1992 NASA and the National Science Foundation (NSF) formed a joint initiative to implement at McMurdo Base in Antarctica a satellite receiving station capable of acquiring and recording data from SAR satellites, and in January, 1995, this installation was complete; the McMurdo Ground Station is now operational and is acquiring data from ERS-1 and -2 and RADARSAT. Two research opportunities for analysis of ASF data have been published, the most recent in 1995, and there are about 175 approved projects for science and applications work with ASF data. Two particularly interesting polar regions mapping projects are planned for RADARSAT in the immediate future: the weekly mapping of the Arctic ice-covered water, which will be further processed into geophysical variables by the RADARSAT Geophysical Processor System (RGPS), and two complete mappings of Antarctica in the RADARSAT Antarctic Mapping Project (RAMP). These activities are discussed further below.

Analysis of SAR Data of the Polar Oceans
Edited by C. Tsatsoulis and R. Kwok

### 9.1.1 ASF Vision and Mission

The ASF is the facility which will fulfill all data system aspects for the United States (US) Government allocation of data from the Canadian Space Agency's RADARSAT. ASF is an end-to-end facility. It compiles and submits requests for data to be taken by the satellite. It schedules and receives data directly at antennas at Fairbanks and McMurdo, as well as receiving data by courier from other network ground stations around the world. All data received either directly or by courier will be archived as raw signal data for a minimum of 10 years after end of mission (or longer depending on mission). All RADARSAT data (as well as data from ERS-1, ERS-2, and JERS-1) are processed to image level (or higher) within the facility. A variety of level-1 products are produced for distribution to users as well as a limited set of higher-level products. ASF also provides software tools for basic manipulation of SAR images; these tools are available through the ASF website (http://www.asf.alaska.edu/daac_documents).

The ASF vision is to advance polar research and earth sciences by supporting the scientific and operational application of remote sensing data in order to better understand the earth as a system. ASF's mission is described as follows:

- Acquire, process, distribute, and archive remote sensing data in support of research, operational applications, and education.
- Participate in programs for scientific validation and applications of the remote sensing data acquired at the facility.
- Support users with acquisition planning, data processing tools, field program data, and resident expertise.
- Implement systems or functions that generate higher-level geophysical data products.
- Preserve and provide access to geophysical data and maps.
- Design and develop systems to accommodate evolving requirements, changing technology, and future sensors.
- Contribute to education at the university, primary, secondary and community levels through student involvement, lectures, curriculum development, electronic information exchange, and tours.

In general ASF data products are to be distributed to US federal agencies and approved science investigators. To become an approved investigator, it is necessary to submit a short (about two-page) proposal to ASF; this proposal is reviewed by ASF and NASA. There are periodic opportunities to gain access to data and research funding through NASA; the most recent of these was the Applications, Development and Research Announcement (ADRO) published by Canada in 1995. As a general rule ASF image data cannot be freely distributed to third parties as intellectual property rights are retained by the flight agencies, but derived products from which SAR backscatter values are no longer obtainable are not included in this restriction.

## 9.1.2 Capabilities

The overall total of ASF capability is extensive and complex. Briefly, ASF is equipped to plan data acquisitions, request data acquisitions from the agencies flying the SAR satellites, receive and record downloaded data from these satellites as they pass through the receiving masks, process the data stream into calibrated SAR images, and distribute these images to users. In the area of data reception, ASF has been built to track and acquire S-band and X-band downlinks from polar-orbiting satellites having the high (ca. 100 million bits per second) data rates characteristic of the SAR satellites. In data processing ASF is capable of processing data from the four currently operational civil SAR satellites, ERS-1 and -2, JERS-1, and RADARSAT, and plans are being formed for known future systems, as shown in Table 1. In the case of RADARSAT, the processing capability amounts to being able to process data from 23 of the 226 beams, the exceptions being the fine beams and the extended low-incidence angle beam, as summarized in Table 2. Higher-level SAR image products are left to the user to generate, and software tools with technical support are supplied by ASF to users so that they can perform geocoding, mosaicking, terrain correction, and similar manipulations on the image data. In future ASF plans to host systems to produce geophysical (higher-level) products from the SAR data.

**Table 1.** ASF SAR MISSION DESCRIPTIONS

| **Instrument:**<br>**Spacecraft:** | AMI<br>ERS-1, -2 | LSAR<br>JERS-1 | SAR<br>RADARSAT | ASAR<br>ENVISAT | VSAR<br>ALOS |
|---|---|---|---|---|---|
| ***SAR*** | | | | | |
| Frequency | C-band | L-band | C-band | C-band | L-band |
| Polarization | HH | HH | VV or HH | VV and/or HH | VV or HH |
| Swath | 100 km | 75 km | 50–500 km | 100–400 km | 70 km |
| Resolution | 25 m | 18 m | 10–100 m | 25–1000 m | 10–100 m |
| Incidence angle | 23° | 35° | 20°–50° | 23°–60° | 20°–55° |
| Orientation | Right | Right | Right[a] | Right | Right |
| Data source | None | 20 min | 20 min | None | 49 min |
| ***Orbit*** | | | | | |
| Inclination | 97.5° | 98.5° | 98.5° | 97.5° | 98.5° |
| Altitude | 785 km | 568 km | 790 km | 785 km | 700 km |
| Repeat | 3, 35, 168 days | 41 days | 24 days | 35 days | 45 days |
| ***Mission*** | | | | | |
| Launch | ERS-1: July 1991 | February 1992 | November 1995 | 1998–9 | 2002 |
| | ERS-2: April 1995 | February 1992 | November 1995 | 1998–9 | 2002 |
| Lifetime | 3 years | 8 years | 5 years | Forever? | TBD |
| Status | Operational | Operational | Integration | Approved | Approved |
| Agency | ESA | NASDA | CSA | ESA | NASDA |

TBD, To be determined.

a Except to map Antarctica in two dedicated missions in which it will look left.

**Table 2.** ASF SAR Products in the RADARSAT era

| Image product | Incidence angles | Swath | Full resolution/ pixel spacing | Low resolution/ pixel spacing |
|---|---|---|---|---|
| ERS-1 and -2 AMI SAR | 19–26° | 100 km | 12 m/12.5 m | 150 m/100 m |
| **RADARSAT** | | | | |
| Standard beams | | | | |
| 1 | 19.6–27.1° | 100 km | 25 m/12.5 m | 150 m/100 m |
| 2 | 24.2–31.3° | | | |
| 3 | 30.5–36.9° | | | |
| 4 | 33.4–39.7° | | | |
| 5 | 36.5–42.2° | | | |
| 6 | 41.5–46.6° | | | |
| 7 | 44.9–49.4° | | | |
| Wide-swath beams | | | | |
| W1 | 20.0–31.3° | 120 km | 35 m/25 m | 300 m/200 m |
| W2 | 30.8–39.6° | 150 km | | |
| W3 | 39.2–45.2° | 150 km | | |
| High incidence beams | | | | |
| EH1 | 48.9–52.2° | 75 km | 25 m/12.5 m | 150 m/100 m |
| EH2 | 50.4–53.7° | | | |
| EH3 | 51.8–54.8° | | | |
| EH4 | 54.5–56.9° | | | |
| EH5 | 55.9–58.3° | | | |
| EH6 | 57.3–59.5° | | | |
| ScanSAR beams | | | | |
| 1 (W1,2,3,S7) | 20.0–49.4° | 500 km | 50 m/100 m | 600 m/400 m |
| 2 (W1,2,S5,6) | 20.0–46.7° | 500 km | | |
| 3 (W1,2) | 20.0–39.5° | 300 km | | |
| 4 (W2,S5,6) | 30.8–46.7° | 500 km | | |
| Low-incidence beam[a] | (Deferred) | | | |
| El1 | 9.9–22.8° | 170 km | 50 m/12.5 m | 300 m/200 m |
| Fine-resolution beams[a] | 37–49° | 45 km | 10 m/6.25 m | |

a Deferred images available from RADARSAT International Inc. but not from ASF at this time

The two flight missions on the right-hand side of Table 1 are future missions. The role of the USA in these missions is not determined as of this writing, but there are good reasons for negotiating to supply data from these SAR instruments to US scientific and operational users, and ASF will be joining with representatives of the user community in seeking access to these data.

## 9.1.3 Reception-Area Masks

ASF archives, processes, and distributes data from the two US SAR ground stations, at Fairbanks, Alaska, and McMurdo Base, Antarctica, and is directly responsible for data reception and recording at Fairbanks, while the NASA Wallops Flight Center manages the McMurdo facility for NSF and NASA. The reception areas for these facilities, called station masks, are shown in Fig. 1. The satellites now flying are in orbits tangent to

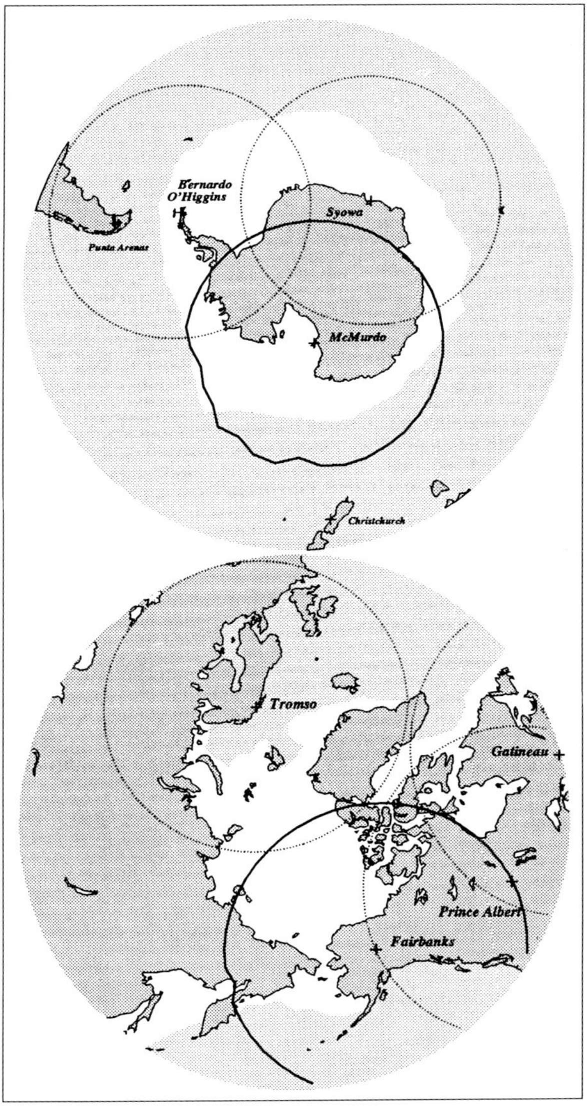


**Fig. 1. a,b.** Reception masks indicating the approximate geographic limits for a satellite in order for real-time data to be acquired or recorded data downlinked from it. *White areas* are those covered by sea ice at the seasonal maximum (March in the north and September in the south). *Heavy lines* indicate the ASF and McMurdo masks; *lighter lines* indicate other-station masks in the polar regions. Other high-latitude stations are in the planning process, e.g., at Svalbard. Note that for the Arctic Ocean a small region in the Laptev Sea is not covered by any of the established stations. (Images courtesy of John Crawford of Jet Propulsion Laboratory).

approximately 82° north and south, and they are oriented with the SAR antenna on the right side, which is to say with the antenna pointed north at both the northern and southern extreme of the orbits. RADARSAT is capable of flying with the antenna on the left side, pointing south at the orbit extrema, and there are plans to orient the spacecraft in this fashion for a few weeks in order to map Antarctica.

### 9.1.4 McMurdo Ground Station

In 1991 representations were made to NSF and NASA regarding the desirability of implementing a SAR-capable ground station at Antarctica, and a Science Working Group was appointed to look into the scientific needs and opportunities in this area; the Group recommended that McMurdo Ground Station be implemented and indicated that the SAR data from this station would find application in studies of the dynamics and mass balance of the Antarctic ice sheet, of air-sea-ice interactions in the Ross, Bellingshausen, and Amundsen Seas, of biological habitat in the ice-ocean environment, and of geomorphology, tectonics, and volcanoes in exposed-rock areas (see Jezek and Carsey 1991). In 1992, NASA and NSF agreed that it would be beneficial and feasible to construct a ground station capable of receiving satellite SAR and other high-data-rate signals at NSF's McMurdo Station in Antarctica.

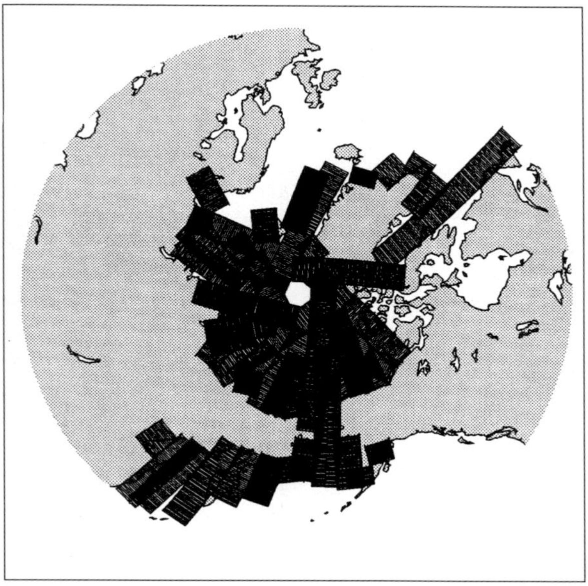

**Fig. 2.** The Arctic Snapshot. *Dark bands* are SAR-covered swaths planned for acquisition by ASF from the Canadian-US RADARSAT in a mode called ScanSAR, in which a swath of 500 km width with a surface resolution of 50 m is acquired. This total coverage is called the Arctic Snapshot and current planning calls for it to be acquired every 6 days. The resulting data set is intended for operational use by the National Ice Center and for scientific use as input to the RADARSAT Geophysical Processor System, discussed by Kwok (this volume). The programmatic duration of the Snapshot is not defined and it is intended for as many years as possible, with a (hoped-for) minimum duration of 5 years, given the known large-scale variability of the sea ice cover. The image and higher-level data from this program will be available through the ASF. (Image courtesy of John Crawford of Jet Propulsion Laboratory).

NASA Wallops Flight Facility, with the assistance of the NSF, installed the ground station and radome at McMurdo Station on Ross Island, Antarctica (77.8°S, 166.7°E) during the 1994/95 Antarctic summer season. In January, 1996, McMurdo became an official ERS ground station, although considerable testing data had already been collected, and in the same month the first RADARSAT data acquisitions were made.

The SAR reception at McMurdo Ground Station will mean that real-time downlink data from SAR satellites is now possible for all of Antarctica, as shown in Fig. 1, because of the coverage afforded by three receiving stations, one at O'Higgins Station on King Island, operated by Chile and supporting a German station, one at the Japanese station at Syowa, and McMurdo. Only McMurdo Ground Station will be operational for 12 months per year, but most of the austral winter data will of necessity be stored at McMurdo until it can be removed in spring, according to current data-handling capabilities. The SAR signal data cassettes will be flown to ASF where they will be processed into images and distributed to users.

## 9.2
## ASF in Polar Regions Science

The disciplines addressed by RADARSAT data focus on polar issues, and they contribute collectively to earth science. Brief scientific motivations for the different areas follow.

*Polar Oceans.* The polar oceans are regions where strong air-sea-ice interactions occur that contribute to the heat and brine balance of the world ocean. Furthermore, the polar oceans and their adjoining subpolar gyres are biologically the most active regions of the world oceans. Mineral resources are also known or presumed to exist in the sedimentary rocks of their continental shelves. Finally, the ice in the polar oceans acts as an impedance to shipping across otherwise highly advantageous marine trade routes.

*Ice Sheets.* The immense ice sheets of Antarctica and Greenland exert important radiative and thermal influences on both regional and global climate, and in addition are vast storehouses of freshwater, thus serving as an important reservoir for sea level changes.

*Glaciers and Lake Ice.* The extent of glaciers and their ablation zones as well as lake ice cover serve as indicators of the seasonal climate cycle and are also significant elements of regional hydrology.

*Boreal Forests.* The global forests are storehouses of carbon in both standing and dead biomass, and they are also critical elements of the hydrological cycle.

*Boreal Wetlands.* The wetlands, primarily boreal in nature, and the permafrost zones are sites of gas flux into the atmosphere and are valuable regional climate indicators.

*Seasonal Snowcover.* Snowcover is critical to the radiative balance and in many regions is the controlling element of soil temperature and moisture.

*Geology and Geophysics.* Geology studies include the study of seismic and volcanic deformation, especially the effects of large explosive eruptions on Earth's radiation balance and hence its climate, the continental record of past climate changes, the history of recent fault movements, and the evolution of the continental crust.

## 9.3 Applications of SAR in Sea Ice Research

Sea ice is important because of its roles in oceanic and atmospheric dynamics, oceanic productivity, and operations. In fact, sea ice and polar oceans studies are the primary motivation behind the RADARSAT program. At this time the central scientific issues for the polar oceans are those relating to the interactions of the high-latitude seas with the world ocean. These issues are: vertical mixing including oceanic convection and shelf processes, freshwater fluxes in the form of ice, and ice-edge-driven dynamics such as upwelling. Also of ongoing interest are the roles of ice in the radiation balance and as a surface flux control in synoptic processes. Understanding these processes calls for data on ice extent, motion, thickness, and concentration. SAR data are uniquely useful in making these determinations on fine scales.

Scientifically, RADARSAT will contribute to a number of specific determinations. The local and global heat and brine budgets are strongly influenced by the dynamics and thermodynamics of polynyas and shelf processes in both the Arctic and the Antarctic, and by the processes occurring during the melt season (Winebrenner et al. 1994). The changes observed in the surface morphology of the ice pack resulting from deformation can be used to estimate changes in regional values for air-ice-ocean momentum transfer, thereby contributing to the more effective utilization of such values in ice dynamics models. Detailed observations of ice movement and deformation have already stimulated the development of more realistic models of the rheology of sea ice (Stern et al. 1995), and this should be extended for its variation with location, season, ice thickness, and ice type. At the coastlines of ice-covered seas the interactions of ice with land are important to the behavior of the ice pack itself and to processes such as erosion and sediment transport in the near-shore region.

A key variable in polar seas is the simple matter of the presence of ice. C-band and L-band SAR systems are generally able to resolve the ice edge to within a few hundred meters. There are situations of ambiguity in this determination, notably when winds roughen the sea adjacent to old ice or first-year ice rubble fields; this situation is the largest source of error in the automatic determination of ice extent, and algorithms are under development to correct that problem. At the ice edge forming ice can occur as "grease ice" or "pancake ice," and each of these has a unique backscatter level and appearance in SAR data, although, again, algorithms for their automatic classification are currently in development.

Ice type is an involved matter. The World Meteorological Organization recognized 13 stages of development of ice, including icebergs, and 12 forms of ice. Various scientists have their own glossaries, often of subsets of the WMO list (WMO 1970). It is likely that no system of observations, including in situ visual sighting, will produce a reliable classification of these stages and forms, but there have been efforts to accomplish a usable classification of sea ice from satellite data, and these have met with various measures of success (Carsey 1992). A common goal of these efforts is to successfully determine the areal concentrations of areas of thin new ice (ice of less than about 50 cm thickness), thicker first-year ice (ice that has not been submitted to a summer melt period), and old or multiyear ice (ice that has survived a summer melt) (Kwok et al. 1992). Some programs additionally address the classification of pancake ice and brash ice, which often appear at the ice margin. Finally, ice type issues are confusing in summer conditions when the thinner ice species disappear and the thicker change in form and surface character (Gogineni et al. 1992; Holt and Digby 1985).

Ice concentration, that is, one minus the fraction of the ice-covered seas which is actually open water, is a variable often mentioned and rarely if ever effectively defined or conclusively determined. The expected behavior of the open water fraction has not been seriously examined. Using SAR data it is possible to determine some kinds of open water areas, e.g., a freezing lead in which Langmuir cells are organizing the newly formed ice bits, but there are areas of ambiguity, e.g., smooth open water looks the same as smooth new ice within the noise limits of most SAR systems. Various ambiguities of open water determination are characteristic of all satellite systems including visible-light, infrared, and passive microwave. An advantage of sequential SAR observations is that open water production due to deformation and divergence can be resolved with good accuracy if repeat data of sufficient timeliness can be obtained.

Ice motion is the key sea ice variable obtainable from SAR data (Holt et al. 1992), second in overall importance only to ice extent as a polar ocean surface variable. The method is simple in principl: features in the backscatter field in the SAR image are tracked in images acquired with an acceptably short time interval so that the motion of interest is resolved. At ASF the ice tracking concept has had a trial using a dedicated processor called the Geophysical Processing System (GPS); results from it have been discussed at length in Stern et al. (1995). The GPS addressed ice motion in the Beaufort Sea, primarily in winter; there were the conditions for which good quality algorithms were available (Holt et al. 1992).

RADARSAT SAR data are expected to increase the capability to monitor sea ice through an increase in coverage arising from the wider swath in the ScanSAR mode (see Table 2) and the use of the onboard recorder. Additionally, improvements in the analysis algorithms have extended the ice tracking capability seasonally and spatially. Consequently, the RADARSAT Geophysical Processing System (RGPS) aims at determining and analyzing ice motion for all the Arctic Ocean ice for all seasons (Kwok et al. 1995), although summer ice edge kinetics are still problematic. The RGPS will convert Lagrangian ice motion data into derived fields of ice age and thickness (for first-year ice) for the entire Arctic Basin; this aggressive program is the first use of satellite radar data to obtain a large-scale geophysical product on a routine basis.

The attempts so far to resolve ice motion in the Southern Ocean have been restricted to special periods and locations, due both to tracking algorithm challenges and to data availability. These constraints are also being removed for the Southern Ocean, however, and in coming years it is likely that an Antarctic RGPS (ARGPS) will be implemented. The ice velocity and deformation environment in the Southern Ocean are appreciably different from those in the Arctic, however, and the required data products for ARGPS will be different from those of RGPS.

The ice-covered seas are locations of operations including shipping, fishing, and petroleum exploration and production (see, e.g., Mulherin et al. 1996). In all there areas ice is a key impediment and often a hazard. To operate safely and efficiently in these regions it is important to have information on ice conditions in the immediate area of operations, several-day ice forecasts for a larger region, and general forecasts for basin-wide areas. This information is best acquired through data gathering and modeling of ice and ocean-atmosphere conditions, and SAR observations have proved quite useful in the identification of ice features that are of operational importance, such as ridges and leads, and for determinations of ice motion, which can lead to dramatic changes of conditions. ASF supports operations in ice-covered seas through the generation of "Quick-Look" data for NOAA and other users. At present this data support is limited to the region of the Alaska mask (see Fig. 1), but it will eventually be extended to include most of the ice-covered seas of the northern and southern hemispheres.

## 9.4 ASF Plans in Sea Ice Science and Operations Support

Sea ice observations continue to be a prime science objective of ASF. Among the recently named Applications, Development and Research teams are 13 sea ice and 2 lake ice projects (out of a total of 60 projects). In addition, operational requirements from the National Oceanic and Atmospheric Administration (NOAA) and NASA call for extensive SAR mapping of sea ice. While the relative importance of sea ice and polar ocean work has declined due to the growth of other SAR applications, the SAR applications to sea ice remain the central task of ASF, and this situation is not seen as likely to change in the near future. The highest priority at ASF is the support of sea ice data for science and operations.

The top long-term priority of ASF is obtaining mapping coverage of the Arctic Sea ice on approximately a weekly basis; this coverage is called the "Arctic Snapshot," as shown in Fig. 2. To produce a predictable data set, the Arctic Snapshot will be acquired four times every 6 days per 24-day RADARSAT orbit cycle (An orbit cycle is the time required for the orbit to repeat; for RADARSAT the orbit of day 1 is identical to that of day 25.) This pattern means that data takes for every 24-day period are identical. The Arctic Snapshot is a high-priority data product for ASF, but regular production must await the working out of data transfer details. It is hoped that the Arctic Snapshot will be acquired and archived by the start of calendar 1997 or sooner.

ASF currently plans to operate RGPS and archive both the input and product data streams after operation begins sometime in 1997 or 1998.

Marine operations support in the northern hemisphere consists largely of timely sea ice data from the Arctic Snapshot. In the part of the Southern Ocean that is in the McMurdo station mask as shown in Fig. 2, there is no commercial shipping at this time. However, research and support vessels are in these waters quite regularly, and these ships require good information on ice conditions for the safety of vessel and crew, and ASF anticipates that this support will become a part of its regular activities in future. The job of getting SAR data from McMurdo to ASF for processing and back to McMurdo, and to the ships plying the Ross Sea and neighboring waters, is substantial, and a start on that job has begun with the installation of high-speed satellite communications equipment at Black Island near McMurdo. The remainder of the communications linkages to supply SAR ice data to ships will require a little more time to develop.

## 9.5 Long-Term View and Future Instruments

Successful SAR-based research projects have been conducted with ERS-1 and -2 and JERS-1, other projects are in advanced planning for RADARSAT, and experienced SAR users are now beginning to anticipate the next generation of SAR flight instruments for future strengthening of their programs and data products. If international agreements allow, there will be significant benefits to the US in the utilization of the planned SAR data coverage of the continental US and EEZ and the polar regions. These benefits accrue in operations, development, science, and technology assessment. Areas of probable benefit include ship operations in lake, sea, and river ice, coastal processes and monitoring, flooding, agriculture, meteorology and severe weather, hydrology, wild-fire monitoring, fishery research and management, and allied areas.

Notable among the SAR flights of the future are the European ENVISAT satellite flying Advanced SAR in 1998 or 1999 and the Japanese ALOS flying in 2002; in each case the instruments flown have the same carrier wavelength as their antecedents – C-band for ESA and L-band for NASDA – with augmented capabilities. These SAR systems of the future offer valuable opportunities for US SAR users in science, applications, and operations. Consequently a national review is required of the purposes to which these future data streams could be put, the requirements imposed by these purposes, the technological maturity of the user communities, the cost implications in providing data products, and a general approach to meeting user needs in a timely and cost-effective means. ASF would support such a review.

The ESA instrument, Advanced SAR (ASAR) on ENVISAT, will fly soon. ASAR is a C-band SAR with variable incidence angle, selectable polarization, and an intriguing wide-swath, low-resolution mode which is planned to be operational whenever no other mode is operational. That is, ENVISAT ASAR will always be "on," and coverage of a survey nature essentially routinely available, analogous to the data stream from the NOAA Polar Orbiter series. For either scientific or operational sea ice data needs, ENVISAT coverage becomes a highly plannable resource. ASAR is expected to be of greatest interest in oceanography and sea and lake ice observations; the use of these data in coastal research and operations is particularly promising.

The ALOS now under development by NASDA carries a fully polarimetric L-band SAR called VSAR (see Table 1). Its longer wavelength will result in applications involving volume scattering, and its polarimetric data will provide an excellent description of ice type and possibly even ice thickness of first-year ice (Drinkwater 1995). Hence land processes will be the key application of ALOS, but sea ice application will be included.

In the immediate future ASF will participate in higher-level data products in two ways: in RGPS operation and in the support of user software. It is the objective of ASF to enlarge that portfolio in ways consistent with the overall ASF mission and vision, and to support higher-level data production at other institutions whenever possible and appropriate, e.g., by supplying SAR images to the RADARSAT Antarctic Mapping Project at Ohio State University.

ASF and its associated McMurdo reception station are poised to serve the sea ice science and operations communities with SAR image and higher-level data products for the foreseeable future. The use of the data products in the past has been directed at some of the key problems in polar ocean surface process determinations by researchers across the US and in other countries. ASF is a national resource in this area; the science and operations communities should continue to utilize its capabilities and place demands on its resources.

## References

Carsey FD (ed) Microwave remote sensing of sea ice, Geophys Monogr 68, American Geophysical Union, Washington, DC

Drinkwater M (1995) Applications of SAR measurements in ocean-ice-atmosphere interaction studies. In: Ikeda M, Dobson G (eds) Oceanic applications of remote sensing. CRC Press, Boca Raton, pp 381–396

Gogineni S, Moore R, Grenfell T, Barber D, Digby S, Drinkwater M (1992) The effects of freeze-up and melt processes on microwave signatures. In: Carsey FD (ed) Microwave remote sensing of sea ice, Geophys Monogr 68

Holt B, Digby S (1985) Processes and imagery of first-year ice during the melt season. J Geophys Res 90: 5045–5062

Holt B, Rothrock D, Kwok R (1992) Determination of sea ice motion from satellite images. In: Carsey FD (ed) Microwave remote sensing of sea ice, Geophys Monogr 68, pp 343–354

Jezek K, Carsey F (eds) (1991) McMurdo SAR Facility. Report of the Ad Hoc Science Working Team, BPRC Tech Report 91-01, Ohio State University

Kwok R, Cunningham G, Holt B (1992) An approach to identification of sea ice types from spaceborne SAR data. In: Carsey FD (ed) Microwave remote sensing of sea ice, Geophys Monogr 68, pp 355–360

Kwok R, Rothrock D, Stern H, Cunningham G (1995) Determination of the age distribution of sea ice from Lagrangian observations of ice motion. IEEE Trans Geosci Remote Sens 33:392–400

Mulherin N, Eppler D, Proshutinsky T, Proshutinsky A, Farmer D, Smith O (1996) Development and results of a northern sea route transit model. CRREL Rep 96-5, USA CRREL, Hanover, New Hampshire

Stern H, Rothrock D, Kwok R (1995) Open water production in Arctic sea ice: satellite measurements and model parameterizations. J Geophys Res 100: 20601–20612

Winebrenner D, Nelson E, Colony R, West R (1994) Observation of melt onset on multiyear Arctic sea ice using the ERS-1 Synthetic Aperture Radar. J Geophys Rev 99: 22425–22441

World Meteorological Organization (WMO) (1970) WMO sea-ice nomenclature. World Meteorological Organization Rep No 259, Geneva

CHAPTER 10

# Polar SAR Data for Operational Sea Ice Mapping

C. Bertoia, J. Falkingham, F. Fetterer

## Contents

## 10.1 Introduction

Many polar nations, including Canada, the United States, Russia, and the Baltic countries, support shipping and other maritime activities in ice-encumbered waters. Visual reconnaissance from aircraft was initially used to map sea ice distribution but this method suffered from extremely limited range and subjective interpretation techniques. Imaging radar became available in the early 1970s when real aperture side-looking airborne radars (SLAR) were deployed to complement and extend the range of traditional visual observations. Canada was the first nation to routinely deploy synthetic aperture radar (SAR) for operational ice surveillance when it commissioned an X-Band SAR on a high-flying jet aircraft. The European Space Agency's ERS-1 was the first satellite to provide SAR data on a sufficiently regular basis for operational ice surveillance. Canada's RADARSAT was specifically designed with operational ice surveillance as one of its major objectives.

Within the operational community to date, SAR has mainly been used as an imaging tool. The images created from the SAR data have been interpreted manually by human analysts to create maps portraying the distribution of sea ice concentration, type and floe size. The data volumes that will be provided by RADARSAT, however, will preclude sole reliance on manual techniques. Close cooperation with the science community has led to the development of several ice classification and ice motion algorithms suitable for operational implementation.

## 10.2 Operational Requirements for Ice Information

The presence of sea ice affects virtually every aspect of Arctic operations. Since the earliest days of Arctic exploration, sea ice has affected the ability of a ship to easily reach its destination. A captain's ice product requirements vary with the ship's ice operating capabilities or limitations. Though accurate spatial depiction of ice conditions is important to the mariner, temporal accuracy is of much greater importance. Generally, a ship's captain requires interpreted rather than raw images, since expert ice interpreters are found at the various national ice centers and generally not aboard ships. The challenge to the ice centers is to process the raw data, interpret it, and disseminate ice products to their customers within 3–12 h of data acquisition. Captains of non-ice-strengthened vessels require ice edge and iceberg limit information in order to plan their routes to avoid all known ice. Captains of vessels with some degree of ice capability require information detailing ice thickness and its associated concentration distribution. Even the most capable icebreakers use information on leads and polynyas to choose the path of least resistance in order to achieve greatest fuel economy.

In order to effectively describe the sea ice cover, nations with Arctic interests have developed a set of common terminology to describe the nature of sea ice and its behavior. The first compendium of internationally accepted ice terminology and symbology was developed by representatives from the USA, USSR, Federal Republic of Germany, and Denmark. A much broader collection of terms was compiled as interest in Arctic operations and exploration increased; this set of revised and expanded terminology was adopted by the World Meteorological Organization in 1968. This terminology was compiled into a volume including illustrative sea ice photographs and was issued as a publication entitled *WMO Sea-ice Nomenclature* (Publication No 259) in 1970. It remains the accepted standard for ice mapping terminology and symbology today.

## 10.3 Ice Information Services Throughout the World

### 10.3.1 United States of America

Interest in developing accurate methods for tracking and forecasting of sea ice conditions in the US began in 1952. US Navy interests in the Arctic ice pack conditions were fueled by the need to establish and maintain distant early warning stations scattered across the high Arctic. In 1951, 30 ships of the 33 vessel group operating under the code name *Operation Bluejay* were damaged by harsh ice conditions while attempting to navigate along the west coast of Greenland to establish the distant early-warning station and air base at Thule, Greenland (McDowell 1990). The identification of new requirements for accurate sea ice observation and forecasting resulted in the 1952 establishment of an ice division at the US Naval Oceanographic Office. In 1958, the first ice mission under the North Pole was accomplished by the submarine *USS Nautilus*. A 1961 *USS Sargo* deployment resulted in a collision with an ice keel and damage to the submarine's sail (McDowell 1990). These events resulted in greater interest in developing capabilities to analyze ice conditions and led to the development of the Naval Polar Oceanography Center in 1972. After a merger with the National Oceanic and Atmospheric Administration (NOAA) in 1976, the organization became the Navy/NOAA Joint Ice Center. This center was tasked with providing global sea ice and Great Lakes ice data to military, government, foreign, and commercial customers (Benner and Bertoia 1992). The organizations were joined by the US Coast Guard in 1995 to form the US National Ice Center (NIC).

The NIC has global responsibility for providing ice information for all sea-ice-covered waters, and lake ice mapping of the Great Lakes. Ice analysis is accomplished in a digital workstation environment using data from polar-orbiting satellites, meteorological models, drifting buoys, aerial ice reconnaissance, and ship and shore station reports. The primary data source for ice analysis in the early years was observation from aircraft. In 1972, the launch of polar orbiting satellites by NOAA led to ice analysis derived from visible and infrared imagery. The NIC currently uses the sensors aboard both the NOAA TIROS-N and the Defense Meteorological Satellite Program (DMSP) platforms. Ice analysis from visible and infrared imagery is highly subjective; images have fairly coarse resolution (0.55–1.1 km) and are subject to lack of solar illumination during the polar night. Climatologically, extensive cloud cover obscures the ice pack nearly 80% of the time during the summer months (Benner and Bertoia 1992). These limitations have led to the increased reliance on sensors from the microwave portion of the electromagnetic spectrum.

The first sensor to provide day/night, weather independent, operationally available imaging of sea ice was the DMSP Electronically Scanning Microwave Radiometer (ESMR), which flew from 1973 to 1976. This sensor was replaced by the Scanning Multichannel Microwave Radiometer which flew from 1978 to 1987. The microwave radiometer currently in use at the NIC is the Special Sensor Microwave Imager (SSM/I). Data from the SSM/I sensor are routinely processed for the NIC at NOAA's National Centers for Environmental Prediction (Grumbine 1996) using the National Aeronautic and Space Administration (NASA) Team ice concentration algorithm (Cavalieri 1992) and incorporating a weather filter (Gloersen and Cavalieri 1986).

Although they provide an all-weather sensing capability, the passive microwave sensors are of very coarse resolution; sea ice products are routinely produced with 25-km pixels. Despite this relatively poor resolution, NIC analysts have historically relied heavily on microwave data due to the extensive cloud cover in the Arctic, which is particularly prevalent in the marginal ice zones (MIZ). NIC analyses were based on microwave data for approximately 30% of the ice analysis chart information content in the summer and 65% in the winter for the period 1972–1983 (Knight 1984). This reliance on passive microwave was reduced to approximately 25% in winter and 35% in summer in the 1990s, when the NIC began to receive digital visible and infrared Operational Linescan System (OLS) data.

The NIC began its investigation into improved capabilities in the microwave band with the initial receipt of SAR imagery from the European Space Agency's (ESA) Earth Remote Sensing satellite (ERS-1) in March of 1992. The NIC routinely receives imagery from three station masks to provide nearly complete Arctic coverage. Images are processed at each of the three Arctic ground stations within 6 h of receipt and transferred to the NIC via the Internet. The Alaska SAR Facility (ASF) processes images at 30-m full resolution and 240-m low resolution; the low resolution products are preferred at the NIC and are simple 8x8 averages of the full-resolution images. The quick-turnaround products from the Tromsø, Norway, and Gatineau, Canada station masks are processed to ESA fast-delivery product standards and received at the NIC at 100-m resolution.

The original SAR workstation hosted at the NIC was derived from the system designed by the Jet Propulsion Laboratory for the Alaska SAR Facility (Kwok et al. 1992). Adaptations to the system, including interactive graphics capabilities required to produce ice products, were developed by the Naval Research Laboratory (NRL), Remote Sensing Applications Branch, at the Stennis Space Center, Mississippi. The ice classification functions allowed operator-supervised or automated labeling of ice types for winter SAR images in the high Arctic. The automated classification algorithm used backscatter-based lookup tables to label four ice types: multiyear ice, two categories of first-year ice, and new ice/open water.

The NIC currently hosts all satellite imagery and digitized ice reconnaissance information on SUN SPARC and Ultra workstations using the NRL Satellite Image Processing System (NSIPS). NSIPS is a PV-WAVE-based image processing package that enables display and manipulation of the suite of satellite data sources required to produce an ice analysis. The NIC uses NSIPS to warp all images to a common polar stereographic projection; images and associated graphics retain latitude and longitude information (Fetterer et al. 1993). DMSP OLS images and SSM/I-derived ice products are analyzed in conjunction with ERS-1/-2 SAR and NOAA Advanced Very High Resolution Radiometer (AVHRR) visible and infrared images. RADARSAT images from Gatineau, Prince Albert, Tromsø, and the ASF are ingested and geocoded into NSIPS. Lines drawn on these images are transferred to a base map where GIF files are created for distribution via the World Wide Web. Analysis graphics files are reformatted into various standard ASCII message files and are sent into ArcInfo, a Geographic Information System

(GIS) software package, where vector and raster ice products are produced. During the 1995–96 Great Lakes ice season, the NIC produced the first completely digital ice products using this production methodology. Work has since expanded to include completely digital analysis of the Arctic (Fig. 1).

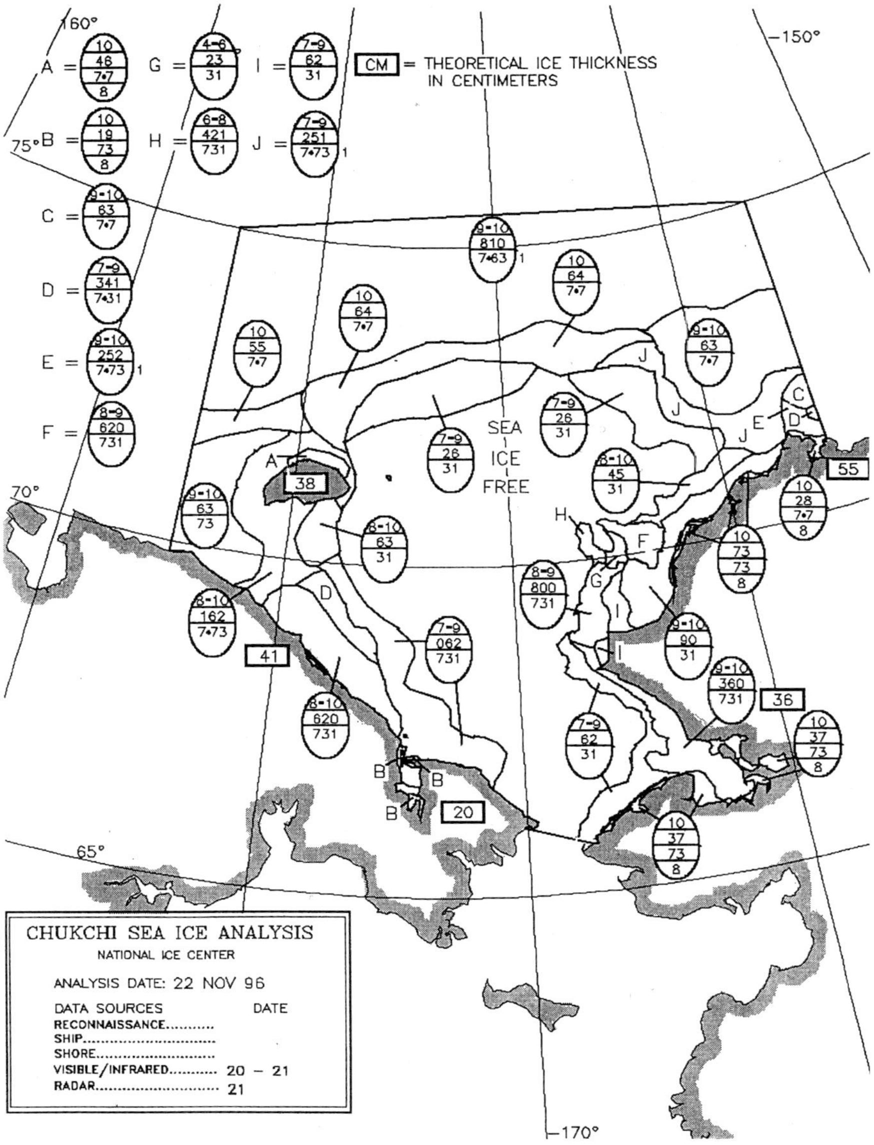


**Fig. 1.** Chukchi Sea ice analysis chart produced at the US National Ice Center using AVHRR, OLS and RADARSAT imagery and aerial reconnaissance. (Available at http://www.natice.noaa.gov)

### 10.3.2 Canada

Canadian ice reconnaissance can trace its roots back to the 1920s when aircraft first flew over the ice-covered waters of the Arctic. The Royal Canadian Air Force maintained responsibility for monitoring ice conditions over the next 25 years, occasionally working in conjunction with the US Navy and the Canadian Department of Transport. In 1954, the responsibility for ice services in Canada was transferred to the Meteorological Branch of the Department of Transport. This organization oversaw the ice services required for the development of ports, trade, fishing and other resources in areas throughout the Canadian Arctic, ranging from the St. Lawrence River to Nova Scotia and Newfoundland. Ice reconnaissance services were also provided for Arctic probes such as those by the tanker/icebreaker *SS Manhattan* in 1969/1970 and *CCGS Louis St. Laurent* in 1972. When the Meteorological Branch was moved into the Department of the Environment at the Atmospheric Environment Service in 1971, the Canadian Ice Service's main data source continued to be aerial ice reconnaissance. DC-4s gave way to Lockheed L-188 Electras in the 1970s. In 1972, the ice service awarded a long-term contract for the service of two Electra L-188-C aircraft dedicated to ice reconnaissance missions. A new technological era began with the installation of the first side looking airborne radar (SLAR) sensor aboard the Electra C-GNDZ. By 1989, the Electras had been retired in favor of a de Havilland Dash-7 and a Challenger jet. In 1995, the Challenger was released from service as satellite data became sufficiently reliable.

The Canadian Ice Service (CIS) considers its area of geographic responsibility to include the Great Lakes, the St. Lawrence River to the Gulf of St. Lawrence, the eastern Canadian seaboard, Hudson Bay, and the Canadian Arctic Archipelago from western Greenland to the Beaufort Sea. Ice products focus on areas where ships are operating, in order to support safe and efficient maritime operations in sea ice. Thus, winter areas of support are found along the Canadian east coast and the Great Lakes. During the summer, focus shifts to the high Arctic regions (Fig. 2).

Radar has been used for operational ice reconnaissance since the early 1960s when search radars were installed on ice reconnaissance aircraft of the Canadian Ice Service. These radars were forward-looking, nonimaging, sweep radars. While these radars had limited capability to determine ice parameters, they were useful for the accurate measurement of range and bearing to the ice edge. The opportunity to employ an imaging radar was presented when the US Coast Guard offered to loan a Motorola APS-94D SLAR, recently returned from military use in Vietnam, to Canada in 1978. With an extensive airborne ice reconnaissance program already in place, Canada was seeking a means to extend the range of visual observations in order to make this type of reconnaissance more cost-effective.

The APS-94 SLAR boasted a ground range of 100 km and an azimuth resolution of 37 m. Because it was a real-aperture system, the range resolution increased from 37 m at the near edge of the swath to about 400 m at the far range. Operating in X-band with very shallow incidence angles, the SLAR was so successful in mapping ice conditions that a second unit was added in 1980. In 1984, the Motorola SLAR was replaced by the CAL-200 SLAR, an instrument that remains in operation in 1996.

Following on the heels of an extensive research program at the Canada Centre for Remote Sensing, airborne SAR was introduced for ice reconnaissance in Canada in the

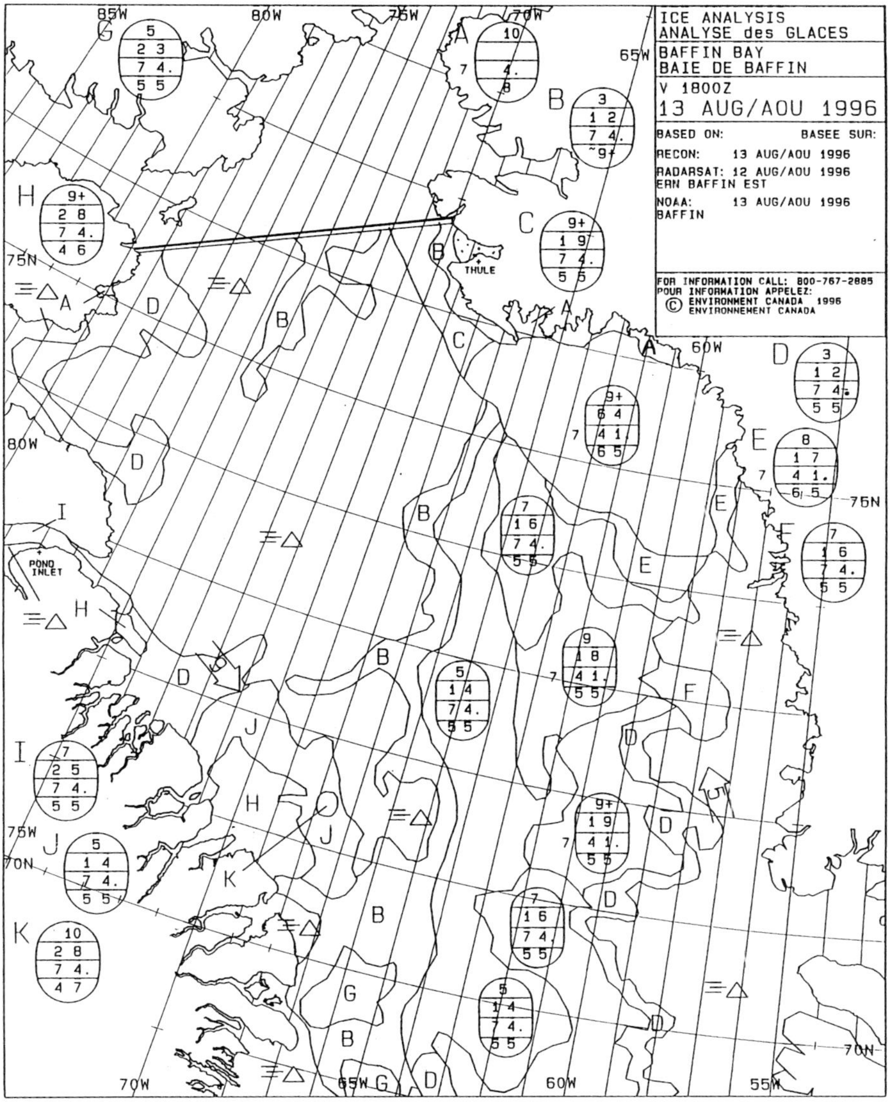


**Fig. 2.** Canadian Ice Analysis Chart. This chart shows the ice conditions in Baffin Bay on August 13, 1996, as analyzed by the Canadian Ice Service based on aircraft reconnaissance, RADARSAT and NOAA AVHRR imagery. The symbols conform to the WMO ice charting standard. (Courtesy Canadian Ice Service, Environment Canada)

mid-1980s. Initial use was for site-specific work by commercial companies working in the Beaufort Sea. The major advantages of SAR over SLAR were that the resolution did not deteriorate with distance from the radar and that the radiometric balance across the range of the radar imagery could be closely controlled, providing uniform gray-scales. The early SARs suffered from very narrow swath widths (on the order of 50 km) but by the end of the 1980s, 100-km swath SARs were developed. This allowed them to become practical for broad scale strategic ice mapping. In 1989, Canada introduced a routine airborne SAR ice reconnaissance service using a pair of X-band SARs mounted on a Challenger jet aircraft. Each radar had a 100-km-wide swath with 25-m resolution. They were mounted on the aircraft to look at opposite sides of the flight track, effectively providing a 200 km wide swath. The aircraft normally flew at an altitude of 35000 feet, well above any inclement weather, to provide an extremely reliable, stable platform. The imagery from this system is still used as a benchmark against which newer imagery sources are compared. Unfortunately, escalating costs and cheaper alternatives forced the termination of this system on March 31, 1995.

After a lengthy development that had its roots in the SeaSat program, Canada launched RADARSAT on November 4, 1995. It is the world's first radar satellite specifically designed for sea ice monitoring. Of primary significance for this role is its ScanSAR mode, which provides a 500-km-wide swath with 100-m resolution. This wide swath provides a high repeat imaging capability, essential for operational sea ice monitoring. By virtue of this wide swath, RADARSAT can image every point on the earth's surface north of 65°N latitude at least once each day. North of 45°N, the entire globe can be imaged in 3 days or less. Imagery on demand, where and when required, is a basic requirement for supporting operational activities where timing and location are dictated by events, whether economic, military, or environmental, far beyond the control of the monitoring manager.

The CIS maintains a communications network that allows RADARSAT SAR data acquired at both the Gatineau and Prince Albert station masks to be processed, geocoded, and delivered to a digital workstation environment called Ice Services Integrated System (ISIS). This system merges ERDAS image processing software with ArcInfo based GIS software. The ISIS software allows analysts to merge RADARSAT SAR with aircraft SAR observations as well as NOAA AVHRR, ERS-2 SAR and DMSP SSM/I imagery to produce complete ice charts. Ice charts and SAR images are transmitted to Coast Guard icebreakers on demand using a digital satellite communications system that employs significant data compression to keep data transfer times within reason. Commercial ships receive ice charts via scheduled radio facsimile broadcasts as well as on demand via Inmarsat facsimile.

### 10.3.3 Baltic Nations

Finland's ice service began in the 1890s but was not considered operational until World War I, when ice charting began in order to secure marine transportation into Finland's many ice-covered ports. During the November to May Baltic ice season, Finnish harbors average some 20000 port calls, of which approximately 5000 require icebreaker assistance.

The Finnish Ice Service operates under the Finnish Institute of Marine Research in Helsinki. Daily ice charts are issued by facsimile and radio broadcast during the severe portion of the ice season. Satellite data sources include NOAA AVHRR and ERS-1/-2 SAR images, which are supplemented with weekly visual aerial reconnaissance flights from October through January. Observations are also received from 30 fixed ice thickness stations, icebreaker-based helicopters, and merchant vessels.

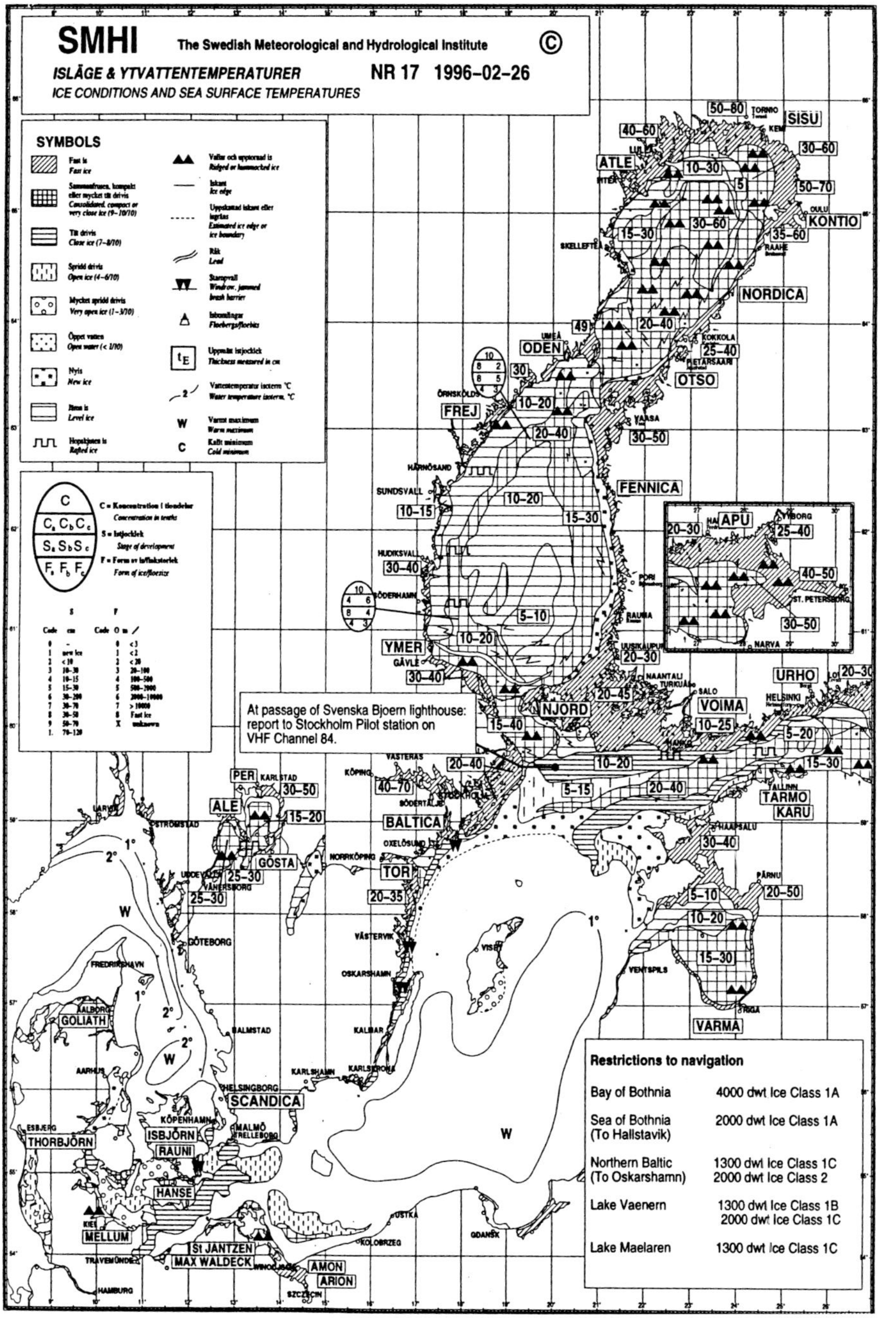


**Fig. 3.** Swedish Ice Service chart depicting ice conditions in the Baltic Sea at maximum ice extent for the 1995/1996 ice season. (Courtersy Swedish Ice Service)

The Swedish Meteorological and Hydrological Institute has issued ice reports in text and ice code since 1928. Ice information is required by a Swedish national policy to keep all Swedish harbors open throughout the year. Ice support products (Fig. 3) are based primarily on NOAA AVHRR imagery, but are occasionally supplemented by ERS-1/-2 SAR, aerial reconnaissance, and ground based observations.

The German Ice Service is based in Hamburg and operates under the auspices of the Federal Maritime and Hydrographic Agency. Its roots are over 100 years old, dating back to a governmental mandate to the Deutsche Seewarte requiring that seafarers be informed of ice conditions and their effect on navigation conditions in German harbors and coastal waters. Since 1927, a network of ice observers has reported local ice conditions in the international Baltic Sea Ice Code. After World War II, responsibility for sea ice services shifted to the German Maritime Institute, which was replaced in 1946 by the German Hydrographic Institute. The Hydrographic Institute was renamed Bundesamt für Seeschiffahrt und Hydrographie (BSH) in 1990. In 1991, BSH assumed responsibility for the region of the former German Democratic Republic, creating one national German Ice Service. The BSH maintains a digital ice chart archive for the period from 1960 to the present in addition to providing near-real-time ice charts covering the German coastal areas and the entire Baltic Sea. Upon request, German vessels may obtain special reports and forecasts of global ice conditions.

Ice analysis software developed at the Finnish Technical Research Centre was hosted on HP 9000 workstations at the three Baltic Ice Service Centers (Finnish, Swedish, and German) in the winter of 1992/93. This versatile image processing workstation allows analysis of NOAA and ERS satellite images in conjunction with climatological ice charts for the production of a new standard ice product. Ice algorithms for SAR interpretation include an ice/no ice discriminator; an algorithm depicting levels of ice roughness was recently added.

### 10.3.4 Japan

Ice observation in Japan began in January of 1892 at local observatories in Abashiri and Nemuro on the island of Hokkaido. With the addition of satellite data in 1966 the first sea ice charts were issued by the Japan Meteorological Agency. Severe ice conditions in the southern Sea of Okhotsk crushed a fleet of fishing vessels in the Kuril Islands in 1970, initiating the routine issue of sea ice information by radio facsimile. Ice products (Fig. 4) are currently produced under the sponsorship of the Japan Meteorological Agency using manual analysis techniques on PC and Unix-based workstations. Images from the Geostationary Meteorological Satellite (GMS), NOAA, Marine Observation Satellites (MOS-1), LANDSAT, and DMSP satellites are supplemented by visual observations from ship and aircraft and shore-based millimeter band radars along the north coast of Hokkaido.

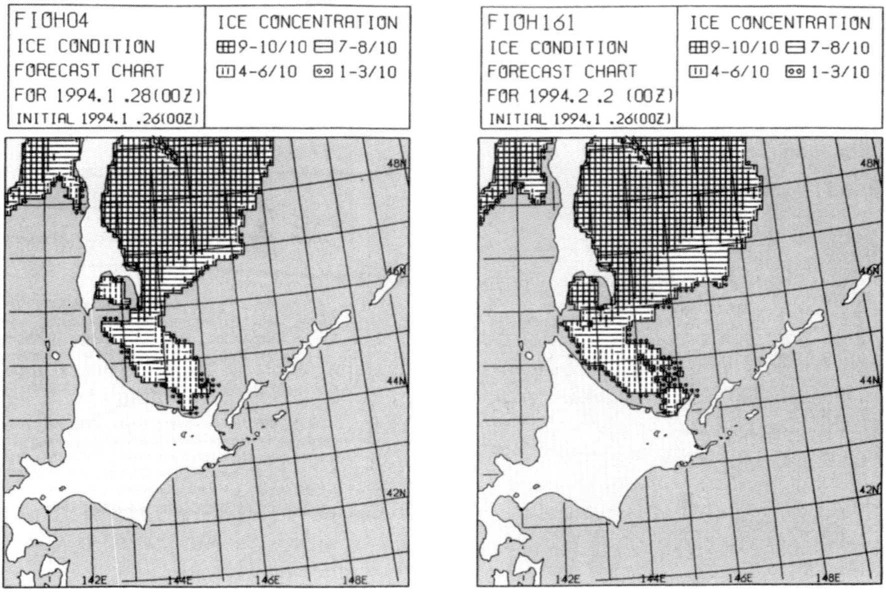


**Fig. 4.** Numerical sea ice forecast chart issued by the Japanese Meteorological Agency. Two-day (*left*) and 7-day (*right*) forecasts. (Courtesy Japanese Meteorological Agency)

## 10.3.5 Russia and Former Soviet Republics

The Russian Ice Service operates under the Russian State Service for Hydrometeorology and Monitoring of the Natural Environment. It has been in existence since 1932. The ice service provides support for navigation through the Northern Sea Route seas, including the Barents, Kara, Laptev, East Siberian and Chukchi Seas, as well as the seasonal ice in the White, Baltic, Bering, Okhotsk, Japan, Azov, Caspian and Antarctic Seas. In the early years, ice analysis was accomplished primarily through the use of visual and SLAR data obtained by aerial ice reconnaissance. Routine aerial surveillance was discontinued in 1992, and is currently conducted only in support of specific operations and expeditions.

Present day analysis is accomplished using a number of satellite sources. These data are combined on PC-based workstations with aerial reconnaissance data as well as visual observations from ship and shore stations. Satellite receiving stations at regional centers in Moscow, Novosibirsk and Khabarovsk receive NOAA AVHRR and Meteor visible and infrared images and OKEAN radar images in high resolution picture transmission (HRPT) mode. Eight areal centers (Archangelsk, Murmansk, Amderma, Dickson, Tiksi, Pevek, Magadan, Yuzhno-Sakhalinsk and Petropavlovsk-Kamchatski) receive and analyze imagery in the lower resolution automatic picture transmission (APT ) mode. Algorithms developed at the Arctic and Antarctic Research Institute

(AARI) are used to automatically delineate ice concentration and ice floe size from visible imagery and to estimate ice thickness from wintertime infrared imagery. ArcInfo GIS software is used to assimilate the data sources and produce the digital ice products. Composite ice charts and associated ice forecasts are issued every 7–10 days.

## 10.4 Radar Monitoring of Ice

### 10.4.1 Ice Reconnaissance: Aircraft SAR and Beyond

Radar has been used for operational ice reconnaissance since the early 1960s, when search radars were installed on ice reconnaissance aircraft of the Canadian Ice Service. In 1978, the United States introduced the first radar images from space when it launched SEASAT, carrying an L-band SAR. Follow-on satellite SAR programs that were utilized by operational ice centers included the European Space Agency's ERS-1 C-band SAR and the Japanese L-band JERS-1. After a lengthy development cycle, Canada launched RADARSAT, the world's first satellite specifically designed for sea ice monitoring. This satellite, and its follow-on clone, RADARSAT-2, with their wide swath and fine resolution, will be the backbone of operational sea ice monitoring work for the next decade.

### 10.4.2 Processing and Communications

The major distinction between "operational" programs and other uses of remote sensing (such as research) is the requirement for timely delivery of data for operational decision making. This decision-making process is often time-critical and must consider many disparate factors. For example, as a ship enters an ice-bound passage, the captain must decide immediately which route will be least hazardous. With the existing potential that advected ice may damage rig equipment, the offshore drilling rig manager must decide when to begin the lengthy process of hauling up a thousand meters of drill pipe to move off-site. Ice is constantly in motion and even a few hours can dramatically change its distribution in a local area. It is critical that the delay between imaging the ice, creating an analysis or forecast, and putting the information into the decision-makers' hands be short and within the timeframe that decisions are made. Remote sensing provides a tool that can help in the decision-making process, but it must perform to demanding time standards.

In the early days of imaging radar use for ice reconnaissance, radars produced images on plastic film aboard the aircraft. In order to provide the information to decision-makers in a reasonable period of time, experts on board the aircraft analyzed the film and plotted their analysis of the ice conditions on paper charts. These charts were transmitted by radio facsimile to users operating in the ice below. This process resulted in a tremendous loss of information since the interpretation could not convey all of the information gathered by the radar. In a few critical situations, the aircraft actually dropped radar film onto the ice in the vicinity of a ship so that the navigators could retrieve the film and the detail imaged by the radar. However, open cracks in the ice, poor visibility, and polar bears, made this procedure much too risky to be employed routinely. Obviously, this data-download technique was not a procedure extensible to space-borne radar, so the development of high speed data downlinks became essential for space-based radar remote sensing of ice in the operational environment.

Canada deployed radar data downlinks in conjunction with its SLAR in 1983, and the Soviet Union soon followed suit. Receivers and film printers on board icebreakers allowed navigators to view the radar image in real time as the aircraft flew along its projected track. By the end of the decade, SLAR and SAR imagery were routinely downlinked to computer display terminals on Canadian icebreakers, and an elaborate network of ground stations with satellite communications had been established. This Ice Reconnaissance Data Network (Falkingham 1991) transmitted SLAR and SAR imagery from ice reconnaissance aircraft flying over the Canadian Arctic or east coast to the Ice Center in Ottawa. Analysts then transferred the essential ice information onto ice analysis charts for broadcast within a few hours.

Operational use of satellite SAR presented yet another challenge: how to get satellite data processed into image form quickly enough for use by decisionmakers. When Canada's SAR aircraft was introduced in 1989, the SAR processor was just fast enough to keep up with the 600 km/h speed of the aircraft, which allowed image data to be broadcast directly from the aircraft. Processing SAR data in realtime at the 20000+ km/h speed of satellites was beyond onboard capabilities, so it was necessary to do the processing on the ground sometime after it was collected by the spacecraft. For example, ERS-1 transmitted "raw" data to a worldwide network of ground receiving and processing stations.

Fast-delivery data processors were installed in the ground stations at Fairbanks (Alaska), Gatineau (Canada) and Tromsø (Norway) specifically to provide imagery for the demanding operational ice reconnaissance requirements. Tromsø has been the most successful at meeting the time constraints, often providing ERS-1 images to icebreakers in the Baltic Sea within 2 h. Tromsø provides a similar capability for RADARSAT data.

The RADARSAT ground processing system in Canada was specifically designed to provide fast turnaround of SAR data for operational ice monitoring. A maximum delay of 6 h was specified for receipt of imagery anywhere within the Canadian economic zone, a vast area spanning 100 degrees of longitude and 50 degrees of latitude. Two ground stations in eastern and western Canada feed raw data to a single processor at Gatineau via high-speed satellite communications links. Another high-speed link delivers images from the processor to the Canadian Ice Center in Ottawa (Fig. 5). Commercial marine communications satellites deliver images and ice analyses from the Ice Center to ships anywhere in the world. At the time of this writing, images are usually received at the Ice Center within 1-2 h and are made available to the ship communication system within another hour, exceeding the specified time requirements. Depending on the size of the image and the resolution required, delivery to the ship can be accomplished within minutes.

The United States has developed a similar capability for its RADARSAT receiving station at the Alaska SAR Facility in Fairbanks. After data have been fast-turnaround-processed to image form, they are transmitted via a dedicated communication link to the US National Ice Center in Washington, DC for analysis of ice conditions. In the true spirit of international cooperation, Ottawa and Washington regularly exchange RADARSAT data across a dedicated communication link (Fig. 6).

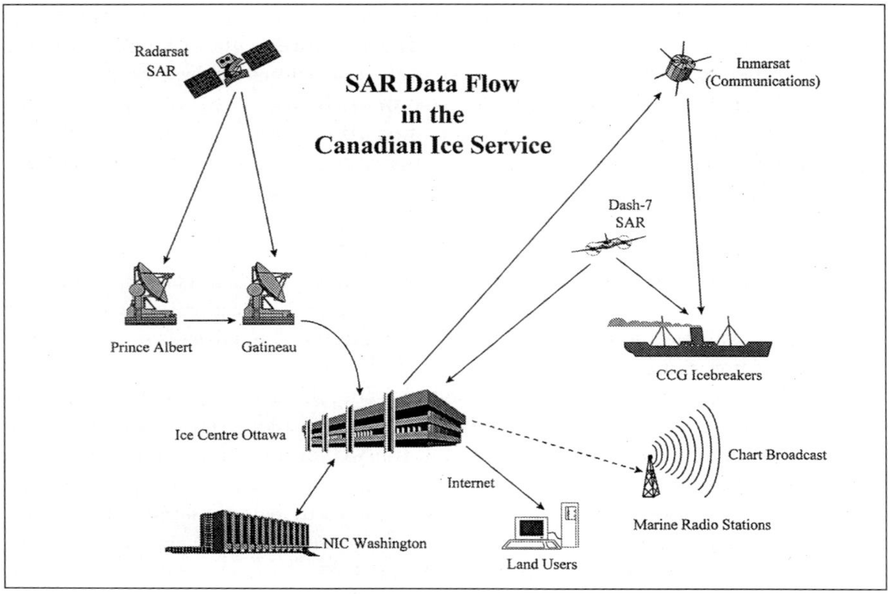


**Fig. 5.** Communication pathways for SAR data into the Canadian Ice Service. SAR data originates from RADARSAT and from the Dash-7 ice reconnaissance aircraft. *Solid arrows* depict the flow of SAR data (both raw and image forms). The *dashed arrow* depicts the flow of interpreted information in the form of ice charts. (Courtersy Canadian Ice Service, Environment Canada)

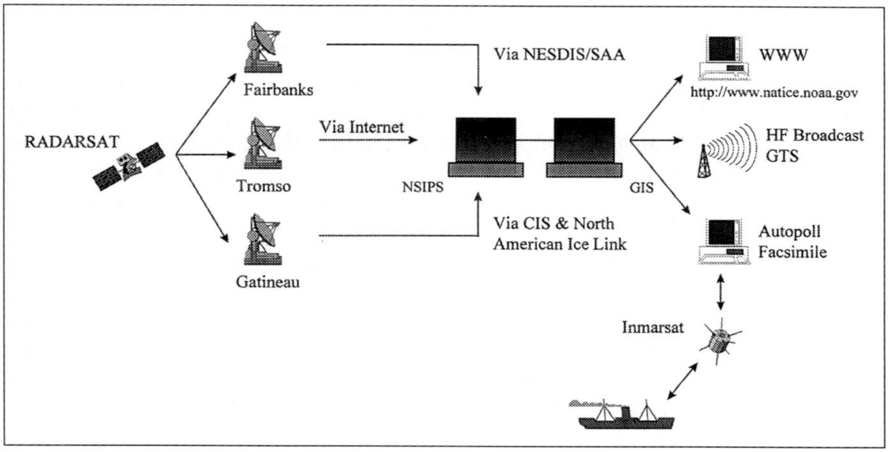


**Fig. 6.** Communication pathways for RADARSAT from the three Arctic ground stations into the US National Ice Center. Block averaged ASF images are received using a T1 link to NOAA's National Environmental Satellite Data and Information Service's (NESDIS) Satellite Active Archive in Suitland, Maryland, then are transferred to the NIC via a wide area network. Block averaged and compressed Tromsø images are transferred via Internet. Gatineau SAR is transferred to the NIC from the Canadian Ice Service over the North American Ice Link, a dedicated communications link designed for the transmission of sea ice information. (Courtersy National Ice Center)

With the rapid and continuous advances in the speed of signal processors, future generations of satellites may feature onboard, real-time processing of SAR data with the potential capacity for direct broadcast to users. This would allow the satellite to simulate aircraft ice reconnaissance broadcasts, providing image displays on ships and offshore platforms in real-time as the satellite passes overhead. The technological advances required to permit this are substantial and they are not expected to be realized before the year 2010.

## 10.5 Manual Analysis Techniques

SAR imagery is the only high-resolution data source capable of penetrating the perpetual cloud-cover conditions and illumination limitations unique to polar regions. Operational demonstrations using SAR imagery from the ERS-1/-2 satellites at the US, Canadian, and Baltic Ice Centers (Bertoia and Carrieres 1994; Ramsay et al. 1994; Herland and Berglund 1995) have proven the feasibility of using SAR images operationally for the creation of ice classification, concentration, and motion products. The techniques developed have led to improvements in a wide variety of ice information and decision-making products.

Operational support requirements place the most emphasis on defining conditions in the MIZ. The geophysical parameters that have been deemed to be most significant to the operational sea ice community are ice concentration, ice type, age and thickness, lead size and orientation, floe size, topography, ice drift, melt state, and convergence/divergence conditions. Interpretation of SAR imagery can provide information describing most of these parameters. Operational experience, however, has proven that SAR imagery alone is of much less value than when it is used in conjunction with other data sources and/or ground truth observations.

Noise associated with SAR imaging is of concern to the ice analyst. Every SAR has a noise floor or noise level associated with the instrument's detection capability. Any received signal which falls below this noise floor cannot be distinguished from the background signal. This can be significant for low return targets such as water and new ice. The ERS-1 noise floor is sufficiently high to allow differentiation between new ice and smooth open water, although early results from RADARSAT have shown this may not be true in the far range of the ScanSAR wide images.

Speckle noise is an inevitable byproduct of image formation in a coherent radar system (see Livingstone 1994). Speckle causes a grainy appearance in SAR images that can make image interpretation difficult. Spatial averaging is the simplest method for reducing image speckle and for this reason the operational ice centers generally use SAR image products at less than the maximum resolution possible for the sensor. For example, the Alaska SAR Facility produces its ERS quick turn around products at 240-m resolution, acquired through simple 8×8 averaging of the 30-m full resolution product. The ESA Fast Delivery Product produced by Norway's Tromsø Satellite Station has a 100-m resolution, as does the product prepared by Canada's Gatineau Station.

Calibration is the process of removing radar sensor-dependent parameters from the radar images so that the digital number of the image intensity accurately represents the imaged surface features. In a properly calibrated system, the same surface feature will consistently return the same pixel value. For the purposes of qualitative visual interpretation of image features, absolute image calibration is not required. In fact, quantitative ice analysis from SAR is not be considered viable until the advent of the Intera STAR-1 radar in Canada. This system, the first that could be calibrated, began service as a sea ice imaging radar in 1986, but operational data communications procedures made absolute calibration, or even relative calibration, of this system infrequent (Livingstone 1994).

### 10.5.1
### Backscatter for Ice Classification

Although ice thickness is not directly derivable from single frequency/single polarization SAR, techniques have been developed that make classification of ice into stages of development possible. The stages of ice development have been defined into standard categories by the World Meteorological Organization based on ice thickness and ice age (Table 1). Sea ice backscatter at the common frequencies for ice analysis (C- and X-band) is a function of the physical properties and surface roughness at the ice/atmosphere interface.

At first glance, an X-band radar image appears very much like a black and white aerial photograph, and in the early years of radar ice analysis it was an easy parallel to draw. For example, open water appears dark while ice takes on various shades of gray and white. Experienced ice observers quickly noted problems with this analogy, however, most notably in the smooth fast ice along coastlines that appeared as black as open water. Knowing that fast ice was present, analysts were able to detect small texture changes in the radar imagery, such as hairline "cracks" which represent the return from ridge lines parallel to the coast. It is now understood that, because of the shallow incidence angle of the airborne radar and the X-band frequency, the smooth fast ice, like open water, reflected most of the radar energy away from the receiver.

**Table 1.** WMO sea ice terminology and symbology [a]

| Stage of Development | Symbol | Thickness | Comments |
|---|---|---|---|
| New | 1 | < 10 cm | Recently formed ice, includes frazil, grease, slush and shuga |
| Nilas | 2 | < 10 cm | May be subdivided into dark nilas (< 5 cm) and light nilas (5–10 cm) |
| Young | 3 | 10–30 cm | May be subdivided into gray and gray-white ice |
| Gray | 4 | 10–15 cm | Usually rafts under pressure |
| Gray-white | 5 | 15–30 cm | Likely to ridge under pressure |
| First-year | 6 | 30 cm– 2 m | Sea ice of not more than one winter's growth |
| First-year thin/white | 7 | 30–70 cm | |
| Medium first- year | 1· | 70–120 cm | |
| Thick first-year | 4· | > 120 cm | |
| Old | 7· | None defined | Sea ice which has survived at least one summer's melt, typical thickness up to 3 m |
| Second-year | 8· | None defined | Old ice which has survived only one summer's melt, typical thickness up to 2.5 m |

a After World Meteorological Organization (1970)

As experience with radar imagery for ice analysis has expanded, radar analysts have become adept at determining ice concentrations and types from subtle clues in the imagery. The strong return from wind-roughened open water that confused early ERS-1 analysts has become an analysis tool with the knowledge that new ice forming in the water dampens the small surface waves, reducing the radar return dramatically. The distinctive wispy shapes depicting ice have become well recognized, even when tone is most definitely non-ice-like.

### 10.5.2
### SAR Image Clues to Ice Stage of Development

New ice begins to form when sea water temperatures reach approximately -1.9 °C. Remote detection of new ice had been nearly impossible from visible and infrared sensors in use at the various ice centers prior to the advent of SAR. It has been proven that SAR is capable of detecting grease ice, the first form of new ice (Fig. 7). The formation

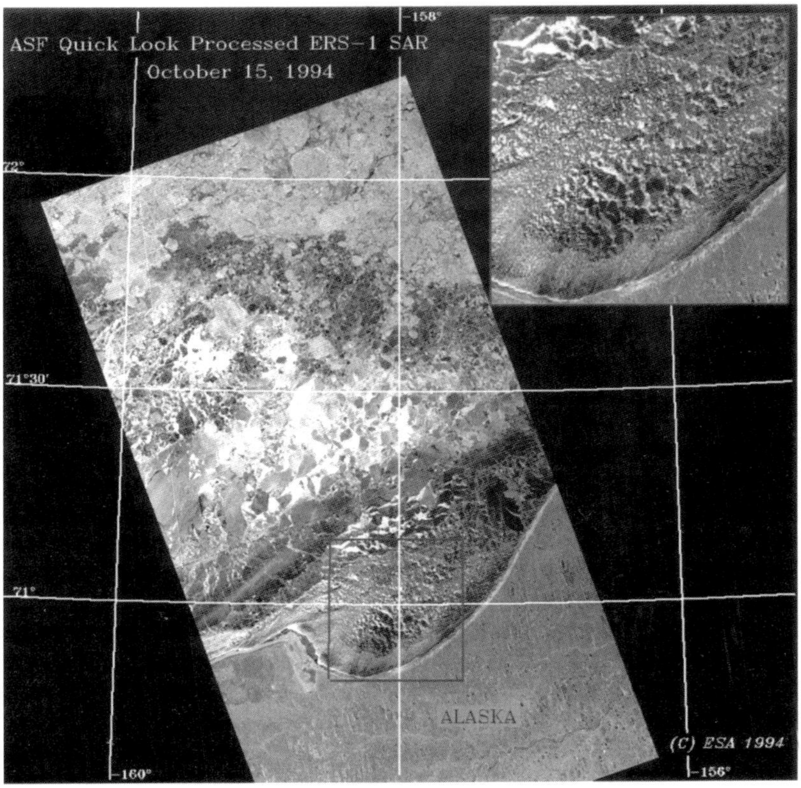


**Fig. 7.** ERS-1 SAR images (ASF image IDs: 154713Q0101, 154714Q0101) from October 15, 1994, showing new ice formation along the north slope of Alaska. The *bright* areas of radar return are the characteristic response of the ERS-1 SAR to high wind conditions. New and young ice flatten the water and result in areas of *dark* return. (ERS-1 imagery copyright European Space Agency)

of this soupy mixture of ice crystals on the ocean surface suppresses Bragg waves and reduces backscatter to a level sufficient to allow detection of this very thin ice (Wohl 1995).

As grease ice thickens into nilas, characteristic finger rafting and tearing patterns are visible; the relatively low resolution of the visible and infrared imagery at the ice centers had in the past masked these features. Nilas exhibits a tone between light and dark gray but has no clear ridge lines. The analyst must use the texture in the image in conjunction with tone to identify this ice type. With continued cold temperatures, new ice freezes into young ice (by definition 10–30 cm thick). Gray ice, at 10–15 cm, will continue to exhibit finger rafting, but it will retain small random convection holes, areas where the relatively warm water has prevented ice from forming. These small ice free areas are sometimes evident in SAR imagery as splotchy areas of dark return within the lighter-toned gray ice area. Gray-white (15–30 cm) ice is the first form of ice to exhibit ridging. Ridges will be evident as narrow radiometrically bright features, although gray-white ice floe forms will remain mostly undefined.

First-year ice is in the 30–200 cm thickness range; generally, undeformed ice thicker than 200 cm is not found. First-year ice can exhibit a significant range in brightness but is normally brighter in tone than new ice and darker than multiyear ice. The most characteristic feature of first-year ice is its distinctive floe shapes. First-year floes are angular and well defined. Ridge lines are bright and long and run completely through the floes. Floe perimeters can appear bright due to raised edges produced by inter-floe dynamic interaction. Extensive very bright rubble fields may be evident in areas which have undergone or are undergoing dynamic activity.

Multiyear ice, or old ice, is ice that has survived at least one melt season and normally exceeds a thickness of 3 m. Lower salinity and greater thickness make old ice the most dangerous ice type to shipping. Prior to the advent of SAR imagery, the only available source for distinguishing first-year and multiyear fractions within the ice pack were brightness temperatures derived from passive microwave. With active systems, the low salinity of multiyear ice allows microwave penetration into the ice; thus, return from this ice type is dominated by volume scattering. The brine pockets and bubbles characteristic of multi-year ice have dimensions in the 1–6 cm range, matching the scale of the C- and X-band radars now in use. The combination of these features makes a bright return characteristic of multiyear ice, particularly in winter conditions. Multiyear ice floes are distinctly rounded and have a mottled texture in SAR imagery due to the prevalent drainage pattern which develops during the summer melt season (Fig. 8). Multiyear floes are often conglomerate, composed of many smaller floes frozen together. Second-year ice (ice that has survived only one summer's melt) can sometimes be distinguished from older ice through identification of a curvilinear pattern of fractures.

The preceding discussion summarized typical ice features found under ideal conditions. In the real world of operational ice analysis, environmental conditions may significantly alter or mask expected SAR signatures from sea ice. SAR signatures are significantly affected by surface moisture, produced by either melting ice and snow or wet precipitation. Wind speeds over 10 cm/s can produce very bright returns from open water, sometimes masking lighter concentrations of ice. Experienced analysts will not usually confuse the bright return from windy open water with that of multiyear ice due to the textural differences between these two ice features.

**Fig. 8.** Mosaic of RADARSAT images covering the western portion of the Canadian Arctic during mid-February 1996. In midwinter, when the ice surface is cold and dry, differences between multiyear ice (*light*) and first year-ice (*dark*) are readily apparent. (Courtesy Canadian Ice Service, Environment Canada, and Canadian Space Agency)

### 10.5.3 Synergistic Use with Other Data Types or Information

Ice type is more accurately determined when viewed at more than one wavelength in the electromagnetic spectrum. More information is also gleaned from a time series than from a single observation. For this reason, operational ice centers attempt to combine data from various sources prior to labeling an ice map. There are two possible approaches to data fusion:

1. Analyze the visible, infrared, in situ observations, passive microwave, and SAR data separately, then mesh the geophysical labels.
2. View all sources concurrently, making decisions about ice characteristics with all sources on screen.

Despite the digital workstation environment present in today's ice centers, some data remain available only on paper (usually in situ observations from ships and aerial ice reconnaissance), requiring the centers to implement a combination of these two approaches.

Ice analysts are trained to start their analyses with the finest-resolution data available. Thus, in situ observations are considered first, followed by interpretation of SAR imagery and then visible and infrared imagery. Finally, passive microwave is used for all cloud covered regions not covered by SAR. The temperature and albedo of ice features derived from infrared and visible imagery is used to supplement the topography and floe shape information provided by SAR, giving analysts a better decision base for ice classification. For example, NIC ice analysts used a combination of DMSP OLS and ERS-1 SAR to characterize ice in the Beaufort Sea (Fig. 9).

The first challenge of this new digital environment at most ice centers is to map all satellite data to the same projection. This is accomplished by the ingest functions of the ice workstations. Rather than map all data to the same resolution, as would be required for automated data fusion techniques, the ice centers have chosen to map imagery to the optimal resolution for ice interpretation (in the case of SAR) or the maximum sensor resolution (in the case of visible and infrared imagery). The next step in the analysis process is to compensate for the temporal differences in images which have been acquired since the last analysis period, from many different sensors. This is accomplished using a combination of ice drift forecast/hindcast models and surface wind and temperature analyses. Time series analysis has proven extremely valuable in discriminating between clouds and ice (in the case of visible and infrared imagery) and between ice and windy open water (in the case of SAR imagery). The final step includes labeling ice features, compensating for resolution and geolocation accuracies for each instrument. At all stages, of course, ancillary information such as season, location, and oceanographic and meteorological conditions is incorporated in the ice analysis.

**Fig. 9.** NSIPS mosaic of ERS-1 SAR (174853Q0101–174861Q0101) and DMSP OLS (F8.95193.0145) imagery, June 12, 1995. The bright return characteristic of windy open water allows the analyst to easily identify open water (with no new ice present) in the visible OLS image. The multiyear floes are easily identified in both the SAR and visible image near the ice edge but become cloud- and surface-water-contaminated toward the inner ice pack. (ERS-1 imagery copyright European Space Agency)

## 10.6 Automated Ice Classification

Producing ice type and concentration estimates automatically from SAR data has been a longstanding goal of both the operational and scientific communities. Scientists desire a stable, long-term data record of ice type and concentration estimates with known accuracy in order to address questions concerning ice mass balance and heat flux. The most feasible way to analyze the volume of SAR imagery required for a climatology of ice characteristics is to use an automated algorithm.

Operational centers require automated algorithms for the same reason: the volume of SAR imagery received precludes manual analysis. At the same time, the high resolution of SAR and the distinctiveness of forms of ice (floe size, shape, and deformation) in SAR imagery give the analyst more operationally useful information than any other satellite sensor. In the US, the science community has benefited from the operational requirements for and Department of Defense (DoD) funding of SAR ice signature and automated image analysis research. In Canada, the Canadian Centre for Remote Sensing (CCRS) and CIS have directly funded ice type algorithm development work.

The operational community has requirements that the science community does not share. For instance, algorithms must run quickly (NIC anticipates receiving about 3 Gb RADARSAT imagery per day), and data cannot be analyzed retrospectively (although past images may aid in classifying a current image). While many studies have shown that multifrequency or multipolarization SAR data can improve classification (e.g., Drinkwater et al. 1992; Rignot and Drinkwater 1994; Collins and Livingstone 1996), the ice centers are not likely to have access to operationally useful multichannel data for decades. Another difference is that the ice centers are required to map ice features. Even if it could be produced in near-real time, the RGPS Arctic Snapshot product (Kwok, this volume) would be of limited use because it does not show the location of ice type features, providing only a distribution of ice type concentrations within 5-km-grid cells.

Operational algorithm development is focused on rapid segmentation of an image so that each pixel is accurately assigned to an ice type or to water. A much more difficult task is automatically extracting and labeling features such as floes, leads, and ridges. Extensive work in this area has been done by Banfield and Raftery (1989) and Vesecky et al. (1990). Data blending, or combining SAR with other satellite data and meteorological analyses to assess ice conditions, is another area which promises operationally useful results. As pointed out Sect. 10.5, analysts currently blend data subjectively and manually. This blending can potentially be performed automatically, using a formal framework such as an expert system or probabilistic inference (in which evidence given by data from different sources, or degree of belief in data, dictates the decision concerning the surface type present). Collins (1992) gives a summary of these and other methods of data blending for sea ice analysis. Currently, all satellite data and meteorological analyses are available as digital data on workstations at both CIS and NIC – the first step in operational data blending.

Beyond the problem of simply blending coincident ice information from different sources is the problem of optimally assimilating and interpolating information which is received at the ice centers irregularly in space and time and at resolutions ranging from 25 km for passive microwave data to 100 m for SAR to a point measurement for buoy data. A form of data blending in which ERS-1 estimates are used for "calibration" tie points for SSM/I ice type estimates was proposed by Fetterer et al. (1993). Further investigation showed that high variability in SSM/I multiyear ice signatures meant that the calibration could not be extended much beyond the specific area of the SAR pass. With the exception of Kalman filtering for blending buoy data with passive microwave data (Thomas and Rothrock 1993), data blending has yet to be demonstrated on an operationally viable scale.

Much of the ground-laying research on SAR ice signatures was performed using aircraft SAR and SLAR data (e.g., Ketchum 1977; Onstott et al. 1982). Ice features such as ridges are apparent in aircraft SAR imagery but are not always distinguishable in satellite SAR imagery, although if present these features contribute to overall pixel backscatter. With satellite imagery, more than one ice type often contributes to an image pixel. These factors hamper the ability to distinguish surface types. Based on previous aircraft SAR and ground-based scatterometer measurements, it was reasonable to expect to be able to separate different types of younger ice from multiyear ice in ERS-1 imagery (Kwok et al. 1992). However, it was discovered that it is not generally possible to distinguish between new, young, and first-year ice using backscatter alone. New and young ice display significant signature variation, and small scale inhomogeneities serve to smear the backscatter distributions for these classes across one another (Fetterer et al. 1994).

## 10.6.1 Automated Ice Classification with ERS-1 at NIC

An ice type classification algorithm was developed to produce ice type maps at the Alaska SAR Facility (ASF). The US Navy funded the adaptation of this ice algorithm for NIC, a major step toward establishing automated ice analysis on a larger scale with RADARSAT data. Some problems encountered in applying the algorithm are described below.

The algorithm [described by Kwok et al. (1992)] is based on a look-up table of backscatter distributions for four ice types: multiyear (MY), first-year rough (FYR), first-year smooth (FYS) and new ice/open water (NI/OW). The algorithm clusters a sample of pixels from an image. The centroid of the brightest cluster, if within 2 dB of -10 dB, is assumed to give the mean MY backscatter for the image. The look-up table distributions are then fixed using the mean for MY ice. The mean for FYR ice falls 4 dB below that of MY, and the means of FYS and NI/OW are lower still. This "sliding scale" feature makes the algorithm robust to errors in absolute calibration. Bayesian maximum likelihood classification is then used to assign each pixel to a class based on the look-up table distributions.

The performance of the algorithm in the Beaufort Sea is assessed by Fetterer et al. (1994). This assessment was extended to the East Siberian Sea by Gineris and Fetterer (1994). Algorithm classification results were compared to those of a supervised maximum likelihood classification using manually selected training sets on an image-by-image basis. This method has two drawbacks: manually classified images are not validated with surface observations, and manually supervised classification with training sets is not as accurate as strictly manual interpretation in which an analyst outlines areas of different ice types using contextual as well as backscatter clues.

The number of images (68 in the Beaufort, 86 in the East Siberian Sea) in the assessment made it necessary to use supervised classification. Therefore, the analysis is actually a measure of how well the look-up table distributions match the actual backscatter distributions for the four ice types. This is valuable information because any between- or within-class variability (that is, changes in the means or standard deviations of the backscatter classes) increases algorithm error.

It was understood at the outset that the algorithm would fail to identify new ice and open water because these classes have a backscatter signature which can overlap that of all other classes. Fortunately, these classes make up only a small percentage of the area covered in the algorithm domain (winter, far from the MIZ). Other errors revealed by the assessment included:

**Inability to distinguish forms of ice younger than MY.** While analysts could usually distinguish between two distinctly different types of FY ice (one brighter overall, one with lower apparent ridge density), the backscatter for these classes overlapped to such an extent that automated classification failed. Backscatter distributions for images of winter ice in the Beaufort Sea tended to be bimodal, with one peak for MY and one peak for all other ice.

**Overestimate of FY ice concentration at high latitudes.** The algorithm erroneously classified MY ice pixels at the low end of the MY backscatter distribution as FY, unless all MY pixels were above the FY/MY decision threshold of about -12 dB. The error became more apparent at high latitudes, where the concentration of MY is usually greater than 95%. When FY ice is present in sufficient quantities (generally at lower latitudes), this error tends to be counteracted by a corresponding misclassification of FY ice as MY, so that the percentages of the FY and MY for the image as a whole are about right.

**Misclassification of FY ice in the MIZ.** Large expanses of FY ice in the Chuckchi Sea had a backscatter of about -12 dB, leading to misclassification as MY ice. At 69°N, this area is south of the stated algorithm domain, but the reason why this ice was brighter than "average" FY ice is not known.

**Degradation of performance in the East Siberian Sea.** Regional variability in the backscatter of MY ice is reported by Kwok and Cunningham (1994). In the Beaufort Sea data set, Fetterer et al. (1994) found that changes in MY backscatter over latitude were matched by corresponding changes in FY ice backscatter. Therefore, the separation between types remained consistent and the sliding scale feature of the algorithm preserved classification accuracy. The root mean square error difference between algorithm and supervised classification for MY/FY separation was only 5%. The backscatter of FY and MY in the East Siberian Sea sector of the Arctic was less well-behaved. Figure 10 displays backscatter samples for different ice types as a function of longitude. Note the increasing variability in MY backscatter from east to west, and the greater overlap between FY and MY ice west of 175°W. Statistics from ice backscatter samples show that the standard deviation of MY backscatter is relatively high in the East Siberian Sea, and the distribution is skewed, with the median lower than the mean. This leads to root mean square error rates that are 6–12% higher than the 5% found in the Beaufort Sea. In the East Siberian Sea, small, rounded MY floes appear embedded in a matrix of mixed ice types. The Beaufort Sea SAR imagery does not exhibit this matrix.

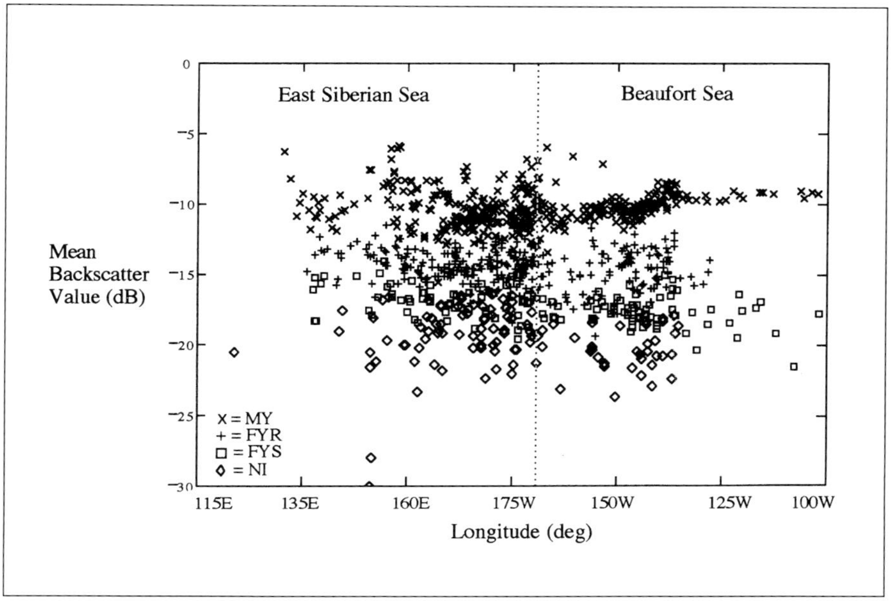


**Fig. 10.** Backscatter of different ice types acquired by manual interpretation and sampling of calibrated ERS-1 imagery during winters of 1991/1992 and 1992/1993. *MY* Multiyear, *FYR*, *FYS* rough and smooth-appearing first-year ice, *NI* new ice. The area covered is roughly the Beaufort, Chukchi, and East Siberian seas to about 82° N. (Figure after Gineris and Fetterer 1994; ERS-1 imagery copyright European Space Agency)

**Summary.** The ERS-1 algorithm does an excellent job of mapping MY ice in the Beaufort Sea during winter as compared to human interpretation. The algorithm was not used extensively at the NIC for two reasons: (1) because routines were not in place to quickly process, mosaic, and display imagery, and (2) because the area covered by available ERS-1 SAR imagery was small enough to be analyzed manually. The algorithm was used to help train new analysts in MY ice recognition, and for quality control of manual analyses.

## 10.6.2 Experience with ERS-1 at CIS

CIS has used manual rather than automated interpretation for ERS-1 imagery, while focusing on developing an automated ice/no ice algorithm for RADARSAT. Noetix Research Inc., under sponsorship by CCRS and CIS, is implementing a tone- and texture-based clustering approach which makes use of ancillary data sets (the previous day's ice chart, SSM/I data, etc.) as well as RADARSAT-specific water signatures to guide the grouping and labeling of derived clusters. All methods have problems in complicated scenes where ice and water signatures overlap. Current work aims to improve results with spatial information in the form of edge maps.

## 10.7 Improving Automated Classification

Based solely on expected backscatter, the single largest impediment to accurate ice classification is overlap of ice type backscatter distributions. The following methods might be expected to improve classification of single-channel SAR data.

### 10.7.1 Dynamic Thresholding

Dynamic or adaptive thresholding divides an image into classes based on local, rather than global, backscatter and a priori thresholds. (The ASF ERS-1 algorithm is an example of a global threshold method.) Wackerman et al. (1988a,b) first suggested dynamic thresholding for ice classification and pointed out that the method does not require strictly calibrated imagery. Haverkamp et al. (1995) use dynamic thresholding to segment an ERS-1 image into three ice types as a front end to an expert system. Thresholds are selected by examining the backscatter distributions for small areas within images. When there is evidence that the local distribution is bimodal, a threshold is selected that divides the populations represented by the two modes. This method becomes complicated because metrics must be devised for choosing the right window size, deciding if a distribution is bimodal, and determining how to interpolate between thresholds in different parts of the image. The resulting classification is sensitive to the scale of ice features relative to the scale of the local area used to determine the local threshold. After the image is segmented, the classes must be labeled. In the implementation of Haverkamp et al., the number of classes is predefined, but, alternatively, a clustering method could be used to determine the number of classes supported by the data.

Theoretically, dynamic thresholding reduces error due to signature variability since the class thresholds are determined based on local statistics. [This was remarked on, but not quantified, by Haverkamp et al. (1995)]. Dynamic thresholding is promising for ScanSAR RADARSAT data because it can potentially adjust for changes in backscatter with incidence angle and across antenna beam patterns. It is also promising for segmenting imagery in the MIZ, where surface types (open water, grease ice, new ice) exhibit varying but locally distinct backscatter distributions.

### 10.7.2 Texture

Ulaby et al. (1986) provide a precise definition of texture in SAR imagery and mathematically separate texture caused by intrinsic spatial variability of backscatter from texture caused by the presence of speckle. Adding parameters given by measures of intrinsic scene texture to measures of tone (i.e., mean pixel backscatter) was shown to significantly improve (by 20% or more) classification of forest types in SeaSat imagery. Results for sea ice have been less definitive. Since the initial use of texture for ice type discrimination by Holmes et al. (1984), numerous papers have been published which attempt to clarify the merits for classification of first-order statistics (measures such as mean, variance, and skewness from an image or subimage) versus the discriminating ability of second-order, or texture, parameters. These are measures based on a gray-level co-occurrence matrix (Haralick et al. 1973), which describes the spatial arrangement of gray tones in an image or subimage. Some conclude that first order statistics are sufficient (e.g. Nystuen and Garcia 1992), while others conclude that second order statistics offer improved discrimination (e.g. Shokr 1991). The lack of consensus arises because of differences in speckle treatment, radar parameters, gray-level co-occurrence matrix derivation parameters, and because most studies draw conclusions based on the analysis of a single image.

These studies evaluated texture based on samples of pure types. Practically, as with dynamic thresholding, there is a "windowing problem" with texture use in an algorithm; texture must be measured within small areas of a larger image and results directly depend on how these areas are selected. Smith et al. (1995) have a workable approach to this problem; first they segment an image without assigning types by filtering the image histogram to sharpen it, then they assign pixels to the same area based on closeness to histogram peaks. Once an image is segmented, texture samples can be extracted without danger of drawing from two distinct areas. Texture then becomes another parameter, along with mean backscatter, with which to classify areas. However, Smith et al. (1995) found that second-order parameters added no discrimination capability beyond first order parameters.

### 10.7.3 Expert Systems

Increasing the dimensionality of data by using multipolarization SAR or by data blending will not be operationally practical in the near term. Algorithms that use texture are slow, and they are not conclusively better than other algorithms. Using a dynamic threshold to segment imagery is expected to reduce, but not eliminate, the problem of backscatter class overlap. Reproducing manual SAR image interpretation techniques is an alternative solution for achieving better automated classification results.

The first expert system for SAR sea ice imagery was built by Intera Technologies Ltd. for Canadian ice analysts (McAvoy and Krakowski 1989). It was used as a training tool rather than for operational analysis. A user supplied answers to questions regarding the appearance of displayed ice, and these answers led to an ice classification based on how human experts would classify the ice. In 1991, CCRS sponsored a pilot project by Norland Science and Engineering Ltd. to construct a knowledge base for the Canadian Arctic. This rule base was not intended to be comprehensive, and not all of the rules could necessarily be implemented. The project anticipated the limitations that the initial image segmentation would place on the expert system's skill. It also stressed the importance of including an ice climatology in the knowledge base. In 1993, Norland Science introduced a package called IceXpert that also used knowledge-based techniques to train analysts.

Haverkamp et al. (1995) created the first "stand-alone" expert system for ice analysis. This system started with an initial classification map from the ASF algorithm. The subsequent classification was refined by applying classification rules to ice features. Features were defined as contiguous pixels identified by the initial classification as the same ice type. Most rules concerned derived geometrical feature characteristics such as "circularity," or position relationships such as "enclosed by" or "adjacent to." A feature might be reclassified or merged with another feature. For instance, a feature classified as multiyear ice could be reclassified as open water if it was sufficiently elongated. The system has two formalisms: a fact base, consisting of assertions about every feature; and a knowledge base, consisting of inference rules that may add, change or delete facts. Rules "fire" when a match is found in the rule base for the "if" portion. This in turn may cause other rules to fire. For example, a rule that enforces the knowledge that in winter and away from the MIZ, all openings in the pack are linear, and first-year ice may at times have very low backscatter, would be:

```
if     (class feature1 OpenWater)
  and  (shape feature1 not-elongated)
  then (retract (class feature1 OpenWater))
       (assert (class feature1 FirstYear))
```

The system stops when a steady state is achieved, that is, when no more rules fire based on existing facts. The authors noted a 10% change in the classification for two subscenes of the ERS-1 image with which the algorithm was developed, and visually evaluated it to be a 10% improvement.

Emulating the way an expert classifies an image is appealing because manual analysis of SAR is the standard by which other methods are (or should be) measured. Actual implementation of an expert system, though, requires unappealing complexity. Subjective guidelines (for instance, "'round' floes are probably MY") must be encoded as rules, although measures of the degree of a characteristic such as roundness could potentially be assigned and used in decision-making with fuzzy set theory. The expert system of Haverkamp et al. (1995) can merge features, but cannot divide features that were initially wrongly assigned as a single feature. The system is therefore dependent upon a relatively good initial classification. To improve the initial classification, the expert system is united with a dynamic threshold algorithm and a feature extraction algorithm. The dynamic threshold algorithm has the advantages cited above, while the feature extraction algorithm uses the image processing technique of dilation and erosion to improve the definition of floe boundaries when floes are touching. The system was run on about 90 ERS-1 images from both winter and summer. Although results were not quantified, the authors noted many instances in which features were reclassified in a manner that concurred with manual analysis procedures.

Expert systems have been used quite successfully for the analysis of geophysical data [the Prospector system for geological data is an early and model example (Duda et al. 1979)]. One criterion for using an expert system is that the problem domain be well-structured; i.e., a model for classification should already exist (Luger and Stubblefield 1993). There is no such rigid structure for sea ice classification. However, the difficulty of correctly classifying single channel radar data procedurally demands continued efforts to code the manual analysis process.

### 10.7.4 Summary

RADARSAT makes it possible to accomplish both Arctic-wide and regional analysis with the same high level of detail. A successful operational algorithm, however, will quickly process imagery from the central Arctic using averaged data for a less accurate but fast analysis. At the operationally important ice margins, a more sophisticated classifier, such as an expert system, or an experienced human analyst, will take over.

## 10.8 Automated Ice Motion Tracking

### 10.8.1 The Development of Automated Ice Motion Algorithms

Operationally, ice motion vectors are useful because they indicate areas of ice convergence and divergence. Motion vectors are also useful in forecasting ice drift. Compared with ice classification, tracking ice motion using SAR is a relatively straightforward problem because floes or recognizable patterns tend to maintain their appearance in successive images. An algorithm need only identify the same feature and calculate its displacement in two georeferenced images. Fily and Rothrock (1987) first implemented a method for accomplishing this automatically by finding an area in two images separated in time in which the pattern of pixels was highly correlated. Subsequent work (e.g. Collins and Emery 1988; Vesecky et al. 1988) refined the method and documented its limitations. This area correlation method works well for the central Arctic where ice undergoes displacement with little rotation. In the MIZ, however, floes may rotate, and it is necessary to use an algorithm that first extracts individual floes or other distinct features and identifies them with rotation invariant parameters, such as floe outline shape. Daida et al., (1990), Banfield (1991), and McConnell et al. (1991) propose methods for this kind of feature tracking.

The ice tracking algorithm developed for ASF (Kwok et al. 1990) was designed to use area correlation for the central Arctic and feature tracking for ice in the MIZ. If area correlation failed (that is, the number of erroneous vectors became high), the feature tracker estimated rotation and translation. Evaluation of approximately 80 ERS-1 image pairs (R. Kwok pers. comm. May 1996) the feature tracker worked well when a few large floes were present in the scenes. It failed, however, in dynamic areas such as the Bering Sea, where small floe sizes and the changeable nature of ice and water signatures made it impossible for the segmentation scheme to extract consistent feature boundaries for the tracker.

In the central Arctic in winter, the algorithm worked quite well. Currently products from the algorithm are being used to study the deformation characteristics of ice (Li et al., this volume), and the algorithm forms the basis of the RGPS (Kwok, this volume). The performance of both area correlation and feature tracking degrades in summer because ice appears nearly featureless in SAR imagery and the signature of open water is variable. A neural-network-based feature tracker is being developed at the University of Kansas to improve tracking capability in the MIZ and in summer (Silveira et al. 1994).

### 10.8.2 SAR Ice Motion Algorithms at US and Canadian Operational Centers

The NIC has adopted the ASF ice motion algorithm but it was seldom used with ERS-1 data because of the difficulty in rapidly acquiring SAR imagery in which the same ice was viewed on two different occasions. The area coverage with ERS-1 was usually too small to be operationally useful, although in a small case study Fetterer and Gineris (1992) found vectors from SAR to be much more reliable than those from AVHRR images. While AVHRR images provide a wider view of ice conditions (under clear sky conditions), errors due to inaccuracy of geolocation and the coarse sensor resolution were unacceptably high, especially under light wind conditions.

CIS uses an ice tracking algorithm called TRACKER that has similar area-matching lineage to the one used by NIC. A form of this algorithm was used for ice tracking with AVHRR imagery (Heacock et al. 1992). The algorithm has undergone development to address operational issues such as time and cost limitations, varying needs for accuracy, and varying operator skill levels (Marko and Gower 1992). In an operational demonstration, ice tracking in the Gulf of St. Lawrence was attempted with ERS-1 data; however, the satellite three-day ground track repeat period was too long to successfully capture rapidly moving ice. A demonstration within the confines of Lake Superior was much more successful. Research sponsored by CCRS and the Canadian Coast Guard to derive ice pressure products from tracked ice displacement as an aid to navigation is currently underway for the Gulf of St. Lawrence. These products will include a modeled forecast component.

Initial runs of TRACKER at CIS with RADARSAT ScanSAR indicate that the wide swath eliminates many tracking problems; however, both NIC and CIS consider the value of SAR-derived ice motion vectors to be in regional or tactical scale support for maritime operations. For basin scale ice drift, a dynamic/thermodynamic model such as the Polar Ice Prediction System (PIPS; Preller and Posey 1989) for the central Arctic and a free drift model at the ice edge perform sufficiently well (when unencumbered by coastal dynamics). Models also have the advantage of providing the predictive capabilities desired by the operational centers, who are interested in where the ice is going to go, not in where it has been.

## 10.9 Problems and Potential

Operational determination of the extent and thickness distribution of sea ice has been undertaken by several countries since the turn of the century. Since that time, operational sea ice mapping techniques have evolved as sensor capabilities for imaging ice have improved, but these techniques have been based primarily on the reasoning developed for the interpretation of visible and infrared imagery. These techniques will need refinement in order for the ice centers to meet their operational requirements using active radar imagery.

There is enormous potential for the application of SAR interpretation techniques to operational analysis of sea ice in the next few years. At the time of this writing, operational use of data from the Canadian RADARSAT satellite has just begun. The future of spaceborne SAR for monitoring ice conditions looks relatively optimistic with several missions well into the planning stages. Canada plans to launch RADARSAT-2, a copy of its predecessor, around the year 2000. Long lead research for RADARSAT-3, a new-generation SAR satellite scheduled for launch in 2005, has commenced. ESA intends to launch Envisat in 1998 with a C-band "Advanced" SAR (ASAR) that will have a swath width of about 400 km and resolution similar to ERS-1 and RADARSAT. NASA has begun to collect user requirements for an L-band LightSAR with the goal of achieving a 1999 launch.

Challenges lie in the handling of the vast amounts of data associated with a SAR based operation. Expensive communication links must be in place to carry images from a global array of ground stations to the ice centers throughout the world. Ice analysts must be trained to transition from the reasoning techniques developed for the interpretation of visible and infrared imagery to the logic required for analysis of radar data. Due to the high volumes of SAR data expected to be available, accurate ice classification algorithms must be developed which run in a timely manner. All this must be accomplished in a period of declining budgets for research, development and operations.

Despite these challenges, the operational ice centers have already made impressive advances in SAR development and utilization. Cooperation among the Baltic ice centers has resulted in the hosting of compatible ice analysis workstations in Finland, Germany, and Sweden. Canada and the United States have long shared digital data over a dedicated communications link between the two ice centers. Ice centers throughout the world willingly share analysis charts and information, and are working toward accomplishing this in a digital environment.

**Acknowledgments.** The authors would like to acknowledge the contributions by Mike Manore of the Canadian Centre for Remote Sensing to the sections on automated ice interpretation work in Canada . We are also grateful to Klaus Strübing of the German Ice Service, Ari Seina of the Finnish Ice Service, Shigeru Miyauchi of the Japan Meteorological Agency, Mats Moberg of the Swedish Meteorological and Hydrological Institute, and Ivan Frolov of the Russian Arctic and Antarctic Research Institute for their contributions describing operational ice support in their respective nations.

## References

Banfield J (1991) Automated tracking of ice floes: a stochastic approach. IEEE Trans Geosci Remote Sens 29, no 6: 905–911

Banfield J, Raftery A (1989) Ice floe identification in satellite images using mathematical morphology and clustering about principal curves. Dept of Mathematical Sciences, Montana State University, Bozeman, Montana

Benner DA, Bertoia C (1992) Operational satellite sea ice analysis at the Navy/NOAA joint ice center. Preprints of the 6th Conf on Satellite Meteorology and Oceanography. American Meteorological Society, Atlanta, Georgia, January 5–10, pp 395–398

Bertoia C, Carrieres T (1994) Operational use of satellite data for sea ice analysis at the US And Canadian national ice centers. Proc Int Geosci Remote Sens Symp IGARSS '94, August 8–12, Pasadena, California, pp 1228–1230

Cavalieri DJ (1992) Sea ice algorithm in NASA sea ice validation program for the defense meteorological satellite program special sensor microwave imager: final report. NASA Technical Memorandum 104559, pp 25–32

Collins MJ (1992) Information fusion in sea ice remote sensing. In: Carsey FD (ed) Microwave remote sensing of sea ice, 478, American Geophysical Union, Washington, DC

Collins MJ, Emery WJ (1988) A computational method for estimating sea ice motion in sequential SEASAT synthetic aperture radar imagery by matched filtering. J Geophys Res 93, no C8: 9241–9251
Collins MJ, Livingstone CE (1996) On the dimensionality of multiparameter microwave image data from thin sea ice in the Labrador Sea. IEEE Trans Geosci Remote Sens 34, no 1: 114–136
Daida J, Samadani R, Vesecky JF (1990) Object-oriented feature-tracking algorithms for SAR images of the marginal ice zone. IEEE Trans Geosci Remote Sens 28, no 4: 573–589
Drinkwater MR, Kwok R, Rignot E, Israelsson H, Onstott R, Winebrenner DP (1992) Potential applications of polarimetry to the classification of sea ice. In: Carsey FD (ed) Microwave remote sensing of sea ice, 478, American Geophysical Union, Washington, DC
Duda R, Gaschnig J, Hart PE (1979) Model design in the Prospector consultant system for mineral exploration. In: Michie D (ed) Expert systems in the microelectronic age. Edinburgh University Press, Trowbridge, pp 153–167
Falkingham JC (1991) Operational remote sensing of sea ice, Arctic 44, Supp.1: 29–37
Fetterer FM, Gineris D (1992) Evaluating ERS-1 ice motion and classification products. Proc 1st ERS-1 Symp, November 4–6, Cannes, France, pp 73-375
Fetterer FM, Gineris D, Johnson C (1993) Remote sensing aids in sea-ice analysis. EOS Trans Am Geophys Union 74, no 24: 265, 267–268
Fetterer FM, Gineris D, Kwok R (1994) Sea ice type maps from Alaska Synthetic Aperture Radar Facility imagery: an assessment. J Geophys Res 99, no C11: 22443–22458
Fily M, Rothrock DA (1987) Sea ice tracking by nested correlations. IEEE Trans Geosci Remote Sens 25, no 5: 570–580
Gineris DJ, Fetterer FM (1994) An examination of the radar backscatter of sea ice in the East Siberian and Chukchi Seas, Proc Int Geosci Remote Sens Symp IGARSS '94, August 8–12, Pasadena, California, pp 499–502
Gloersen P, Cavalieri DJ (1986) Reduction of weather effects in the calculation of sea ice concentration from microwave radiances. J Geophys Res 91: 3913–3919
Grumbine RW (1996) Automated passive microwave sea ice concentration analysis at NCEP. NOAA Tech Note No. 120
Haralick R, Shanmugan K, Holtzman JC (1973) Textural features for image classification. IEEE Trans Syst Man Cybern 3: 610–621
Haverkamp D, Soh LK, Tsatsoulis C (1995) A comprehensive, automated approach to determining sea ice thickness from SAR data. IEEE Trans Geosci Remote Sens 33, no 1: 46–57
Heacock T, Hirose T, Lee F, Manore M, Ramsay B (1992) Sea ice tracking on the east coast of Canada using NOAA AVHRR imagery. Ann Glaciol 17: 405–413
Herland EA, Berglund R (1995) Demonstration of operational sea-ice monitoring in the Baltic Sea with ERS-1 SAR. Int J Rem Sens 16, no 17: 3427–3440
Holmes QA, Nesch DR, Shuchman RA (1984) Textural analysis and real-time classification of sea-ice types using digital SAR data. IEEE Trans Geosci Remote Sens 22, no 2: 113–120
Ketchum RD (1977) An evaluation of side looking radar imagery of sea ice features and conditions in the Lincoln Sea, Nares Strait, and Baffin Bay. NORDA, NSTL Station, Mississippi

Knight RW (1984) Introduction to a new sea ice database. Ann Glaciol 5: 81–84
Kwok R, Cunningham GF (1994) Backscatter characteristics of the winter sea ice cover in the Beaufort. J Geophys Res 99, no C4: 7787–7803
Kwok R, Curlander JC, McConnell R, Pang SS (1990) An ice-motion tracking system at the Alaska SAR Facility. IEEE J Ocean Eng 15, no 1: 44–54
Kwok R, Rignot E, Holt B, Onstott R (1992) Identification of sea ice types in space borne synthetic aperture radar data. J Geophys Res 97, no C2: 2391–2402
Livingstone CE (1994) Synthetic aperture radar images of sea ice. In: Haykin S, Lewis EO, Raney RK, Rossiter JR (eds) (Eds.) Remote sensing of sea ice and icebergs. John Wiley, New York
Luger GF, Stubblefield WA (1993) Artificial intelligence: structures and strategies for complex problem solving. Benjamin/Cummings, Redwood City, California
Marko JR, Gower JFR (1992) TRACKER II, an improved algorithm for automatic extraction of displacement fields from sequential satellite imagery. Adv Space Res 12: 155–161
McAvoy JC, Krakowski EM (1989) A knowledge based system for the interpretation of sea ice. Proc Int Geosci Remote Sens Symp IGARSS '89, July 10–14, Vancouver, British Columbia, pp 844–847
McConnell R, Kwok R, Curlander JC, Kober W, Pang SS (1991) Y-S correlation and dynamic time warping: two methods for tracking ice floes in SAR imagery. IEEE Trans Geosci Remote Sens 29, no 6: 1004–1012
McDowell JS (1990) The other cold war. Mariner's Weather Log 34, no 1: 3–7.
Nystuen JA, Garcia JFW (1992) Sea ice classification using SAR backscatter statistics. IEEE Trans Geosci Remote Sens 30, no 3: 502–509
Onstott RG, Moore RK, Gogineni SP, Delker C (1982) Four years of low-altitude sea ice broad-band backscatter measurements. IEEE J Ocean Eng OE-7, no 1: 44–50
Preller RH, Posey PG (1989) The polar ice prediction system – a sea ice forecasting system. Naval Ocean Research and Development Activity, Stennis Space Center
Ramsay B, Hirose T, Heacock T, Walton A (1994) Automatic Synthetic Aperture Radar Information Extraction Algorithms for Sea Ice Applications. Proc Int Geosci Remote Sens Symp IGARSS '94, August 8–12, Pasadena, California, pp 505–507
Rignot E, Drinkwater MR (1994) Winter sea-ice mapping from multiparameter synthetic aperture radar data, J Glaciol 40, no 134: 31–45
Shokr ME (1991) Evaluation of second-order texture parameters for sea ice classification from radar images. J Geophys Res 96, no C6: 10625–10640
Silveira PE, Van Dyne M, Tsatsoulis C (1994) Feature matching from SAR arctic data using neural networks. Proc. Int. Geosci. Remote Sens. Symp. IGARSS '94, August 8–12, Pasadena, California, pp 496–498
Smith DM, Barrett EC, Scott JC (1995) Sea-ice type classification from ERS-1 SAR data based on grey level and texture information. Polar Rec 31, no 177: 135–146
Thomas DR, Rothrock DA (1993) The Arctic Ocean ice mass balance: a Kalman smoother estimate. J Geophys Res 98, no C6: 10053–10067
Ulaby FT, Kouyate F, Brisco B, Williams THL (1986) Textural information in SAR images. IEEE Trans Geosci Remote Sens 24, no 2: 235–245
Vesecky JF, Samadani R, Smith MP, Daida JM, Bracewell RN (1988) Observation of sea-ice dynamics using synthetic aperture radar images: automated analysis. IEEE Trans Geosci Remote Sens 26, no 1: 38–48

Vesecky JF, Smith MP, Samadani R (1990) Extraction of lead and ridge characteristics from SAR images of sea ice. IEEE Trans Geosci Remote Sens 28, no 4: 740–744
Wackerman C (1992) Digital SAR image formation. In: Carsey FD (ed) Microwave remote sensing of sea ice, 478, American Geophysical Union, Washington, DC
Wackerman C, Jentz RR, Schuchman RA (1988a) Present status of SAR sea ice algorithms. Environmental Research Institute of Michigan, Ann Arbor, Michigan
Wackerman C, Jentz RR, Schuchman RA (1988b) Sea ice type classification of SAR imagery, Proc. Int. Geosci. Remote Sens. Symp. IGARSS '88, September 12–16, Edinburgh, pp 425–428
Wohl G (1995) Operational sea ice classification from synthetic aperture radar imagery. Photogramm Eng Remote Sens 61, no 12:1455–1462.
World Meteorological Organization (1970) WMO Sea Ice Nomenclature. Secretariat of the World Meteorological Organization Publication No. 259.TP.145, Geneva, Switzerland

# CHAPTER 11

# The RADARSAT Geophysical Processor System

R. Kwok

## Contents



## 11.1 Introduction

The RADARSAT Geophysical Processor System (RGPS) is a data processing system for the analysis of synthetic aperture radar (SAR) data of sea ice acquired by the Canadian RADARSAT satellite (Raney et al. 1991). RADARSAT was launched in November of 1995 into a 24-day repeat cycle, an orbit configuration which provides near repeat coverage of the high latitudes at 3- and 7-day intervals. The SAR sensor has a number of imaging modes, one of which illuminates a wide swath (460 km) designed for large scale mapping of the Earth's surface. The science plan of the Alaska SAR Facility calls for weekly coverage of the Arctic Ocean. With this mode, the coverage of the Arctic Ocean every 7 days (Fig. 1) with 100 m pixels produces approximately 1 GB of image data daily. In an effort to make this large volume of SAR data more useful to the scien-

Analysis of SAR Data of the Polar Oceans
Edited by C. Tsatsoulis and R. Kwok

**Fig. 1.** RADARSAT summer snapshot of the Arctic Ocean using ScanSAR coverage. (Provided by J. Crawford, Jet Propulsion Laboratory)

tific community, the objective of the RGPS is to process this image data into fields of geophysical variables to a spatial and temporal scale suitable for supporting Arctic-scale science investigations.

A similar system for the processing of sea ice and ocean data, the Geophysical Processor System (GPS) (Kwok and Baltzer 1995), was installed at the Alaska SAR Facility shortly after the launch of the European Earth Remote Sensing (ERS-1) satellite in July of 1991. During its 3 years of operation, this system routinely provided products of ice motion and ice types. This system demonstrated the feasibility to produce geophysical products in a semiautomated fashion. The GPS filled part of the need to convert SAR images into geophysical data sets: it produced these products in roughly 100 × 100 km-scene-sized parcels. However, it did not coalesce these data parcels into fields of observations, although some work by investigators moved in this direction. The narrow swath and limited data availability hampered some of these efforts. With the RADARSAT mapping mode and frequency of repeat coverage, it is now possible to convert these high resolution radar data sets into large scale fields.

The scientific goal of the RGPS is to provide data sets to improve our current understanding of the impact of sea ice on global climate. Interactions between sea ice, ocean, and atmosphere in the polar regions strongly affect the Earth's climate. Sea ice growth, movement, and decay affect the energy and mass balance of the polar ocean system. The surface heat and brine fluxes associated with sea ice growth contribute significantly to convection of the ocean and thermohaline circulation. Snow covered sea ice reflects most of the incident solar radiation back into space, while freshwater fluxes associated with melting ice serve as stablizing elements in the circulation of the North Atlantic waters. Processes along the ice margin and coastlines participate in water-mass formation, upwelling, convection sediment transport and other phenomena. Understanding and modeling these phenomena requires information of sea ice motion, thickness, and concentration.

We plan to estimate and monitor five sea ice parameters in the RGPS system: ice motion, ice age/thickness, date of melt onset/freeze-up, open water fraction, and histograms of backscatter. Below, we describe the scientific relevance of these geophysical variables. A fundamental RGPS measurement is that of ice motion, obtained by tracking common features in successive SAR images. In the RGPS, a Lagrangian view of the motion field is produced from a sequence of SAR imagery. From these observations, the ice age and ice thickness histograms of the thin ice fraction of the ice cover in the winter are estimated. In the summer, we estimate the open water fraction from backscatter and area change information. The backscatter histogram of each cell, an areal parcel on the ice cover, is recorded during each observation. The times of melt onset in the spring and freeze-up in the late fall are estimated from changes in the records of the backscatter signature of the ice cover. These fields of geophysical variables will support process studies, model validation and the development of climatologies of sea ice processes. A brief description of the importance of these five parameters and their methods of measurement are provided in the rest of this section. More detailed discussion of the algorithms can be found in Sect. 11.2.

## 11.1.1 Ice Motion

Large scale circulation of sea ice determines the advective part of the ice balance and provides a velocity boundary condition on the ocean surface. Smaller scale processes involve the detailed motion of individual floes, aggregate of floes, and the formation of leads. Small-scale ice motion controls the abundance of thin ice and therefore the many surface processes dependent on thin ice, such as turbulent heat flux to the atmosphere. Lead formation during periods of divergent motion produce open water and thin-ice areas that dominate the heat flux into the atmosphere and salt flux into the ocean. In contrast to this are pressure ridges that form during periods of ice convergence. Small scale motion and deformation of the ice cover under wind and current stress are measurable in SAR data, as has been demonstrated by numerous studies (Curlander et al. 1985; Fily and Rothrock 1987; Kwok et al. 1990). Time-series SAR data provide a small-scale view of sea ice motion obtained by tracking common features in successive observations provided by the high resolution SAR data.

## 11.1.2 Ice Age/Thickness Distribution

Estimation of the thickness distribution is the primary motivation for tracking the ice age distribution. Among the many properties of sea ice strongly dependent upon its thickness are compressive strength, rate of growth, surface temperature, turbulent and radiative heat exchange with the atmosphere, salt content, and brine flux into the oceanic mixed layer (Untersteiner 1986). Our present knowledge of the Arctic ice thickness distribution is derived largely from analysis of sonar data from submarine cruises. More recently, moorings with upward looking sonars have also been used to sample the thickness distribution at fixed locations. These and other remote sensing techniques under development for measurement of the ice thickness are reviewed by Thorndike et al. (1992) and Wadhams and Comiso (1992). These instruments typically provide a one dimensional transect of ice draft (or, equivalently, thickness). These data are suitable for computing the volume transport of ice through a region or could be used for building up a climatology of mean ice thickness over a long period of time. However, they do not provide a spatial picture of the thickness distribution.

The use of ice type as inferred from active (Kwok et al. 1992) or passive microwave (Cavalieri et al. 1984) data sets as a proxy thickness indicator of ice thickness, in substitution for the direct resolution of the thicker ice classes, is at best a poor replacement of direct measurement. Surface processes typically dominate the backscatter and emissivity observed by spaceborne sensors. At higher probing frequencies, atmospheric water vapor and cloud are additional confounding factors in the identification of ice types.

Kwok et al. (1995) demonstrated that the age distribution of sea ice is a fundamental quantity that can be measured using ice motion information by keeping track of the area changes of deforming Lagrangian cells. The age distribution of sea ice specifies the fractional area covered by ice in different age classes as a function of time. In the Arctic Ocean in winter, new ice forms from freezing sea water that is exposed by the opening of leads in the ice cover. These horizontal openings manifest themselves as additions of areas to the local ice cover and are directly observable in time sequences of SAR imagery. The new ice in these leads ages and thickens. Repeated temporal sampling of these elemental areas provides us with an indication of when new areas or leads were created, the length of their existence, and thus a record of their age. The resolution of age is dependent upon the sampling interval provided by SAR observations. With some knowledge of the heat exchange between the atmosphere and ocean, the observed ice age distribution can be used to estimate the thickness distribution of the sea ice in these cells.

### 11.1.3 Melt Onset/Freeze-Up

The location and timing of melt onset and freeze-up on Arctic sea ice play significant roles in a number of geophysical processes. A large decrease in surface albedo accompanies the onset of snow melt during the spring and thus increases the absorption and heating of the ice by the shortwave radiation. The rapid temperature decrease during freeze-up in the fall has the opposite effect on albedo and marks the beginning of the ice growth season and the end of the summer. The length of the melt season is defined by the two transitions described above. Changes in the timing and spatial patterns of these transitions and the length of the melt season are important parameters in polar climate and may also serve as sensitive indicators of climate change in the high latitudes.

Winebrenner et al. (1994 and 1996) have shown that the onset of melt and freeze-up events are clearly detectable as changes in radar backscatter in ERS-1 imagery. During melt onset, the appearance of liquid water in the snow cover on multiyear ice is marked by a steep decrease (almost 9 dB) in the observed backscatter. At the end of the summer, the C-band backscatter from sea ice which survived the summer season increases rapidly over several weeks as temperatures fall below freezing and attains a stable backscatter signature which is characteristic of multiyear ice in the winter (Kwok and Cunningham 1994).

### 11.1.4 Open Water Fraction

During the melt season, fresh water is reinjected into the upper ocean. Meltwater increases ocean stratification and therefore the heat flux between the ocean and atmosphere. The heating of the upper ocean also increases as the total ice concentration decreases over the summer. Thus, the areal extent of open water is an important parameter to monitor for an understanding of the heat and mass balance of the ice cover.

After the onset of melt in the spring, the contrast between first-year and multiyear ice at C-band is lost and there is at present no effective means for ice type classification in the summer time. The summer sea ice cover at C-band has an average range of backscatter that is between -17 dB and -12 dB. At C-VV, open water backscatter is dependent on wind speed and is typically higher than that of the ice cover if the wind speed is above 4–5 m/s. The azimuthal look direction introduces only 1–2 dB modulation of the backscatter at ERS-1 look angles. Comiso and Kwok (1996) used an algorithm which takes advantage of the higher backscatter of wind-roughened open water relative to the ice cover to estimate the open water fraction. In addition to the large range of incidence angle in the RADARSAT data, the SAR is operated at C-HH. For use in the RGPS, this algorithm needs to be modified to take advantage of the response of roughened water at near range and the lower backscatter of open water at higher incidence angles at far range.

### 11.1.5 Backscatter Histograms

The time-series evolution of backscatter contains information on the state of the surface of snow covered sea ice. It may be possible to link the conditions of the ice surface to a variety of energy and mass fluxes based upon the state of ablation of the snow and sea ice (Barber et al. 1994). The RGPS will collect time series histograms of backscatter within each Lagrangian cell to support investigations for relating the evolution of backscatter to geophysically relevant parameters.

The intention of this chapter is to provide a description of the RGPS, the processing algorithms, and the data products. In the next section, we outline the algorithms which are implemented in the RGPS for analysis of the RADARSAT radar data. In Sect. 11.3, the sea ice data products produced by the system are described. The implementation of the system is discussed in Sect. 4. Summary remarks are provided in Sect. 11.5.

## 11.2 Sea Ice Algorithms

Ice motion and backscatter history are the two primary measurements made directly from the time sequence of SAR imagery. The other sea ice parameters are derived from the Lagrangian ice motion observations and backscatter record. The algorithms described herein are strongly dependent on the availability of the time series of ice motion, and estimation of some of these geophysical variables would not be feasible without continuous observations of the ice cover. The specific details of the algorithms described here can be found in the references in the text and elsewhere in this book and will not be given here. Rather, we use examples to illustrate the procedures used in the estimation of these parameters.

### 11.2.1 Lagrangian Ice Motion

The ice motion tracker is based on the procedure described in Kwok et al. (1990). Briefly, the ice tracking algorithm operates on pairs of images separated in time by *n* days, using a combination of area-matching and feature-matching techniques to track an array of points from a "source" image to a "target" image. Initially, a regular array of points is defined on the first image of a series of images covering a region. These points constitute the corners of a regular array of square cells measuring 5 km on a side. The ice features at these points are identified and tracked in each of the subsequent images in the repeat observations using the ice tracking algorithm. This provides a field of ice displacement vectors. Figure 2 shows the field of displacement vectors derived from ERS-1 imagery. Each point acquires its own trajectory, and the array of cells defined by these points move and deform with the ice cover. The deformation of the ice cover as depicted using the trajectory information is shown in Fig. 3. This differs from the current GPS

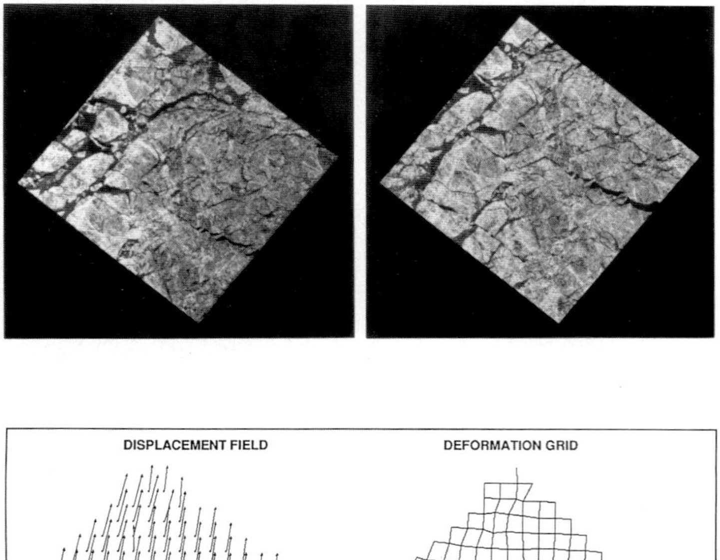

**Fig. 2.** Displacement vectors derived from a pair of ERS-1 SAR images (ERS-1 images copyright ESA 1996).

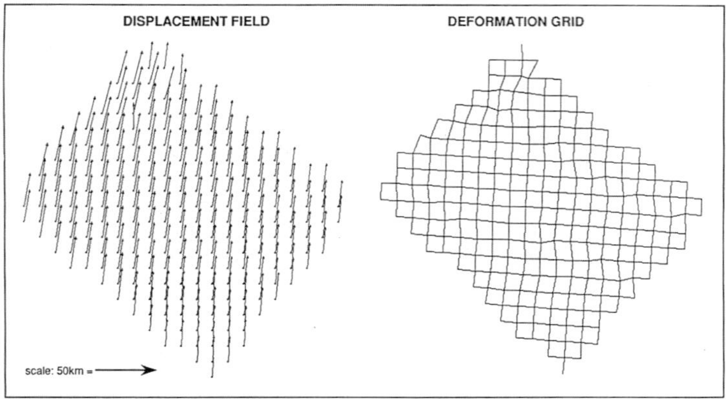

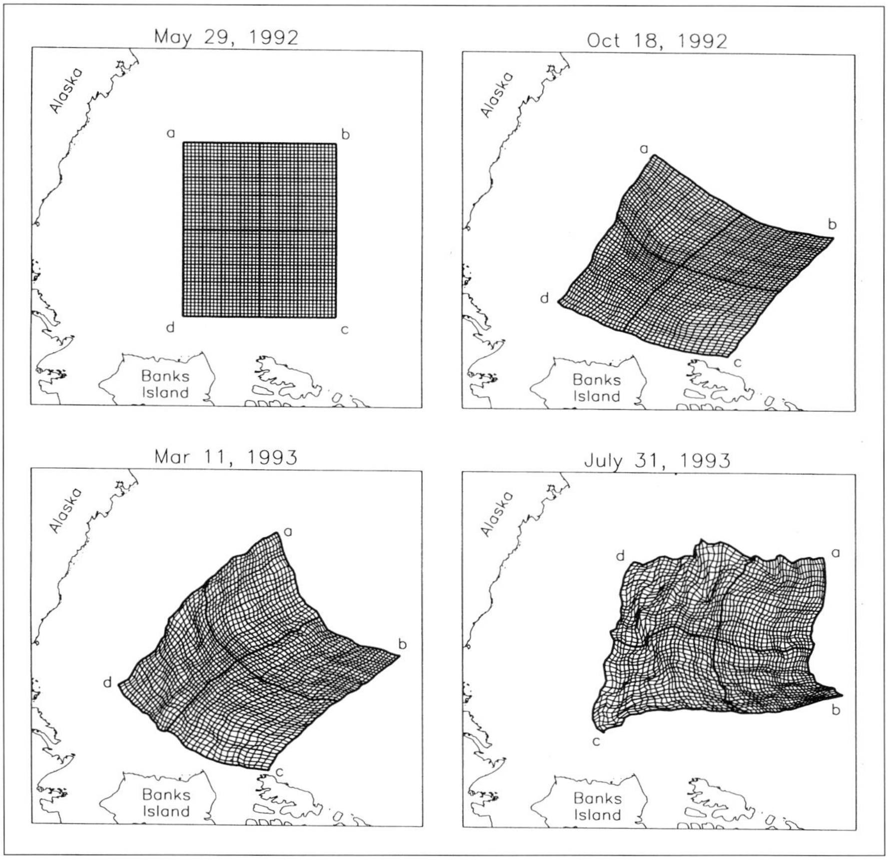


**Fig. 3.** Deformation of the ice cover (May 1992 through July 1993) constructed from Lagrangian ice motion observations: A simulation. (The Lagrangian trajectories are created from AVHRR ice motion observations)

ice tracking strategy, in which the motion between image pairs is referred to an Earth-fixed grid, giving a quasi-Eulerian picture of the displacement field. Here, the RGPS uses this ice tracker to follow the same set of points over an ice season, giving a densely sampled Lagrangian picture of ice motion identical to that derived from drifting buoys.

## 11.2.2
## Ice Age/Thickness Distribution

This procedure allows us to provide estimates of local age and thickness distributions of the Arctic sea ice cover. Again, the age distribution of sea ice specifies the fractional area covered by ice in different age classes as a function of time. We note that the algorithm works only during the winter and also that it provides a fine age resolution of only the young end of the age distribution. We refer to the time interval between sequential images as a time step. During each time step, the cell areas are computed. A positive change indicates that new ice was formed in the cell. A negative change is assumed to have ridged the youngest ice in the cell, reducing its area. The assumption here is that once ridging starts, the deformation tends to be localized in the thinner, weaker ice that recently formed in the lead systems. The age classes are determined by the lengths of the time steps. The area of ice in each age class in each cell is updated at each time step. In this manner, we keep track of the age distribution of the young ice within a cell.

The area of multiyear ice in each cell is also computed at each time step using the ice classification algorithm described in Kwok et al. (1992). The ice type algorithm uses a maximum likelihood classifier and a look-up table of expected backscatter characteristics to assign each image pixel to one of four classes: multiyear ice, deformed first-year ice, undeformed first-year ice, and a low-backscatter type characteristic of smooth, younger ice types and calm open water. Two possible sources of error are from wind-roughened open water and from frost flowers growing on new ice. Both of these physical phenomena exhibit highly variable radar backscatter, causing the classifier to label these pixels sometimes incorrectly as multiyear ice because of overlapping signatures. We use the time series of multiyear ice area for each cell to resolve these ambiguities, resulting in a more accurate classification. Since the area of multiyear ice within a cell should remain constant (no multiyear ice is created in the winter), any anomaly which shows up as a transient spike or hump could be filtered out. The ice classification algorithm is not used to identify different types of first-year ice because the accuracy of the classifier is lower for these ice types. Since the areas of young ice and multiyear ice in each cell are accounted for by the above procedures and the total cell area is known from its geometry, the residual area is simply labeled as first-year ice. For a series of five images with, say, 3 days between successive images, the age classes would be: 0–3 days, 3–6 days, 6–9 days, 9–12 days, first-year ice, and multiyear ice.

With records of the near-surface air temperature, the young end of the age distribution is converted to a thickness distribution using a simple empirical relation between accumulated freezing-degree days and ice thickness (Maykut 1986). We note that this scheme does not provide estimates of the thickness of first-year or multiyear ice. Our present method provides a two-dimensional, potentially basin-wide picture of the thickness of young ice, although it does not give any information about the mean thickness of the ice cover as a whole, since young ice occupies only a small fraction of the total area. The main output product of our age analysis procedure is the thickness distribution of young ice at a fine spatial resolution and the areal fraction of first-year and multiyear ice at regular time intervals. Further details of how ice age is related to cell area changes can be found in Kwok et al. (1995).

We select two examples to illustrate the procedure used for ice age computation within a cell. These cells are extracted from ERS-1 image sequences of the central Beaufort Sea acquired during 1992. The time series spans a period of 12 days from March 18 to March 30. The sampling interval or time step of the sequence is 3 days. The initial cell size is $5 \times 5$ km. We assume, in the two examples shown here, that the initial distribution contains only first-year and multiyear in the cells. Table 1 and Fig. 4 illustrate a few steps in the procedure. The increase in cell area between days 077 and day 086, due to the continual opening of a lead, is evident. The area changes of the cell as a function of time are plotted in Fig. 4b. This cell had an initial area of 2500 units, 1336 of which were classified as multiyear ice. Over the first time interval (day 077 to day 080), the area increased to 3034, giving a young ice class (which is between 0 and 3 days old), an area of 534 units. The remaining 1164 units were assigned to the first-year ice class of undetermined age. The cell area increased to 3205 and 3317 during the second (days 080 and 083) and third (days 083 and 086) time intervals, respectively. This new area of 171 units created during the second time interval replaces the 534 units as the youngest age group. Similarly, the 112 units created during the third time interval replaces the 171 units as having the youngest age. The 534 units created during the first time interval have become 3–6 days and 6–9 days old, during the second and third time intervals, respectively. A closing event (between Day 086 and 089) caused a decrease in cell area from 3317 to 3235, or 82 units. At this time step, the newest age class has zero area, since no new area was created, and the next-youngest class loses $82 \times k/k-1$ units to account for the lost cell area. The factor $k$ accounts for the redistribution of thin ice into ridged ice; we assume that all ridged ice is $k$ times its original thickness and occupies an area of $82 \times k/k-1$. The area of this ridged ice is kept as a separate category in the age distribution record. Here, we use a value of $k$=5 from the ridging study of Parmerter and Coon (1972). Thus, approximately 20 units of ridged first-year ice are created from the area decrease of 82 units. Then 1265 units were classified as multiyear ice, leaving 1235 units of old first-year ice. Note that the area of multiyear ice does not remain constant throughout the 12-day period. This is due to the high backscatter of the open lead, the signature of which overlaps with that of the multiyear ice backscatter, leading to an overestimation of multiyear ice. We discuss a procedure to resolve this classification error below.

**Table 1.** Record of parameters (area changes, age distribution) from time-sequence analysis: example 1

| Record | Time (Day:Hh) | Mean temp., $T$ | Cell area, $A$* | FDD | Area* of age class $j$ 1 | 2 | 3 | 4 | FY | MY | Ridged FY |
|---|---|---|---|---|---|---|---|---|---|---|---|
| 1 | 077:22 | -23 | 2500 | | | | | | 1164 | 1336 | 0 |
| 2 | 080:22 | -20 | 3034 | 65 | 534 | | | | 598 | 1902 | 0 |
| 3 | 083:22 | -20 | 3205 | 126 | 171 | 534 | | | 536 | 1964 | 0 |
| 4 | 086:22 | -19 | 3317 | 166 | 112 | 171 | 534 | | 1046 | 1454 | 0 |
| 5 | 089:22 | -17 | 3235 | 229 | 0 | 10 | 171 | 534 | 1235 | 1265 | 20 |

FDD, Cumulative freezing-degree days; age class 1, 0–3 days old ice; age class 2, 3–6 days old ice; age class 3, 6–9 days old ice; age class 4, 9–12 days old ice; FY, first-year ice; MY, multiyear ice

* 1 pixel = $100 \times 100$m = 10 000 m$^2$ = unit area

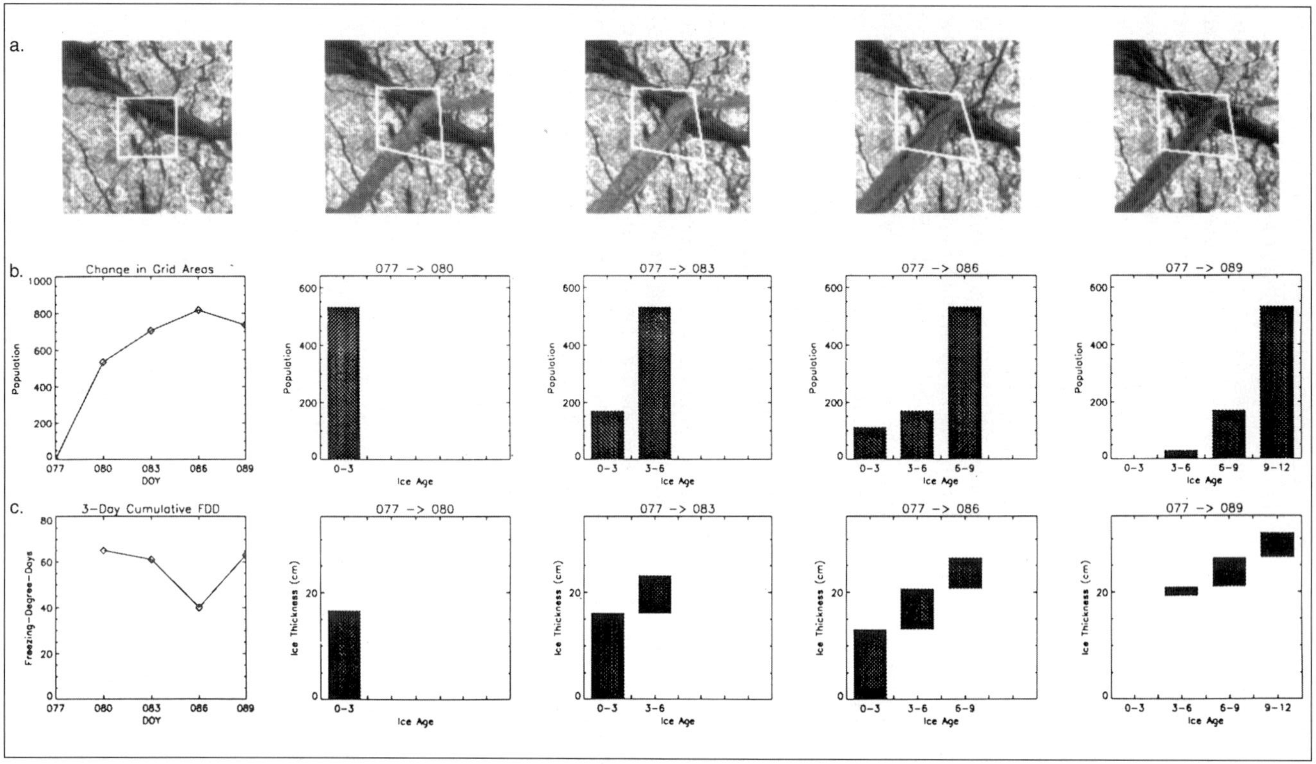


**Fig. 4.** **a-c.** Determination of ice age distribution and thickness distribution in a grid cell extracted from the time sequence. **a** The image sequence showing the deformation of a grid cell. **b** Plot of area changes as a function of time, and age histograms computed from the area changes. **c** Plot of the freezing-degree days and the computed thickness of the ice. (1 Pixel = 100 × 100 m = 10 000 $m^2$ = unit area.) (ERS-1 images copyright ESA 1996)

Table 2 shows the results from the temporal evolution of the grid cell in Fig. 5. The grid cell is extracted from the image time series. This cell increased in area during the first three time intervals, with a decrease in area during the last time interval in the sequence. The total decrease in area of 82 × k/k–1 units during the last time interval is accounted for by decreasing the area of the three youngest age classes. Seventy-one units of area are added to the ridged first-year ice category. The multiyear ice area remained relatively constant during the whole time sequence because the backscatter of the ice in the leads was low in this case.

The block diagram for the entire age/thickness scheme is shown in Fig. 6. The measurements of ice displacements at each node and multiyear area within each cell are computed during each time step. The temperature at each cell is also recorded. The area changes of the cells are then used to update the age distributions as described above. The thickness distribution is then determined using the temperature record accumulated for that cell. Some features of the scheme are described below.

The age distribution needs to be initialized at fall freeze-up. Since we do not know the ice age distribution at freeze-up, observations must be made for a start-up period of time just before the initial conditions or initial age distribution no longer affect the age distribution estimates. This start-up period is determined by the oldest desired ice class (excluding first-year and multiyear) we decide to track. For example, if the oldest desired age class is 30 days, 30 days will be needed for the initial conditions (young ice present at start-up) to grow into the first-year category. Otherwise the initial distribution of ice in the 0- to 30-day range would have "aged" beyond the range of observation. After this period, the initial conditions are determined and the age distribution will be correctly represented by the computational procedure above.

As previously noted, the presence of wind-blown open water or frost flowers on thinner ice could cause the ice classifier to overestimate the area of multiyear ice even though the winter signature of multiyear ice has been shown to be remarkably stable (Kwok and Cunningham 1994). Using the time series of multiyear ice for a particular cell, these misclassification events can be identified as positive spikes or humps. Filtering out these events leaves the background or true multiyear ice area. This may still not be perfectly constant since the cell boundaries (straight lines connecting the corner nodes) are not necessarily material boundaries and these moving edges can cause

**Table 2.** Record of parameters (area changes, age distribution) from time-sequence analysis: example 2

| Record | Time (Day:Hh) | Mean temp., $T$ | Cell area, $A$* | FDD | Area* of age class $j$ | | | | | | |
|---|---|---|---|---|---|---|---|---|---|---|---|
| | | | | | 1 | 2 | 3 | 4 | FY | MY | Ridged FY |
| 1 | 077:22 | -23 | 2500 | | | | | | 692 | 1808 | 0 |
| 2 | 080:22 | -20 | 2736 | 65 | 236 | | | | 575 | 1925 | 0 |
| 3 | 083:22 | -20 | 2807 | 126 | 71 | 236 | | | 612 | 1888 | 0 |
| 4 | 086:22 | -19 | 2981 | 166 | 112 | 71 | 236 | | 540 | 1960 | 0 |
| 5 | 089:22 | -17 | 2697 | 229 | 0 | 0 | 0 | 126 | 588 | 1912 | 71 |

FDD, Cumulative freezing-degree days; age class 1, 0–3 days old ice; age class 2, 3–6 days old ice; age class 3, 6–9 days old ice; age class 4, 9–12 days old ice; FY, first-year ice; MY, multiyear ice

* 1 pixel = 100 × 100m = 10 000 m$^2$ = unit area

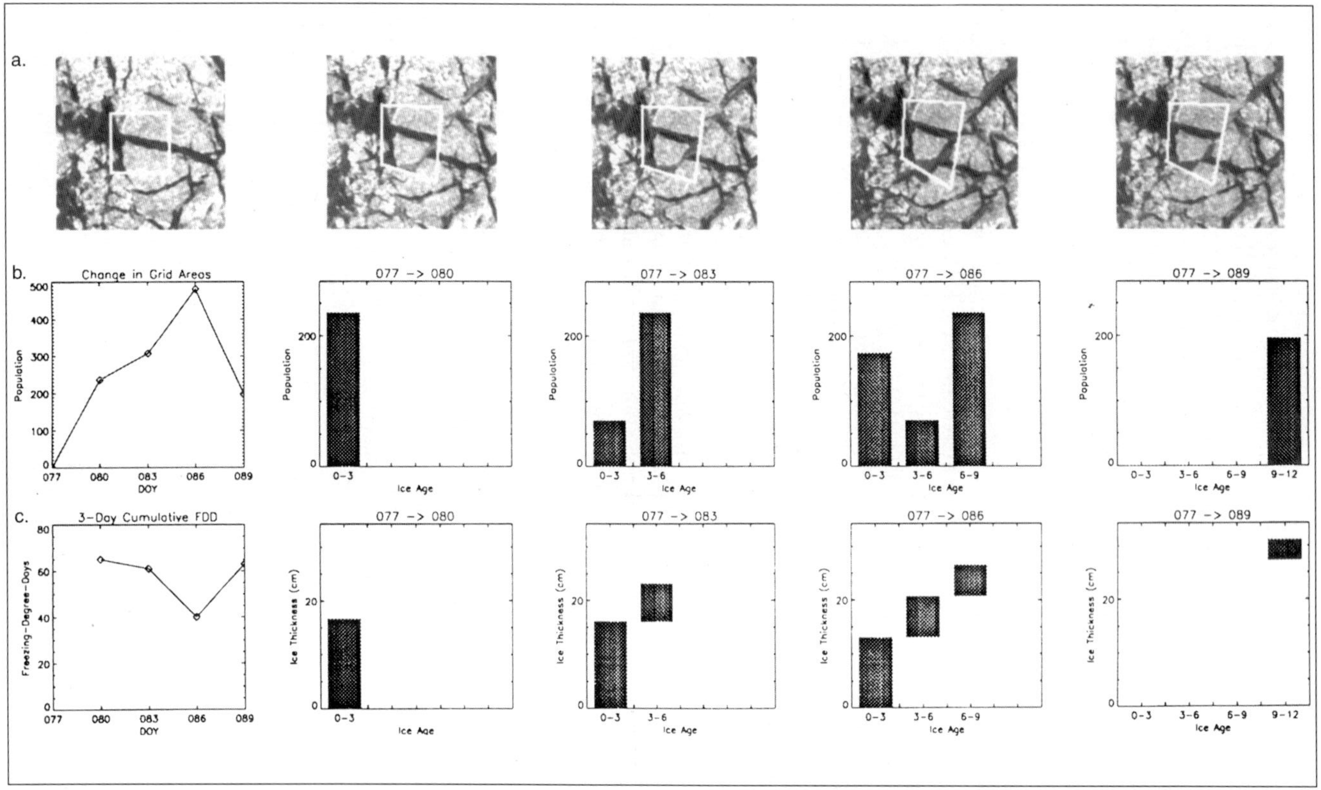


**Fig. 5.** **a-c.** Determination of ice age distribution and thickness distribution in a grid cell extracted from the time sequence. **a** The image sequence showing the deformation of a grid cell. **b** Plot of area changes as a function of time, and age histograms computed from the area changes. **c** Plot of the freezing-degree days and the computed thickness of the ice. (1 Pixel = $100 \times 100$ m = 10 000 $m^2$ = unit area.) (ERS-1 images copyright ESA 1996)

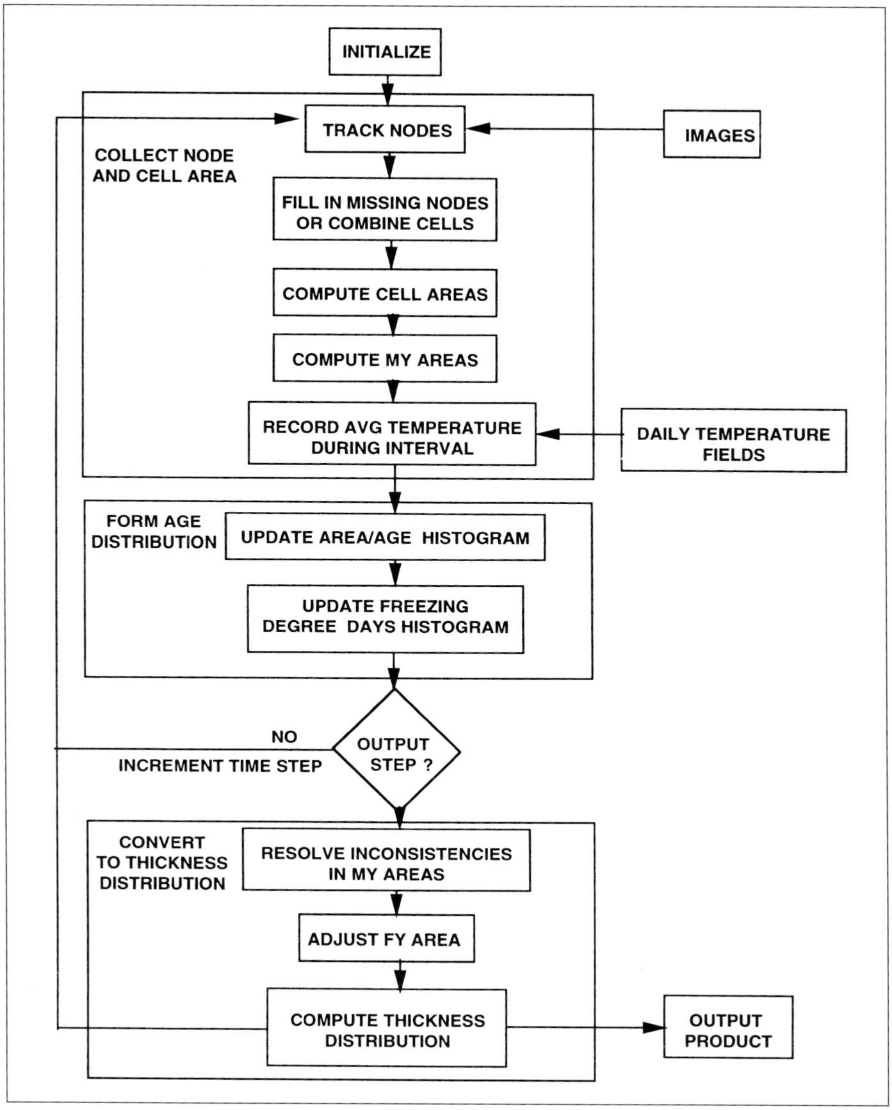


**Fig. 6.** Block diagram showing the general scheme of the ice thickness distribution algorithm

ice to shift into or out of a cell. We use a simple procedure, which uses the assumption that multiyear ice within a cell remains constant within a cell, to estimate the average multiyear ice area (Kwok et al. 1995). Figure 7 shows observed multiyear ice area for the two grid cell examples used in the previous section; the dashed line indicates the estimated multiyear ice area determined with this procedure. In the first example, the new-

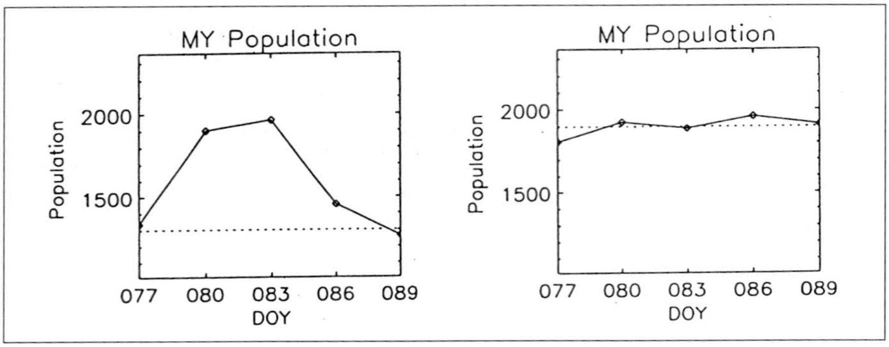


**Fig. 7.** Plot of the area of MY ice in the two grid cells. **a** Grid cell 1 (Fig. 4); **b** grid cell 2 (Fig. 5). *Dashed line* shows the estimated average MY area. (1 Pixel = 100 × 100 m = 10 000 $m^2$ = unit area.)

ly opened lead has a backscatter signature which overlaps with that of multiyear ice and therefore increased the estimated areal fraction of multiyear ice in the grid cell. As the ice in the lead aged, its backscatter evolved toward a more first-year-ice-like signature and appearance, resulting in a decrease in the area of multiyear ice. The multiyear ice area returned to approximately its area before the lead opened. The second example shows that the backscatter of the ice in the leads remained low and therefore did not confound the backscatter-based classifier. In both cases, a reasonable estimate of the multiyear ice fraction was obtained.

It is the freezing rate, not temperature or time alone, that tells the thermal history of each age class. To convert age to ice thickness, we must know the freezing rate. We approximate this rate as being proportional to the number of freezing-degree days (FDD) associated with each age class of each cell. In Table 1, we have recorded the mean temperature over each time interval. It is obvious that the accuracy of the temperature field is important; we discuss the source of the temperature fields we use in the RGPS in the next section. We convert the age distribution to thickness distribution with a simple procedure that utilizes the dependence of thickness, $H$, on freezing-degree days, $F$. For each area of young ice, there is an upper and lower bound on the age of the ice due to the length of each time step: the opening in the ice could be created during the beginning or near the end of the time step. Consequently there are two values of $F$, upper and lower bounds, that apply to each age category. We do not keep track of $F$ for the first-year and multiyear classes. We used Lebedev's parameterization (discussed in Maykut 1986), with $H = 1.33\ F^{0.58}$. This relationship is based on 24 station years of observations from various locations in the Soviet Arctic. Lebedev's parameterization describes ice growth under "average" snow conditions, in contrast to others which describe ice growth with little or no snow cover. The thickness of the snow cover is an important parameter which controls ice growth, but there is at present no routine measurements of snow depth over the ice cover which could be used for better estimation of the growth rate.

The upper and lower bounds of ice thickness for each age class are shown in Figs. 4c and 5c. The high rate of ice growth when the ice is young gives the largest uncertainty in the thickness in this youngest age class. This relative uncertainty improves as the ice

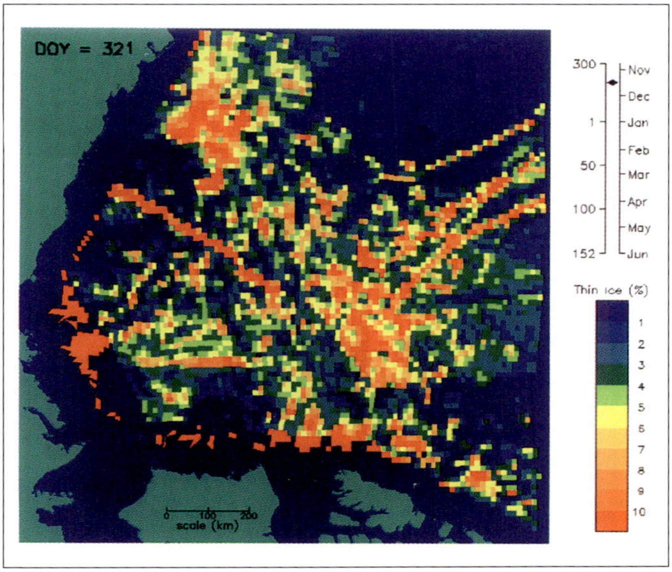


**Fig. 8.** Thin ice fraction (<25 cm) estimated by the ice age/thickness algorithm using the simulated ice motion trajectories used to create Fig. 3

ages and the growth rate decreases. The area occupied by sea ice within a thickness range can be read directly from Figs. 4c and 5c. Certainly, a more sophisticated model could be implemented for computation of ice thickness, and this could easily be incorporated at a later stage. Figure 8 shows a snapshot of the thin ice fraction computed from the trajectories used to create Fig. 3.

## 11.2.3 Summer Open Water Fraction

The summer open water fraction within each cell is determined with an algorithm which uses the wind dependent backscatter characteristics of open water. We first compute the expected backscatter cross-section of open water using the surface wind speed and its direction relative to the radar look direction at each point. A C-HH model function (constructed from current C-VV model functions using the expected copolarized response of open water) is used to provide the backscatter cross-section of open water using the above information. Summer ice (bare ice, ice with wet snow cover) has a very narrow range of backscatter, between –17 dB and –12 dB. At C-HH, depending on the wind velocity and incidence angle, the backscatter of open water could overlap or be above or below that of the ice. We anticipate that a refinement of this algorithm will be made based on the backscatter of open water at C-HH. Also, we do not estimate the open water fraction if the backscatter of water and ice overlap. At points where the open water backscatter is above or below that of the ice, we use backscatter thresholds computed using the model function to determine whether a pixel belongs to the open water category. Figure 9 shows an ice/open water ERS-1 and the corresponding classification map of water and ice. The C-VV model function is used as a guide in this case.

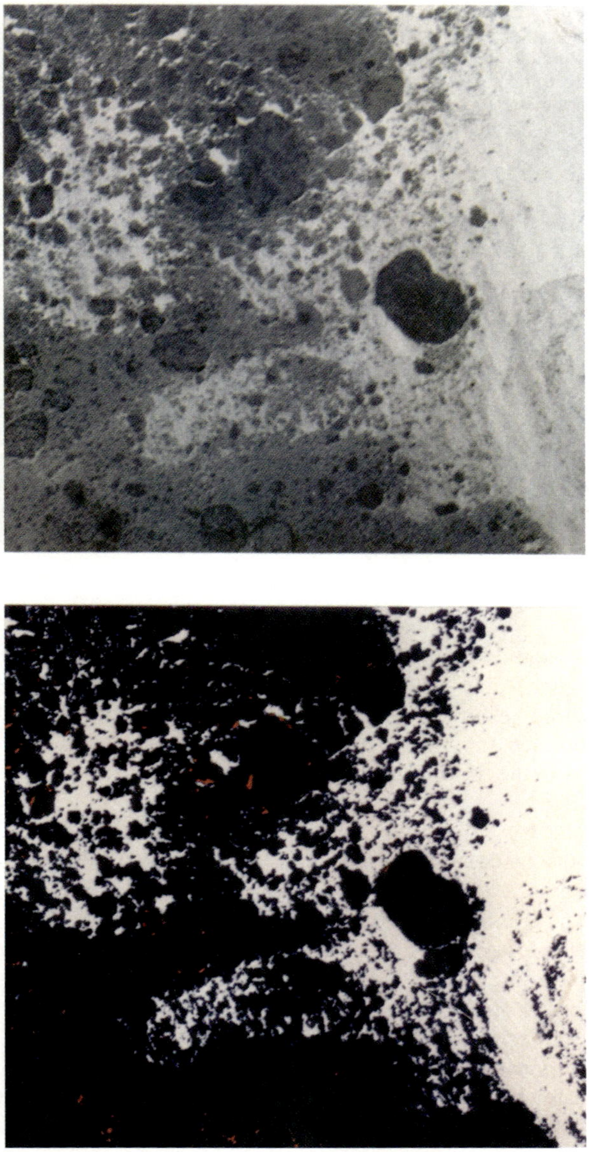

**Fig. 9.** Open water fraction computed using the expected backscatter of wind-roughened water as threshold. (ERS-1 images copyright ESA 1996)

### 11.2.4
### Melt Onset/Freeze-Up

The backscattering cross section of a Lagrangian cell is recorded at each time step. A time series of backscattering cross section histograms of each cell is used to determine the date of melt onset and freeze-up. The melt onset algorithm tracks the peak of the histogram for cells with more than 50% multiyear ice. The detection criterion is when the peak value dips below a predetermined threshold. We estimate the date of melt onset to be the time when the temporally interpolated peak crosses this threshold. The freeze-up algorithm also tracks the peak of the backscattering histogram. When this peak remains above a specified threshold for a period of *N* days, the date of freeze-up is estimated at *M* days prior to the threshold crossing. The details and physical basis for these algorithms are discussed by Winebrenner (this volume).

## 11.3 RGPS Products

In this section, we describe how the time series of sea ice data are stored and how the products are synthesized from the basic observations of ice motion and backscatter. The measurements made at each time step are stored in three main tables (Table 3): the connectivity table, the grid point trajectory table, and the cell attributes table. This information is available to users who need the tables at the lowest level. The size of these tables grows linearly with time. There are a total of 14 RGPS products (Table 4) derived from these tables. Seven are produced directly from the Lagrangian ice motion record; they contain information about individual cells and their properties. Six gridded products (on a 50-km grid), on an SSM/I polar stereographic projection, are summary prod-

**Table 3.** RGPS database tables

| | |
|---|---|
| *Grid point trajectories* | |
| Grid point identifier 1 | Time of observation |
| | Latitude |
| | Longitude |
| Grid point identifier 2 | Time of Observation |
| : | : |
| *Grid connectivity table* | |
| Cell identifier 1 | Number of vertices (*N*) |
| | Grid point identifier 1 |
| | : |
| | Grid point identifier *N* |
| | Birth date |
| | Death date |
| Cell identifier 2 | Number of vertices (*N*) |
| : | : |
| *Cell attributes table* | |
| Cell identifier 1 | Time of observation |
| | Area |
| | Air temperature |
| | Wind u,v |
| | Winter: multiyear ice are |
| | Summer: open water fraction |
| | Incidence angle |
| | Backscatter histogram |
| Cell identifier 2 | Time of observation |
| : | : |

**Table 4.** RGPS data products

| Description | Season | Grid spacing | Frequency |
|---|---|---|---|
| Lagrangian motion trajectories | Winter/summer | Variable | Weekly |
| Gridded ice motion | Winter/summer | 50 km | Weekly |
| Eulerian ice motion | Winter/summer | 5 km | Variable |
| Ice age histogram | Winter | Variable | Weekly |
| Ice age histogram (interpolated) | Winter | Variable | Weekly |
| Gridded ice age histogram | Winter | 50 km | Weekly |
| Ice thickness histogram | Winter | Variable | Weekly |
| Gridded ice thickness histogram | Winter | 50 km | Weekly |
| Area/open water fraction | Summer | Variable | Weekly |
| Gridded open water fraction | Summer | 50 km | Weekly |
| Backscatter histogram | Winter/summer | Variable | Weekly |
| Gridded backscatter histogram | Winter/summer | 50 km | Weekly |
| Melt onset/freeze up | Winter/summer | Variable | Twice yearly |
| Gridded wind/temp./pressure | Winter/summer | 50 km | Daily |

ucts derived from the cell data set. The last product contains gridded sea-level pressure, surface wind, and temperature estimates, which are used in the analysis of the SAR data.

### 11.3.1 Database Tables

**Grid Point Trajectory Table.** At each time step, the geographic location of that grid point is recorded. This table is updated each time a grid point is tracked to a new location. Displacement vectors are computed from differences in their sampled locations.

**Connectivity Table.** This keeps track of the number of points in the polygon used to define a cell. For each cell, the grid point identifiers of the vertices of the polygon are stored. Also, the birth and death dates of the cell are stored. The table is updated when a grid point or cell is added or deleted. The birth date is the date the cell was created, while when a grid point is deleted, the cells which are defined by that grid point are no longer valid, thus causing the death of these cells. A grid point is deleted when that point is advected out of the region of interest or can no longer be identified by the ice tracker.

**Cell Attributes Table.** The properties of each cell, e.g., backscatter histogram, multi-year ice fraction, area, etc., are stored in this table, which is updated at each time step. These properties are listed in Table 3.

### 11.3.2 Products

**Ice Motion (Lagrangian/Eulerian).** Operationally, a regular array of points is defined initially on the first image of a long time series of SAR images and the ice motion tracker finds the positions of those points in all subsequent images of the series. This Lagrangian product contains a record of the trajectories or an array of position measurements of the 'ice particles' that are located on the initial grid which covers a region or the entire Arctic Ocean. The sampling interval is determined by repeat coverage of these points by SAR imagery. With an initial grid spacing of 5 km, there are approximately 500000 points in this array. Over a winter season of approximately 8 months, there are close to 32 observations of each trajectory. The gridded 50-km Eulerian product represents weekly ice motion which is spatially and temporally interpolated to the grid locations from these trajectories. We also produce routine Eulerian observations of ice motion from image pairs for locations like the Greenland Sea and Bering Strait, where the character of the ice motion is such that the deformation does not allow us to maintain the density of points to define a Lagrangian grid. In these special regions, we select a high-repeat coverage frequency.

**Ice Age Histogram.** The ice age distribution of sea ice specifies the fractional area covered by ice in different age classes, within a given region, as a function of time. This ice age distribution is computed from the field of Lagrangian trajectories described above. The algorithm for determining ice age works only in the winter: the assumption is that there is ice growth in all new leads. The process for ice age determination is initialized shortly after fall freeze-up and is operated till the onset of melt. For each observation of a cell, we update the ice age histogram. The resolution of age is dependent on the sampling interval of the area of interest. As an example, for a series of images with, say, 7 days between successive images, the age classes would be: 0–7 days, 7–14 days, 14–21 days, 21–28 days, ..., first-year ice, multiyear ice, and ridged first-year ice. We note that the ice age categories for a given age histogram do not necessarily have uniform age ranges because the sampling intervals can be nonuniform. The multiyear ice fraction is obtained using an ice classification algorithm. At each step, the accumulated freezing-degree days associated with each age class are also recorded. The surface air temperatures for computing the freezing-degree days are extracted from analyzed temperature fields (described below). There are three ice age histogram products. The first ice age histogram product contains records of local ice age distributions and the accumulated freezing-degree days of each age class within each Lagrangian cell. The interpolated ice age histogram product provides uniform age categories (3-day) rather than the nonuniform age categories in the previous product. The last product is a gridded (50-km) product which summarizes the ice age histograms on an Earth fixed grid.

**Ice Thickness Histogram.** The ice thickness distribution specifies the fractional area covered by ice in given thickness ranges within a region as a function of time. The ice thickness distribution is estimated from the ice age distribution using an empirical relationship between freezing-degree days and ice thickness as described above. The highest resolution products provide the ice thickness histograms of the Lagrangian cells. Again, a 50-km gridded product of the ice thickness histograms is produced from the cell observations.

**Open Water Fraction.** Summer ice conditions are characterized by an open water fraction product. The summer open water fraction in a Lagrangian cell is estimated from kinematics and backscatter data. Kinematics gives the expected area change relative to a previous observation and the backscatter distribution provides an estimate of the areal coverage of open water in a cell. We plan to use both measurements to obtain a better estimate of the open water fraction.

**Backscatter Histogram of Lagrangian Cells.** The backscatter histograms of the Lagrangian cells are recorded in this product. The backscatter histogram for each cell is saved in twenty 1-dB bins. The backscatter history will allow the correlation of surface properties of the sea ice to atmospheric and surface conditions.

**Date of Melt Onset/Freeze-Up.** There is a fairly well-defined change in the backscatter of the snow/ice at the onset of melt and freeze-up. The analysis algorithm will detect and estimate the date this seasonal transition using the time series of backscatter product. Different parts of the Arctic go through these transitions at different dates and this product would provide a high spatial resolution of the date of transition. The temporal resolution is, however, dependent on the repeat observations of the Lagrangian cells.

**Daily Gridded Fields of Pressure, Temperature and Geostrophic Wind.** The gridded surface pressure and wind fields are derived by interpolation of the pressure and wind fields obtained from the National Meteorological Center. The analyzed temperature field is a 6-hourly, 2-m temperature data set (Martin and Munoz 1997) provided by the POLES (POLar Exchange at the Sea Surface), a NASA interdisciplinary project. The gridded air temperatures are estimated using an optimal interpolation scheme with temperature inputs from drifting buoys, manned drifting stations, coastal land weather stations, and ship reports. These are air temperature estimates used in the conversion of ice age to ice thickness.

### 11.3.3
### Sampling Issues and Measurement Errors

**Spatial and Temporal Undersampling.** The spatial sampling of the small scale motion field is limited by the density of grid points. In this case, the spatial sampling of the motion field is on the order of 5 km, thus smaller-scale motion details are not resolved and the field is undersampled. The deformation of a region of ice is represented by these deforming polygons, so the geometry of the region is very coarsely represented. We do not know the quantitative effect of this spatial undersampling. The motion field is also temporally undersampled. The sampling rate is dependent on the repeat observation of a region which in our case is constrained by orbit geometry and repeat cycles. Again, the motion details between observations are not resolved and the temporal record is undersampled. The temporal undersampling has a direct effect on the computation of total open water production and ridging. These parameters will be underestimated, as was pointed out by Fowler et al. (1996), because opening/closings of leads probably happen at a time scale which is shorter than the sampling period possible with current SAR observations. Coon et al. (1996) studied the effects of undersampling using buoy data and suggested correction factors to compensate for the weekly sampling issue.

**Measurement Errors.** There are two primary sources of error in measuring ice motion from satellite imagery: (1) the geolocation uncertainty of each image pixel and (2) the tracking error which is the uncertainty in identifying common features in an image pair. A more detailed discussion of these errors can be found in Holt et al. (1992). For ERS-1 imagery, the uncertainty in ice displacement is approximately 0.3 km. Deformation computations are typically not affected by geolocation errors if these errors are positional biases. For SAR imagery, the geolocation errors are biases and the variations are negligible within an image. Stern et al. (1995) estimated uncertainty in divergence computed from the GPS ice motion products to be approximately 0.008. For RADARSAT, these errors need to be reassessed.

## 11.4
## The RGPS System

The mapping approach uses the RADARSAT ScanSAR Wide B mode to provide a close to 6-day repeat coverage of the Arctic Ocean and adjacent Seas. The swath width of this mode is approximately 460 km. The drift of the orbit over 7 days, within the 24 day repeat cycle, is modest and always eastward. In our approach, the passes from days 1,2 and 3 are used for the first map (Fig. 1). This mapping is completed in 72 h. Later maps are made on days 7/8/9, 13/14/15, and 19/20/21. Each map takes less than 107 min radar on-time. In the following 24-day cycle, we repeat the mapping starting with passes from days 1,2, and 3. The interval between the sequential maps is 6 days. This sequence is repeated for all cycles. The maps are constructed predominantly with descending passes and gaps are filled with ascending passes. Over a 24-day repeat period, 423 min ScanSAR data are required to map the Arctic Ocean for the purposes of the RGPS. This represents a daily data rate of close to 14 500 × 460 km ScanSAR images. Some features of the system are described below.

**Throughput/Storage.** The ice motion tracker presents the largest computation load on the system. At a grid spacing of close to 5 km, each Arctic map contains approximately half a million observations of ice motion. The RGPS throughput is designed to slightly exceed the daily rate of data collection, in order to accommodate a certain amount of reprocessing of the data. The computation capability required to sustain this throughput is provided by application accelerators which provide an aggregate floating point performance of approximately 1 Gflops. We store approximately 30 days of image data in the system for efficient pairing of the image frames for motion tracking. In addition, storage space is required to accommodate the database tables described in the previous section. A total storage capacity of 100 GB is available for such purposes.

**Ice Motion Tracking.** The Lagrangian observations are sensitive to errors introduced by the tracker because errors in location propagate through the time sequence. In the RGPS, all results from the automated tracking procedure are visually inspected by an operator before the observations are recorded. To make this process more efficient, the operator is provided with the capability to view areas of large deformation as designated by the tracker and to correct the tracker results where appropriate.

**Other Data Sources.** The system assimilates wind and pressure data produced by the National Meteorological Center and regrids them into a polar stereographic projection. Also, the 2-m temperature field (described previously) is provided to us by the POLES project. These data sets are used in the analysis of the SAR data.

**Archive and Distribution.** The RGPS does not archive and distribute data products. The RGPS products are delivered to the Alaska SAR Facility where they are archived and distributed to users.

**Flexibility.** A goal in system development is to provide the flexibility for ease of incorporation of new algorithms. We envisage that new algorithms and methods to enhance current algorithms will be proposed as the science community develops experience working with SAR data and the current set of geophysical products. The system is designed such that the addition of new algorithms is a relatively simple procedure.

## 11.5 Summary and Discussion

We have described the sea ice algorithms that are used to generate the RGPS data products. The processing system described here serves three purposes: (1) to transform values of radar backscatter into fields of information that are useful for scientists; (2) to analyze images consistently, so that comparisons can be made over wide spatial and temporal scales without biases due to processing; and (3) to handle large volumes of data. The RGPS algorithms give us an opportunity to estimate sea ice parameters (ice age/thickness) previously inaccessible using other data sets. The potential payoffs in geophysical understanding of the polar regions will be very large if the radar coverage and data quality allow us routinely to make basin-scale observations of the Arctic Ocean.

In April of 1994, an RGPS Working Group was established to define the geophysical variables which could be routinely extracted from the RADARSAT SAR data, to specify the data products, and to monitor the implementation process and the evolution of the system. The current membership of this group contributes directly to the development of the analysis algorithms and definition of the data products. This group is also responsible for the verification of the system, the validation of the data products, and processing strategy. We define verification as the process used to confirm that the system and procedures work as designed and validation as the test of the scientific validity of the products. After the initial release of RGPS data products, we envisage a larger working group which includes participation of the broader user community to facilitate exchange of information and ideas amongst users of the products.

At the time of this writing (June 1996), the RADARSAT commissioning phase activities are drawing to a close. These activities include the checkout of the space segment and ground segment, and the calibration of the RADARSAT beams. The release of the SAR data for RGPS analysis will be available shortly thereafter. Based on the current schedule, the implementation of RGPS will be completed in October of 1997. This is followed by a 6-month verification phase, for assessment of the performance of the system and verification of the algorithms, after which the system will start routine processing of the Arctic Ocean data.

**Acknowledgments.** R. Kwok performed this work at the Jet Propulsion Laboratory, California Institute of Technology under contract with the National Aeronautics and Space Administration.

## References

Barber DG, Papakyriakou TN, LeDrew EF (1994) On the relationship between energy fluxes, dielectric properties, and microwave scattering over snow covered first-year ice during the spring transition period. J Geophys Res 99, no C11: 22401–22411

Cavalieri DJ, Gloersen P, Campbell WJ (1984) Determination of sea ice parameters from Nimbus 7 SMMR. J Geophys Res 89, no D4: 5355–5369

Comiso JC, Kwok R (1996) Surface and radiative characteristics of the summer Arctic sea ice cover from multisensor satellite observations. J Geophys Res 101, no C12:28397–28416

Coon MD, Knoke GS, Echert DC (1996) Compensating for daily lead-width motion when calculating open-water production from weekly data. NWRA Report NWRA-CR-96-R158

Curlander JC, Holt B, Hussey KC (1985) Determination of sea ice motion using digital SAR imagery. IEEE J Ocean Eng 10, no 4: 358–367

Fily M, Rothrock DA (1987) Sea ice tracking by nested correlations. IEEE Trans Geosci Remote Sens 25, no 5: 570–580

Fowler C, Emery W, Maslanik J (1996) The consequences of 7-day sampling for RADARSAT ice motions and derived fields. NSIDC Report, Boulder, Colorado

Holt B, Rothrock DA, Kwok R (1992) Determination of sea ice motion from satellite imagery. In: Carsey FD (ed) Microwave remote sensing of sea ice, Geophys Monogr 68

Kwok R, Baltzer T (1995) The geophysical processor system at the Alaska SAR Facility. Photogramm Engd Remote Sens 61, no 12: 1445–1453

Kwok R, Cunningham G (1994) Backscatter characteristics of the sea ice cover in the winter Beaufort Sea. J Geophys Res 99, no C4: 7787–7802

Kwok R, Curlander JC, McConnell R, Pang S (1990) An ice motion tracking system at the Alaska SAR Facility. IEEE J Ocean Eng 15, no 1: 44–54

Kwok R, Rignot E, Holt B, Onstott RG (1992) Identification of sea ice types in spaceborne SAR data. J Geophys Res 97, no C2: 2391–2402

Kwok R, Rothrock DA, Stern HL, Cunningham GF (1995) Determination of ice age using Lagrangian observations of ice motion. IEEE Trans Geosci Remote Sens 33, no 2: 392–400

Martin S, Munoz EA (1997) Properties of the Arctic 2-m air temperature for 1979–present derived from a new gridded data set. J Clim, in press

Maykut GA (1986) The surface heat and mass balance. In: Untersteiner N (ed) The geophysics of sea ice, Series B: Physics, vol 146, Plenum Press, New York

Parmerter RR, Coon M (1972) Model of pressure ridge formation in sea ice. J Geophys Res 77: 6565–6575

Raney RK, Luscombe AP, Langham EJ, Ahmed S (1991) RADARSAT. Proc IEEE 79, no 6: 839–849

Stern, HL, Rothrock DA, Kwok R (1995) Open water production in Arctic sea ice: satellite measurements and model parameterizations. J Geophys Res 100, no C10: 20601–20612

Thorndike AS, Parkinson C, Rothrock DA (eds) (1992) Report of the sea ice thickness worksh. 19–21 November, 1991, New Carrolton, Maryland, Applied Physics Laboratory, University of Washington

Untersteiner N (ed) (1986) The geophysics of sea ice. Series B: Physics, vol 146, Plenum Press, New York

Wadhams P, Comiso J (1992) The ice thickness distribution inferred using remote sensing techniques. In: Carsey FD (ed) Microwave remote sensing of sea ice, Geophys Monogr 68

Winebrenner DP, Nelson ED, Colony R, West RD (1994) Observations of melt onset on multiyear Arctic sea ice using ERS 1 synthetic aperture radar. J Geophys Res 99, no C11: 22425–22441

Winebrenner DP, Holt B, Nelson ED (1996) Observation of autumn freeze-up in the Beaufort and Chukchi Seas using ERS-1 synthetic aperture radar. J Geophys Res 101, no C7: 16401–16419

# Towards Operational Monitoring of Arctic Sea Ice by SAR

A.J. Sephton and K.C. Partington

## Contents



## 12.1 Introduction

This chapter describes work at GEC – Marconi Research Centre (MRC) on the development of image analysis tools for sea ice monitoring based on analysis of ERS-1 SAR data. Issues of product quality, degree of automation and operational performance are considered, with particular emphasis on the SUN-based Ice Pilot Application Project (IPAP) system that has been developed within the Application Visualization System (AVS) software environment at MRC over the period 1992 – 1995, in conjunction with SAIC (Science and Engineering) and Scott Polar Research Institute, and three commercial polar operators (Nunaoil, Canarctic, and BP) with varying interests in the Arctic.

Table 1 lists the main operational requirements for sea-ice products as specified by the commercial operators within the IPAP study. System constraints on orbit geometry, swath width, resolution, and duty cycle clearly limit the degree to which these requirements can be met for any one satellite, and coupled with the need to minimize data volumes to maintain reasonable processing performance all the IPAP products were derived from standard ERS-1, three-look, precision image SAR scenes (aerial coverage 100 × 100 km), averaged to 100-m pixels.

Analysis of SAR Data of the Polar Oceans
Edited by C. Tsatsoulis and R. Kwok

**Table 1.** IPAP Operator Requirements

| Product | Areal coverage (km) | Spatial sampling (m) | Temporal sampling |
|---|---|---|---|
| Ice type and concentration | 150 x 300 | 100 | 12 h |
| Ice edge | 150 x 300 | 100 | 12 h |
| Ice motion | 150 x 300 | 100 | 3 days |
| Iceberg detection | 240 x 240 (centered on rig) | 10 | 12 h |
| Enhanced SAR imagery (speckle-suppressed and contrast enhanced) | 150 x 300 | 100 | 6 h |

**Fig. 1.** Ice Pilot Application Project (IPAP) system level diagram

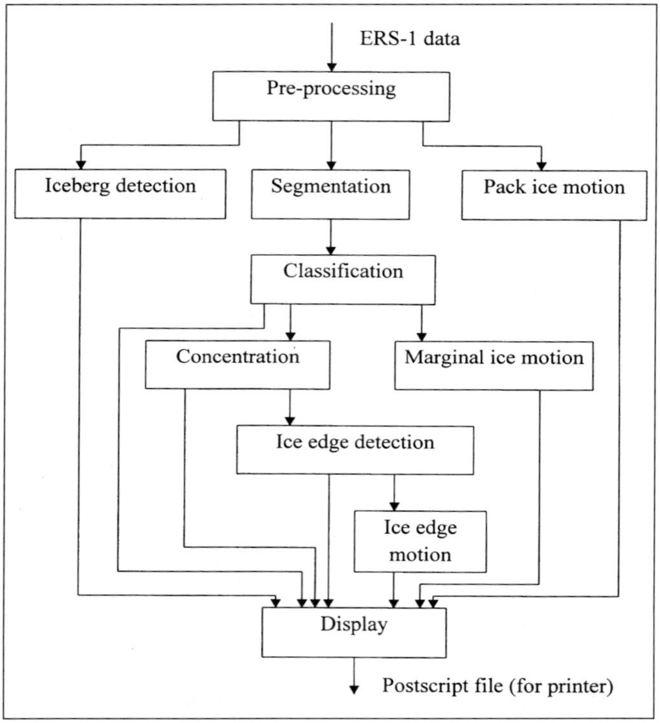


The overall structure of the IPAP system is illustrated in Fig. 1. The preprocessing subsystem enables IPAP products to be georeferenced to UTM or UPS map projection, and manually or automatically land-masked as appropriate. The display subsystem is designed to ensure that the output is in a form that is of direct use to the operators. To date, IPAP products have been transmitted to the operators by facsimile, optionally annotated with a latitude-longitude grid. In the following sections, each of the main product areas within the system will be reviewed.

## 12.2 Ice Classification and Concentration

### 12.2.1 Overview

There are three main approaches to the classification of sea ice in SAR data. The first involves a purely manual classification, perhaps using a customized PC tool for producing ice charts. This approach makes the reasonable assumption that the human interpreter is himself the best image analysis "tool" available. The second approach is to use an automated classification, with image analysis tools classifying images based on a predetermined set of optimized algorithms and algorithm parameters. This technique has the advantage of minimizing processing time and user intervention. An example of this approach is that provided by the Alaskan SAR Facility geophysical processor system, where ERS-derived ice products make use of look-up tables to classify image data (Kwok et al. 1992). A third, hybrid classification involves manual classification of selected regions in an image and automated (but supervised) classification of the remainder of the regions based on their similarity to the manually classified regions. An example of this approach is provided by the IPAP image analysis tools.

An ideal classification tool would be fully automated, to minimize processing time and the requirement for skilled labor, but there are two main reasons why this is difficult to implement at the current time, namely the problems of speckle and inherent variability in backscatter properties.

#### 12.2.1.1 Speckle

SAR data are subject to "speckle," which is noise inherent to the coherent imaging process. For intensity data from a homogeneous region, multilook speckle is gamma-distributed with:

$$\sigma / \mu = 1 / \sqrt{N} \tag{1}$$

where $\mu$ is the mean pixel amplitude, $\sigma$ is the local standard deviation, and $N$ is the effective number of independent looks. The number of looks is determined by the processing bandwidth and any smoothing after image construction, and can be estimated from the window size used to calculate $\sigma$ and the spatial resolution of the data. Speckle will degrade any estimate of the backscatter coefficient and hence our ability to classify an image based on this parameter alone.

If the data are sufficiently smoothed, then speckle is reduced, but this is achieved at the expense of poor spatial resolution (see Sect. 12.2.2). If higher spatial resolution is required, then it is necessary to carry out segmentation of the image data into homogeneous regions of ice and open water prior to classification. By extracting attributes of regions rather than individual pixel intensities, the effect of noise (particularly on estimates of backscatter coefficient) is reduced. Importantly, segmentation also allows attributes other than mean backscatter to be measured, and this can assist in region classification in the (general) case where backscatter coefficient alone is an ambiguous attribute for classification. Segmentation is described more fully in Sect. 12.2.2.

#### 12.2.1.2
#### *Inherent Variability in Backscatter*

The second issue which needs to be addressed in any classification of ice concerns the inherent variability in backscatter within ice types, between ice types, and within areas of open water. A summary of early ERS-1 measurements of backscatter coefficient provided by Partington and Hanna (1994) indicated that ranges of backscatter within a single ice type could be as high as 12 dB (e.g., rough first-year ice), and that the range of backscatter for multiyear ice is covered by that of rough first-year ice. Bright features include ridges, old ice (which has both appreciable radar penetration depth and generally rough surface characteristics), the edges of ice floes (which can produce a multiple reflection geometry with the water surface), frost flowers, and wind-roughened open water. Dark features include smooth water, new ice, frazil ice (millimeter-scale fragments of ice in open water), and areas of new ice frozen into sheets. However, there are also variations in backscatter contrast between different ice types which depend on surface temperature (particularly on whether the ice is melting) and on large-scale geographic variations in the mode and rate of formation of sea ice (e.g. salinity variations, typical compressive pressures, etc.). This problem, together with that of speckle, makes automated classification difficult except under well-understood, restricted conditions. It also makes a pixel-based classification system difficult to justify, despite significant savings in computational time.

The alternative approach, alluded to earlier, is to make use of attributes other than backscatter coefficient in any classification. This suggests a requirement for segmentation and for a classification scheme which makes use of region attributes including backscatter coefficient and measures of image texture. In practice this is what happens in a manual classification, with the human interpreter making use of texture and context as much as mean backscatter coefficient in any classification. In a hybrid classification, such as that described below, segmentation and classification are partially automated (with texture being provided as a set of attributes), but context can be included in a classification through the manual classification of selected 'training' regions. Region-based classification is described in Sect. 12.2.3.

### 12.2.2
### Segmentation

Segmentation is the process of separating an image into physically distinct regions such as ice floes (that is, regions of differing local statistics, though the regions themselves may exhibit continuous variations in the local mean intensity). Local operators for edge enhancement and detection exist in profusion. The optimum statistical test for detecting the presence of an edge between two speckled regions of uniform intensity is Student's $t$-test (Kendall and Stuart 1977). However, this assumes that the boundary between the two samples of pixel values to be compared in the $t$-test coincides with the edge position. In practice, of course, the edge positions of a SAR image are unknown a priori and so this requirement is impossible to satisfy.

Instead, Sephton et al. (1994) have evaluated the performance of the ratio of the local standard deviation to mean, $\sigma/\mu$, calculated within a $5 \times 5$ pixel window and assigned to the central pixel in the window as it is passed across the image. Although $\sigma/\mu$ is a less powerful technique than the *t*-test for edge detection (see Fig. 2a), it is sensitive to the presence of an edge irrespective of the edge direction. For an image corrupted by multiplicative noise, the expectation value of this ratio for a homogeneous area is constant independent of the mean intensity. When the $\sigma/\mu$ window includes the boundary of two regions differing in mean intensity, however, the variation in pixel value is much increased and the value of $\sigma/\mu$ increases accordingly.

The performance of the $\sigma/\mu$ filter (for a $5 \times 5$ pixel window) assuming uncorrelated pixel data has been assessed by Sephton et al. (1994) from simulations of SAR intensity data corrupted by *n*-look multiplicative noise. Defining a 90% detection level as the minimum detectable edge-contrast ratio, it can be seen from Fig. 2a that the $\sigma/\mu$ (also referred to as "sigmu") filter is noticeably poorer than the *t*-test at detecting edges for the same false alarm rate (values shown in parentheses), requiring two to three times the number of looks in order to achieve the same performance.

**Fig. 2.** **a** Performance of SIGMU edge detection filter. **b** Approximate spatial resolution of ERS-1 data

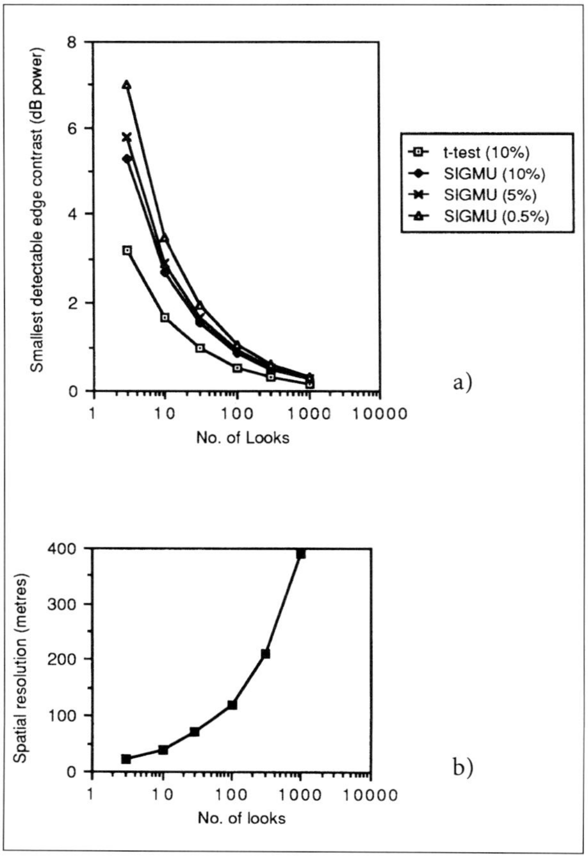

While in principle almost any given level of edge detection can be achieved by smoothing the image to increase the effective number of looks, it should of course be noted that this will in turn produce a degradation in the spatial resolution. For an ERS-1 image, with a spatial resolution of approximately 24 m in range and azimuth for three-look data, the degradation in spatial resolution can be represented as in Fig. 2b.

The values of $\sigma/\mu$ obtained in the edge detection process do not directly provide edge positions in the image as they contain no edge direction information. Instead, the values of $\sigma/\mu$ are used to "bond" neighboring pixels so that pixels are linked by bonds to all other pixels in the same region, but not to pixels in any other region. A typical bonding method used is that described by Oddy and Rye (1983). As a consequence of artificially dividing some regions in the bonding process it may be necessary to include a step of region merging, based for instance on the *t*-test (Sephton et al. 1994), before attributes are derived for each region.

### 12.2.3
### Classification and Concentration

The number of approaches to classification is large: this section describes the philosophy adopted within the IPAP system. We have already discounted a pixel-based approach on a priori grounds. There is a further decision to be made regarding whether supervised or nonsupervised methods are applicable. We have selected the supervised approach, on the basis that we desire manual intervention in assigning the attribute characteristics to a class before beginning classification. The unsupervised approach finds its own class characteristics and members before requiring the user to assign labels to each class, but suffers the disadvantage of requiring the number of ice classes present to be known a priori. Supervised classification options include parametric classification, which requires assumptions about the distribution of samples in a class, and is therefore considered less appropriate than nonparametric approaches, which make no such assumptions. So we reject (for example) maximum likelihood classification in favor of *K*-nearest neighbor (K-NN) and neural network, multi-layer perceptron (MLP) methods.

The K-NN method is based on the assumption that vectors from a given class tend to be close together, so that an unknown vector should be given the same class as some *K*-weighted set of vectors which form the nearest class. For $K = 1$, the error rate is close to twice the Bayes error (which defines the minimum possible error) and as *K* increases, the error rate approaches the Bayes limit asymptotically (Cover and Hart 1967).

The MLP method is based on the principle that sequential Von Neumann computers are exceeded in performance by processors which mimic animal brains through structures which contain heavily interconnected nodes, and that this improved performance extends even to the case where the number of processing units may be tens rather than the many orders of magnitude more that exist in animal brains (Lippman 1987). An MLP typically contains three layers, with nodes in each layer interconnected to nodes in layers above and below. The weights associated with each node allow the predicted class to be matched to the actual class by backwards adjustment of weights within the network following poor matches.

Supervised classification is a two-stage process. In the first stage, selected regions are classified manually. A judgment needs to be made on how many regions will be sufficient to be representative of the population of regions from each particular class. The classification is carried out by displaying the underlying image together with an overlay of region boundaries (from the segmentation). The operator then selects regions and assigns a class (effectively as an additional region attribute). This information is added to the vector database. Once sufficient regions have been selected, the updated vector database is written to disk as a vector training database.

In the second stage of the classification process, the training database is accessed in a classification of the entire image, using the MLP or K-NN algorithm. This is a partially automated stage of the classification process and may involve more than one iteration as the resulting image classification may need to be improved by trying a new classifier or, more likely, by refining the training database.

Typical results of the classification are shown in Fig. 3. From the original ERS-1 SAR image (Fig. 3a) acquired in the Kara Sea in March 1994, a classification product (Fig. 3b) and a total ice concentration product (Fig. 3c) are derived. Other products (not shown) which can be generated by the system include an "old ice" concentration map and a classification confidence map (based on the relative distance of attributes in each region from the cluster center of its assigned class).

The region attributes used in the classification include both texture and mean backscatter attributes, calculated using pixels located away from the region boundaries. The selection is not based on an understanding of the relative value of each attribute, although there is some evidence for benefits in the use of some texture measures including the normalized standard deviation $\sigma/\mu$ and inertia and entropy measures from the spatial gray level dependency matrix (Haralick et al. 1973; Sun et al. 1992). Instead, a range of attributes are included in a black box approach to maximize the probability of including relevant attributes in each particular circumstance. By including the slant range coordinate of a region as an attribute, the classification can also take into account incidence angle variations in the attributes (particularly in mean backscatter). By providing scatter plots for the various attributes, and estimating their separability from manually classified regions, it was found that $\sigma/\mu$ and entropy were useful discrimination measures for winter ice, whilst inertia, uniformity, and dissimilarity also added discrimination capability, but of more marginal value.

The hybrid classification approach advocated here is considered to be a reasonable compromise between an automated scheme which is unlikely to be valid outside very restricted circumstances and a fully manual scheme which fails to make use of any computer intelligence in assisting the user. Table 2 illustrates the division of classification tasks into manual and automated. Note in Table 2 that for the tasks specified as manual, it is not envisaged that the selection of the attributes would need to be performed on every occasion.

Figure 4 shows a comparison of AVHRR-derived ice concentrations and ERS-1 SAR-derived ice concentrations for the Greenland sea, in a campaign carried out in August 1994. The AVHRR-derived concentrations are estimated from Danish Meteorological Institute (DMI) ice charts recorded on the same day as each of eight ERS-1 SAR images. The errors on the AVHRR-derived ice concentrations result from the binning used by DMI in producing the ice charts, and in the SAR-derived ice concentrations represent ±1 standard deviation. It can be seen that open water and pack ice are recognized consistently using the two sensors. At moderate ice concentrations, there is a discrepancy of up to 2/10ths, which to date has not been explained. In the long term, the availability of spaceborne SAR data with dual polarization will reduce the reliance on manual intervention.

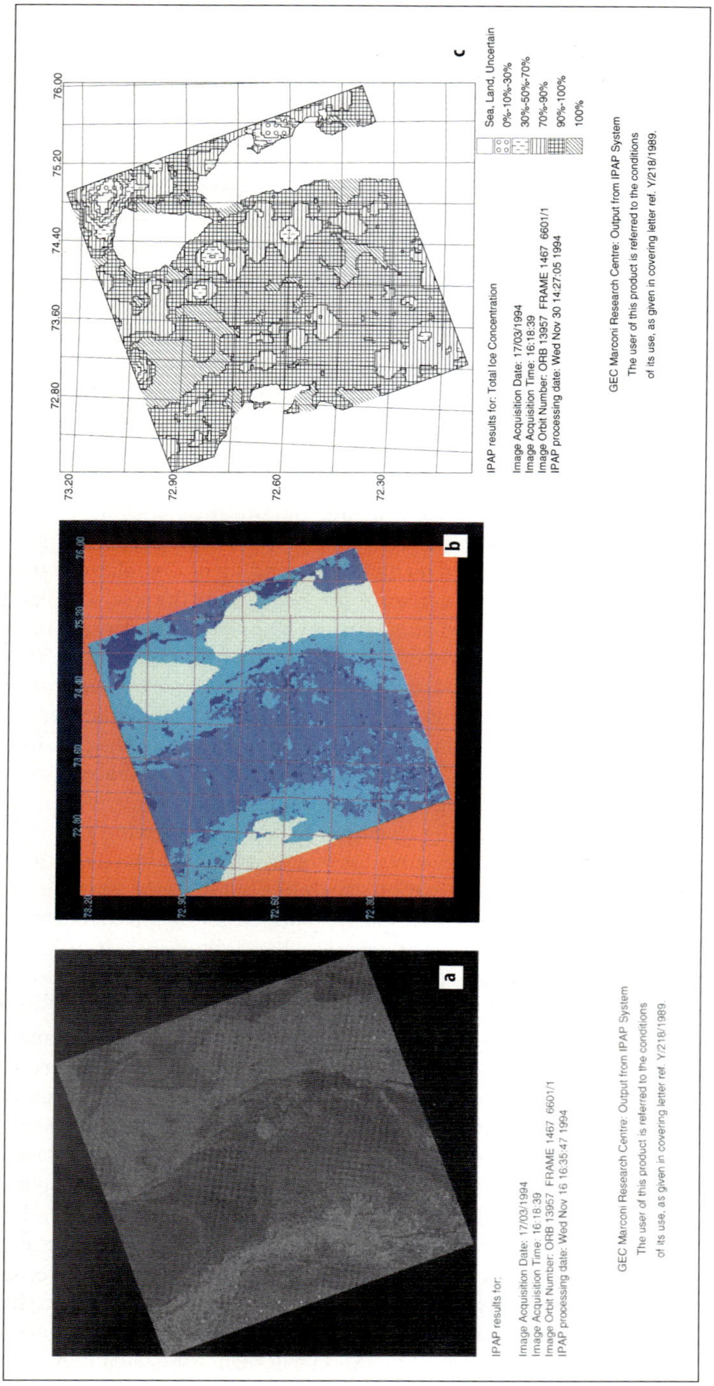


**Fig. 3.** **a** ERS-1 SAR image from the Kara Sea recorded in March, 1994. **b** Classification of data shown in **a** into multiyear ice (*green*), first-year ice (*light blue*), open water (*dark blue*), and land (manually classified in *yellow*). **c** Total ice concentration map derived from data in **b.** (ERS-1 image copyright ESA 1994)

**Table 2.** Generation of ice-type concentration products: Manual and automated tasks

| *Manual* | *Automated* |
|---|---|
| Selection of segmentation parameters:<br>• Smoothing<br>• Edge detection threshold<br>Vetting of result (iteration if required)<br>Selection of region attributes<br>Construction of training dataset | Segmentation |
| Selection of region attributes<br>Selection of K-NN or MLP algorithm | Calculation of region attributes |
| Vetting of result (iteration if required) | Classification of nontraining regions<br>Generation of products:<br>– classification, confidence<br>– ice-type concentration<br>– ice edge chart |

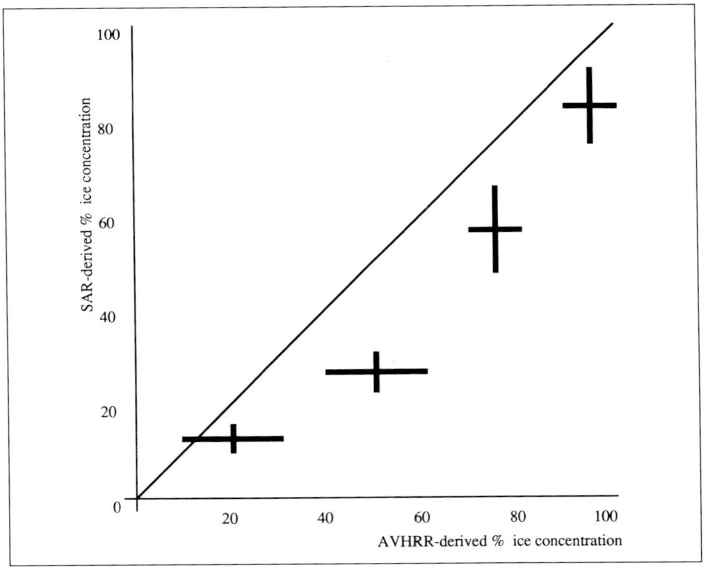


**Fig. 4.** Comparison of SAR and AVHRR-derived ice concentrations, using charts derived from the same day for various dates in August 1994 for the Greenland Sea. Each cross indicates an ice class from the AVHRR-derived ice chart: 1 – 3/10ths concentration, 4 – 6/10ths concentration, 7 – 8/10ths concentration, and 9 – 10/10ths concentration

Subjectively, we have concluded that the approach proposed here is most successful in winter conditions and with large ice floes (of diameter greater than 1 km). In summer conditions, or where floes are very small, segmentation is unable to detect boundaries reliably and so is unable to provide a suitable template for a region-based classification.

## 12.3 Ice Motion

### 12.3.1 Overview

Ice motion is of interest in a number of contexts, identified by Carsey and Holt (1987), as: latent heat advection, oceanic surface stress, passive tracing of currents, production of open water by way of ice divergence and shear, and interaction between icebreaker navigation and ice structure. In addition, ice motion is of importance in an operational context, such as in giving an indication of where the ice is compressing and therefore of danger to a vessel (perhaps forming pressure ridges), where (navigable) leads are likely to form, and for drilling platforms to take evasive action if threatened by the movement of heavy ice.

Sea ice moves in response to forcing by surface winds and ocean currents, and to stresses from and within the ice itself. In the central ice pack and marginal ice zone (MIZ), the complex nature of the relative ice motion under divergence and shear is a very active research area. Convergent stress conditions in the ice cover result in the production of pressure ridges, which contain a large fraction of the total sea ice volume and are operational hazards, whereas divergent stress conditions produce areas of open water (leads and polynyas). Through these areas of open water, significant exchanges of heat and moisture occur. The varying coverage of leads and polynyas also result in varying fluxes of brine in the sea.

In the Arctic, large-scale ice motion is commonly along atmospheric surface isobars, and has been measured by tracking buoys and individual floes (e.g., Colony and Thorndike 1984). The effect of wind and ocean currents causes most of the ice to rotate around the Arctic basin. However, ice is continually swept south through the Fram strait, and the warm West Spitzbergen current flows north, creating an ice-free area to the south-west of Svalbard. The presence of land creates zones of shear, such as that found within the area north of Alaska. Over scales less than 100 km however, ice motion can be quite irregular as floes ranging in size up to tens of kilometers translate and rotate as rigid bodies between other tightly-packed floes.

In order to resolve the spatial structure of this smaller-scale ice motion, a large number of measured trajectories of floes is needed, with a data sampling density of approximately 1 $km^{-2}$. As a means of achieving such sampling rates over large areas, the use of sequential satellite images is now generally considered to be the most effective method of studying small-scale ice motion, the resolution of the imaging system being the limiting factor in the size of feature that can be identified. Estimates of ice motion made from satellite image data have revealed a mean motion of some 5 km / day in the pack ice, and approximately 15 km / day in the MIZ (Curlander et al. 1985). Ito and Muller (1982) have reported observing ice velocities in satellite imagery of up to 34 km / day, whilst velocities up to 80 km / day are not unknown.

Two main categories of automatic and semiautomatic methods have been developed in recent years to track the motion of sea ice in remote sensing data between image pairs, covering the two main types of ice condition, i.e., pack ice and marginal ice. In pack ice, floe boundaries are generally indistinguishable but the overall motion is mainly translational (over short time scales); in marginal ice, floe boundaries are generally distinguishable but independent rotation of floes is also likely. Algorithms that perform well with one type of ice are generally not best suited for determining motion in other ice conditions. Following the example of the Alaskan SAR Facility (Kwok et al. 1990; Rothrock et al. 1992), it was therefore decided to adopt a hybrid solution for the IPAP system: i.e., to develop one algorithm of each type to be selected as best appropriate to the prevailing ice conditions.

The main methods that have been developed which are appropriate for pack ice are area correlation (Fily and Rothrock 1986, 1987), matched filtering (Ninnis et al. 1986; Collins and Emery 1988), and optical flow methods (Yan 1992). Of these algorithms, the area correlation method has been the most widely used, and as one of the algorithms implemented under IPAP is fully described in Sect. 12.3.2 below. Since the main drawback of the method is in terms of computational efficiency, the matched filtering technique (in which the correlation is performed in the Fourier domain) would seem to hold out significant potential. In particular, the cross correlation function of the power spectra might be used as a first step to rotationally correct the images prior to applying a conventional area correlation, the power spectra being related by a rotation which is the same as the relative rotation between the two images in the spatial domain [Fily and Rothrock (1987) estimate that rotations beyond 7° will start to affect the correlation values obtained]. For the optical flow method, in which the components of ice velocity are calculated at each image pixel in terms of partial derivatives (temporal and spatial) of the image intensity, the assumptions made regarding stationary image intensity and gradient are likely to be valid only for small velocities of the order of a few pixels, and so this method may not be appropriate for large-scale ice displacement.

The main methods that have been developed which are appropriate for marginal ice are tie-point extrapolation (Zhang et al. 1989), edge-feature matching (Vesecky et al. 1988; McConnell et al. 1991), postsegmentation region matching (see Sect. 12. 3.3), invariant moment matching (Zhang 1992), and stochastic approaches (Banfield 1991). For the tie-point extrapolation method, the degree of manual intervention makes it too time-consuming for routine use within an operational environment. The other approaches can all be considered as variations on a common theme, i.e. matching common features (generally floes) between images based on some measure of shape. For each of these techniques, the primary requirement is the input of a good, repeatable segmentation. Assuming that there is no noise in the boundary determination, then the best representation of the region shape will be obtained from one of the edge-feature matching techniques which characterize region boundaries at the pixel level. At the other extreme, the invariant moment matching method is based on a parameterization of the region shape in terms of its moments and, whilst being less sensitive to boundary noise, would not be able to cope with tracking regions that split or merge. The stochastic approach represents an intermediate solution in that floe boundaries are approximated by principal curves from which edge-pixel probability distributions are derived. The postsegmentation region matching scheme also offers an intermediate solution in that segmented regions are approximated to by polygons (of user-defined maximum deviation from the true boundary position), but also offers the flexibility of incorporating tests based on other attributes such as classification label. This algorithm was that selected for implementation as part of the IPAP system (see Sect. 12.3.3).

### 12.3.2
### Pack Ice Motion

The area correlation techniques which have been used for assessing pack ice motion in remote sensing data (see Sect. 12.3.1) are based on the cross-correlation of blocks of pixels in two images, the displacement vectors between matching points being taken to be between centroid positions of the blocks with the highest correlation. In order to reduce the required degree of processing, a hierarchical scheme has been adopted in the implementation at MRC, similar in principle to that developed by Fily and Rothrock (1986, 1987). In this method, the correlation computation is facilitated by applying the correlation to an image-pyramid data structure (one pyramid for each image), where cross-correlation peaks corresponding to motion vectors are first found in highly averaged images, and repeated stepwise (up to three iterations in the algorithm implemented at MRC) in images of finer and finer resolution. Providing that the measured correlation peaks are statistically significant (at the 1% level for a correlation window size of 16 pixels), the "best match" positions derived from the coarser resolution images are taken as the starting point for the correlation at the next level in the pyramid.

At the end of each smoothing iteration, a filter may optionally be applied (in the MRC implementation) in a user-defined number of iterations in order to correct "spurious" displacement vectors at odds with their neighbors. This helps to prevent errors in the top levels of the pyramid propagating down to the lower levels. In the first filter iteration, if the displacements in both the $x$ and $y$ directions for a particular pixel corresponding to the second best correlation differ less than the $x$ and $y$ displacements corresponding to the best correlation from the mean $x$ and $y$ displacements of the pixel's neighbors, then the displacements for that pixel are updated by the displacements associated with the second best correlation. In subsequent filter iterations, if the displacements in either of the $x$ and $y$ directions for a particular pixel differ from the mean $x$ and $y$ displacements of the pixel's neighbors by more than a user-defined threshold (in standard deviations), then the displacements for that pixel are updated by the mean $x$ and $y$ displacements.

The ice motion displacements are displayed corrected for any geographic offset between the images, either in the form of a grid representation for all points in the data or as arrows showing those displacements with the highest correlation. Up to three classes of vector are displayed dependent on the strength of the associated correlation. The mean vector displacements of ice between the two images can also be generated, averaged over a grid of user-specified dimensions.

Evaluation of the area correlation algorithm based on three pairs of ERS-1 images each recorded 3 days apart off the east Greenland coast in March 1992 shows that for pack-ice there is a high degree of consistency in the measured vector displacements both within and between images, as well as with manual interpretation and with the known patterns of ice motion in this area. Figure 5 shows by way of example the results of the correlation obtained using a pair of images recorded on March 24 and 27, 1992. The image of March 24 is shown overlaid by the grid representation of the total correlation space in red and by arrows showing the vector displacements of ice in the later image with respect to the earlier image, yellow arrows corresponding to displacement vectors which have an associated correlation coefficient between 0 and 1 standard deviations above the user-defined threshold of 2 standard deviations above the mean, and

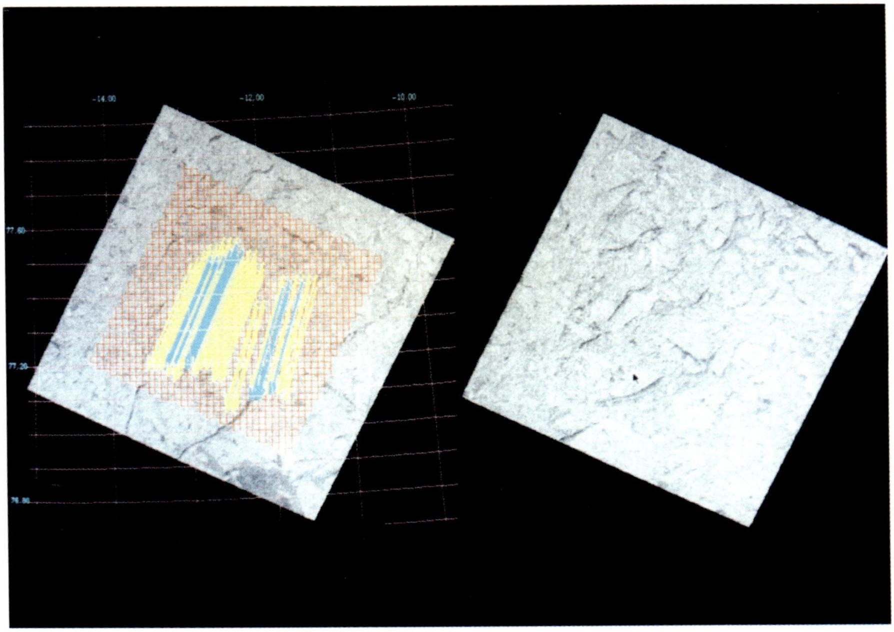

**Fig. 5.** Area correlation of pack ice for a pair of ERS-1 images recorded 3 days apart off east Greenland in March 1992. See text for further details

turquoise arrows corresponding to displacement vectors which have an associated correlation coefficient between 1 and 2 standard deviations above the user-defined threshold. The effect of applying this correlation threshold is that only those vectors associated with the highest degree of correlation (and therefore confidence) are displayed. The image of March 27 is shown next to the image of March 24 for comparison. In each case, the data have been georeferenced to UTM map projection. Figure 6 shows the resulting ice motion product for transmission to the operator in the field, obtained by averaging the same vector displacements as in Fig. 5 over a grid of size 5 intervals.

### 12.3.3 Marginal Ice Motion

The post-segmentation region matching technique was originally developed at MRC as part of the ESPRIT II Multi-Sensor Image Processing (MuSIP) project funded by the European Commission, the main aim of which was the combination ('data fusion') of region attributes from multitemporal land imagery. For each segmented region in an image (see Sect. 12.2.2), the boundary position is stored in the form of Freeman code, specifying for each boundary pixel the direction of the boundary with respect to neighboring boundary pixels. Region shape is characterized by a polygonal approximation to the region boundaries.

**Fig.6.** Same vector displacements as in Fig. 5, averaged over a 5 × 5 grid, in the form of an operational IPAP product

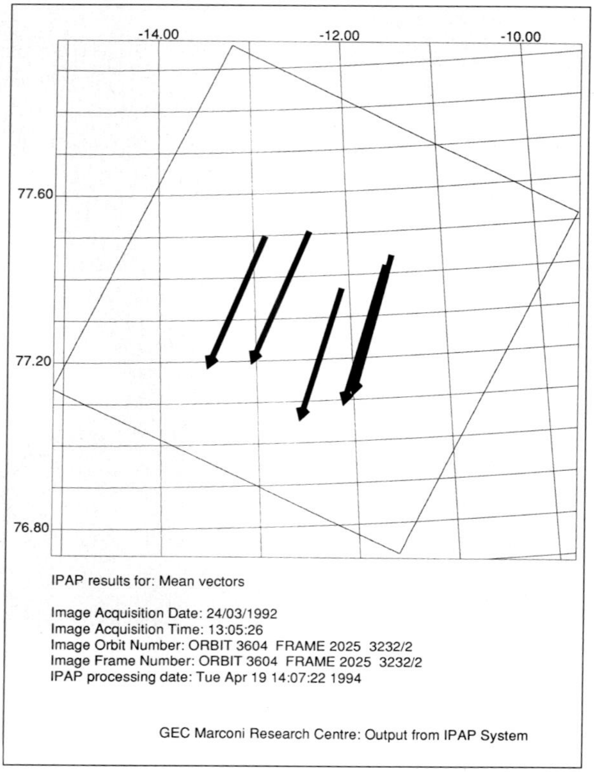


Qualifying regions from one image are selected, based on suitability for matching, and for each such region a defined search area in the other image is used to find regions which are considered as candidates for matching. The boundary-matching test is based on a comparison of the length, orientation, position, and order of region boundaries, and matches can be either complete or partial. The matching process optionally uses the classification of the regions as well as their shape in order to determine whether the regions correspond to the same ice floe and whether the shape, position, and orientation of the floe has changed. Detection of partially matching regions enables the system to track floes which have either split or merged between the two images.

Evaluation of the algorithm has been performed with ERS-1 data of marginal ice off the east Greenland coast recorded in August 1994 as part of an operational demonstration (see Sect. 12.5) in which IPAP products were directly faxed to a survey ship operated by Nunaoil. For this ERS-1 image pair (Figs. 7, 8), the image frames are not completely overlapping.

The results of the region matching are shown in Fig. 8, the displacements of floes being indicated by the length and direction of the arrows superimposed on the later image of the pair (the arrow tails corresponding to the floe positions in the earlier image). Five regions from the two images are matched, corresponding to five large multiyear ice floes in the images (region matching was here performed only on the multi-

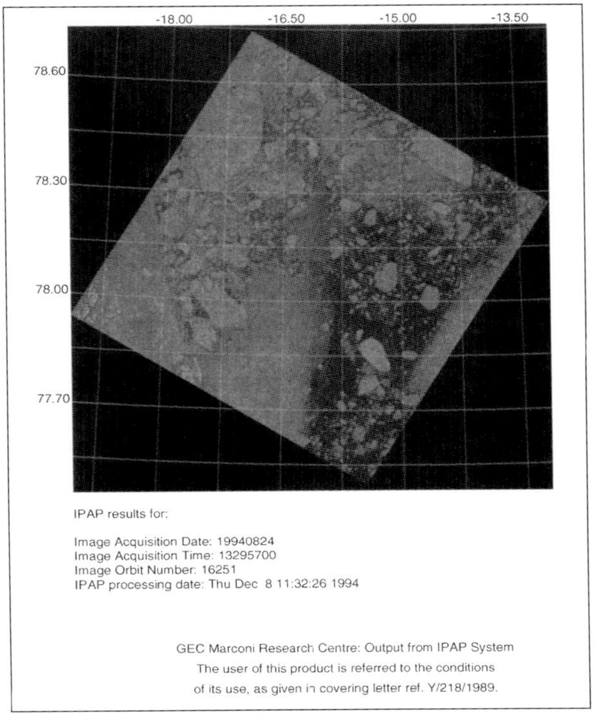


**Fig. 7.** ERS-1 SAR image of marginal ice recorded off east Greenland on August 24, 1994

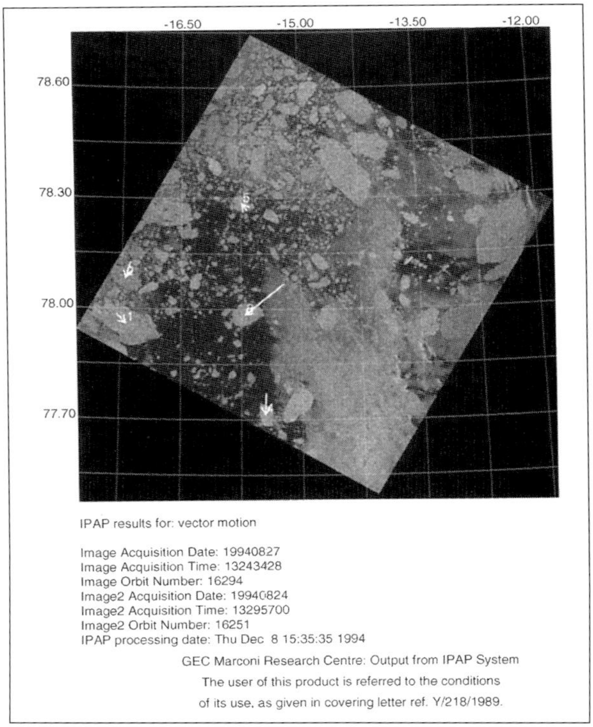


**Fig. 8.** Displacement of ice floes with respect to data in Fig. 7, determined from region matching algorithm, superimposed on later image recorded on August 27, 1994

year ice floes as these are the most distinctive floes in the image pair). By comparison with Fig. 7 it can be seen that the algorithm has accurately tracked the motion of each of the five floes, without generating any spurious matches, despite variations in region boundary caused by the segmentation process.

In comparison, for ERS-1 data of primarily marginal ice recorded in the Gulf of St. Lawrence in March 1992, the region matching algorithm does not perform well. This is mainly due to the higher concentration of ice (above 8/10ths) in which the floe boundaries are not so distinct, and the much smaller average floe size (below 1 km in dimension) and hence less distinctive floe shape. Where groups of ice floes are moving together, then increased matching accuracy might be obtained by using contextual information (relations to and attributes of neighboring regions such as the direction and extent of adjacency, degree of "surroundness," and region area) to assist the matching of individual floes.

## 12.4 Iceberg Detection

Drifting icebergs can be found many kilometers outside the ice edge in areas such as Baffin Bay, the Labrador Coast, the Grand Banks, and the North Atlantic shipping lanes. Because they are extremely hard (consisting of nearly pure water crystals), and can be of enormous size and mass (drafts of 300 – 400 m are not uncommon, whilst an iceberg of 13 m diameter will typically weigh some 8000 tons), they represent a major hazard to oil rigs and shipping.

The spatial resolution and relatively steep (23°) incidence angle of ERS-1 means that less than 100% detection rates are achievable, though useful information can potentially still be derived for planning and as an adjunct to marine radar data. It is also demonstrated here that although the icebergs can be small compared to the imaging spatial resolution, their effect on the local environment, and hence their signature size, can be much larger. An example is shown in Fig. 9 where an iceberg of approximately 100 m diameter is shown in a full-resolution ERS-1 SAR subscene. The large signature size is a result of the iceberg shadow, which in this case extends to approximately 500 m.

There are three characteristics of an iceberg which can be used in its identification:

- The iceberg itself (generally bright)
- The iceberg "shadow" (dark)
- The iceberg wake (linear, if visible)

The iceberg itself is generally brighter than surrounding open water and ice, as a result of strong internal scattering from air bubbles and, in some cases, double-bounce scattering from the sides of the iceberg and surrounding water. The "shadow" is a dark region adjacent to the iceberg which may be caused by fresh water damping of capillary waves or locally lower wind speeds (in the lee of the iceberg). A wake may be present, depending on wind conditions, iceberg size, and currents. Certainly in areas of sea ice, the wake can show up as strongly scattering linear regions of brash ice (Thomas and Roth 1996). In open water, a wake is rarely visible to the eye. The detection characteristics of icebergs are described in more detail by Willis et al. (1996).

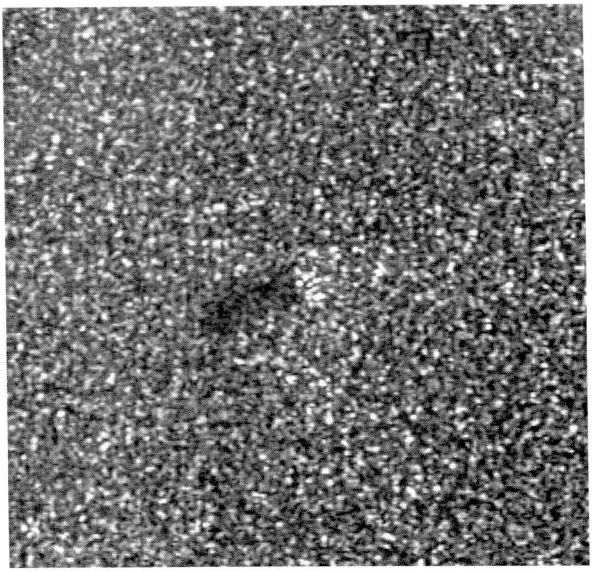

**Fig. 9.** Iceberg signature in ERS-1 SAR data, showing the bright response of the iceberg and the associated "shadow." (Copyright ESA and Int J Remote Sens)

Segmentation, as described earlier, is not generally applicable to icebergs because of their small size. Tabular icebergs, as found in the South Atlantic, would almost certainly be detected by such methods, but the utility of spaceborne SAR lies in being able to detect icebergs close to the spatial resolution of the instrument. As part of the IPAP system, we have implemented an algorithm which detects icebergs in open water based on their neighboring bright and dark signatures (indicating the iceberg and its shadow). The image is first normalized so that detection thresholds are independent of local means. The bright and dark pixels in the image are then detected (using thresholds specified by the operator), and bright and dark pixel 'clusters' identified by a process of morphological erosion and dilation. Adjacent bright and dark pixel clusters are connected by a 'gap bridging' algorithm where appropriate, and the object is at this point labeled as an iceberg and its center of gravity used to provide its position.

An assessment of the performance of this algorithm has been undertaken using International Ice Patrol observations of icebergs in the Grand Banks region of Newfoundland, with an example result shown in Fig. 10. One hundred percent of icebergs over 120 m in diameter were detected, 66% of those in the size range of 60–120 m, and 40% of those between 15 and 60 m in diameter.

The algorithm is clearly successful in terms of true detection, but is currently hampered under a range of conditions through its high false alarm rate which can range from below 10% in calm sea conditions to 50% or more in conditions where the ocean has major textural features or is very rough. It may be possible to improve the false alarm rate through using Hough transforms to detect iceberg wakes in open water and matching these detections to those of the iceberg and its shadow, but to date these have not been investigated. However, in comparison to ice type concentration products, the algorithm is very quick to run and has a strong possibility of being automated should the key technical issue of the high false alarm rate be addressed satisfactorily.

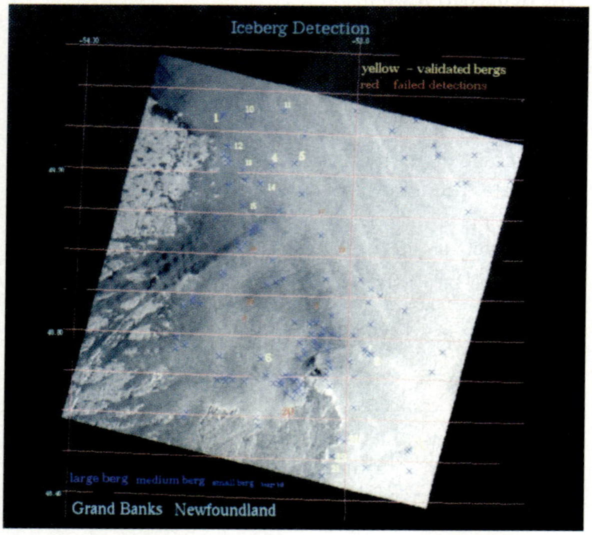


**Fig. 10.** True and false detections of icebergs from the Newfoundland coast, validated using International Ice Patrol observations. True detections are indicated by *yellow* numbers, whilst undetected icebergs are indicated by *red* numbers. The size of the number indicates the size of the iceberg, as indicated. Unvalidated iceberg detections are indicated by *blue*

## 12.5 Operational Implementation

The IPAP system has been operationally demonstrated in near to real time using ERS-1 data recorded off the east Greenland coast in summer 1994 and sent to a survey ship (MV *Thetis*) being operated by Nunaoil in support of seismic exploration for oil and gas reserves. The ERS-1 data were received by West Freugh ground station in Scotland where they were first processed by the Rapid Information Dissemination System (RAIDS) developed by Matra Marconi Space UK, before generation of the high-level ice products at MRC. Under optimum conditions, the products were received onboard the survey ship within about 3 h of data acquisition by the satellite.

A strategic assessment of the operational utility of the IPAP products undertaken as part of this demonstration, and by the two other commercial operators in the IPAP team, has revealed the following key observations:

- Facsimile transmission of image or enhanced image data can result in poor quality reproduction, especially where no recognizable ice features are apparent, emphasizing a requirement for digital transmission of the products.
- The SAR-derived products will not generally be used in isolation. Information obtained via remote sensing is generally regarded as an adjunct to (say) marine radar, and the facility needs to be provided for the fusion of data from the different sources.
- The resolution of the products needs to be carefully selected as a trade-off between detail and clarify of presentation.
- A facility to overlay certain products on the original SAR data should be provided.

The demonstration proved a key learning experience, not without problems as can be seen, which enabled a strong view to be taken of the requirements for an operational system. These can be listed as follows:

**Product requirements**

- Coverage of greater than 200 × 200 km (ideally more), without major gaps
- Update frequency between 12 h and 72 h (depending on operations)
- Appropriate pricing structure, based on subscription
- Relatively small file size to aid transmission times and memory requirements
- Flexibility in algorithms and scope for manual intervention

**Transmission requirements**

- Product delay under 4 h (SAR acquisition to product reception). Implies broad bandwidth link
- Digital reception at destination
- Transmission of other relevant information (e.g. meteorological data) via the same program, preferably from a single bulletin board source

**On-board manipulation**

- Coregistration of different products and ingestion of shipborne information (position, bearing, marine radar, meteorological products, AVHRR-derived ice charts, etc.)
- Facility to display coverage of new products as a means of selecting and ordering new products.

Meeting these requirements is a basic prerequisite for the routine use of spaceborne SAR in tactical ice monitoring. To integrate spaceborne SAR data firmly within the infrastructure of ice monitoring will require cooperation from existing meteorological and ice services, who will themselves be best placed to provide the bulletin board for ice products and to convince the end users of the value of the data.

Ongoing activities with the IPAP system include installation of the software at the Defence Research Agency Farnborough for evaluation in Royal Navy trials using ERS-2 SAR data (in "strip mode" format) towards the end of 1996. However, it is important to note that the IPAP system has been developed primarily as a flexible research tool to support algorithm research and development, rather than as a dedicated operational workstation running in near to real-time and with the minimum of operator intervention. The commercial viability of such a system has to date also been seriously limited by the constraints put on the operator requirements given in Sect. 12.1 by the use of ERS-1 data. The most important recent development in the capability of satellite SAR systems for sea ice observation has been the launch in 1995 of RADARSAT, a C-band HH polarization SAR with the ability to collect data over a number of variable incidence angles (20° – 50°) and swath widths (50 – 500 km), at a variable resolution of 30 – 100 m. This degree of flexibility with a nominal repeat period of 24 days greatly enhances the sensor's coverage and thus its operational utility. It is intended to adapt the IPAP system to handle RADARSAT data at a future date.

Regarding future satellite SAR instruments, the C-band ASAR instrument onboard ENVISAT-1 (currently due to be launched in 1999) will also have the flexibility of a variable incidence angle in the range of 15° – 45° and swath width up to 400 km. It will additionally have a dual polarization (HH and VV) capability, which in its alternating polarization mode (half of the looks in a scene acquired with horizontal polarization, and the other half with vertical polarization) is likely to enable significant improvements in ice type and iceberg discrimination performance.

**Acknowledgments:** The IPAP study was undertaken with funding from the British National Space Centre and was performed in collaboration with SAIC (Science and Engineering), University of Cambridge (Scott Polar Research Institute), Matra Marconi Space UK, Nunaoil, Canarctic Shipping, and BP. Acknowledgment is also made to DRA Farnborough who managed the study, and to the European Space Agency for the provision of ERS-1 SAR data.

## References

Banfield J (1991) Automated tracking of ice floes: a stochastic approach. IEEE Trans Geosci Remote Sens 29, no 6: 905–911

Carsey F, Holt B (1987) Beaufort-Chukchi ice margin data from Seasat: ice motion. J Geophys Res 92, no C7: 7163–7172

Collins MJ, Emery WJ (1988) A computational method for estimating sea ice motion in sequential seasat synthetic aperture radar imagery by matched filtering. J Geophys Res 93, no C8: 9241–9251

Colony R, Thorndike AS (1984) An estimate of the mean field of Arctic sea ice motion. J Geophys Res 89, no C6: 10623–10629

Cover TM, Hart PE (1967) Nearest neighbor pattern classification. IEEE Trans Inf Theory 13: 21–27

Curlander JC, Holt B, Hussey KJ (1985) Determination of sea-ice motion using digital SAR imagery. IEEE J Ocean Eng 10, no 4: 358–367

Fily M, Rothrock DA (1986) Extracting sea-ice data from satellite SAR imagery. IEEE Trans Geosci Remote Sens 24, no 6: 849–854

Fily M, Rothrock DA (1987) Sea-ice tracking by nested correlations. IEEE Trans Geosci Remote Sens 25, no 5: 570–580

Haralick R, Shanmugan K, Dinstein I (1973) Textural features for image classification. IEEE Trans Syst Man Cybern 3: 610–621

Ito H, Muller F (1982) Ice movement through Smith Sound in Northern Baffin Bay Canada observed in satellite imagery. J Glaciol 28: 129–143

Kendall MG, Stuart A (1977) Advanced theory of statistics, 4th edn. Griffin, London

Kwok R, Curlander JC, McConnell R, Pang SS (1990) An ice-motion tracking system at the Alaska SAR Facility. IEEE J Ocean Eng 15, no 1: 44–54

Kwok R, Rignot E, Holt B, Onstott R (1992) Identification of sea ice types in spaceborne synthetic aperture radar data. J Geophys Res (Oceans) 97, no C2: 2391–2402

Lippman RP (1987) An introduction to computing with neural Nets. IEEE ASSP Magazine April Issue, 4: 4–22

McConnell R, Kwok R, Curlander JC, Kober W, Pang SS (1991) Y-S correlation and dynamic time warping: two methods for tracking ice floes in SAR images. IEEE Trans Geosci Remote Sens 29, no 6: 1004–1012
Ninnis RM, Emery WJ, Collins MJ (1986) Automated extraction of pack ice motion from advanced very high resolution radiometer imagery. J Geophys Res 91, no C9: 10725–10734
Oddy CJ, Rye AJ (1983) Segmentation of SAR images using a local similarity rule. Pattern Recogn Lett 1: 443–449
Partington KC, Hanna M (1994) Modelling radar sea ice backscatter in support of ERS-1. EARSeL, Adv Remote Sens 3, no 2: 9–23
Rothrock DA, Carsey FD, Curlander JC, Holt B, Kwok R, Weeks WF (1992) Ice tracking techniques, implementation, performance, and applications. Adv Space Res 12, no 7: 141–147
Sephton AJ, Brown LMJ, Macklin JT, Partington KC, Veck NJ (1994) Segmentation of synthetic aperture radar imagery of sea ice. Int J Remote Sens 15, no 4: 803–825
Sun Y, Carlstrom A, Askne J (1992) SAR image classification of sea ice in the Gulf of Bothnia. Int J Remote Sensing 13, no 13: 2489–2514
Thomas M, Roth R (1996) Monitoring of lead formation and heat exchange in the Weddell Sea. Proc 2nd ERS Applications Worksh, London, 6–8 December 1995 (ESA SP-383), pp 229–234
Vesecky JF, Samadani R, Smith MP, Daida JM, Bracewell RN (1988) Observation of sea ice dynamics using synthetic aperture radar images: automated analysis. IEEE Trans Geosci Remote Sens 26, no 1: 38–48
Willis CJ, Macklin JT, Partington KC, Teleki KA, Rees WG, Williams RG (1996) Iceberg detection using ERS-1 synthetic aperture radar. Int J Remote Sens 17, no 9: 1777–1795
Yan S (1992) Ice Motion retrieval from SAR imagery in terms of intensity derivative. Proc Int Geosci Remote Sens Symp IGARSS '92, May 26–29, Houston, Texas, pp 585–587
Zhang H, Pedersen LT, Gudmandsen P (1989) Ice Motion in the Greenland Sea determined from satellite remote sensing data. Proc EARSeL Symp, Espoo, Finland, June 27 – July 1, pp 129–130
Zhang H (1992) Automatic tracking of ice floes from satellite imagery via invariant moment matching. Proc Int Geosci Remote Sens Symp IGARSS '92, May 26–29, Houston, Texas, pp 582–584

# Author Citation Index

# Subject Index

# Springer and the environment

At Springer we firmly believe that an international science publisher has a special obligation to the environment, and our corporate policies consistently reflect this conviction.
We also expect our business partners – paper mills, printers, packaging manufacturers, etc. – to commit themselves to using materials and production processes that do not harm the environment. The paper in this book is made from low- or no-chlorine pulp and is acid free, in conformance with international standards for paper permanency.

Springer

Printing: Saladruck, Berlin
Binding: Buchbinderei Lüderitz & Bauer, Berlin